Chemistry Experiments for Instrumental Methods

Chemistry Experiments for Instrumental Methods

Donald T. Sawyer

Professor, Department of Chemistry
University of California
Riverside, California 92521

William R. Heineman

Professor, Department of Chemistry
University of Cincinnati
Cincinnati, Ohio 45221

Janice M. Beebe

Associate Professor, Department of Chemistry
Frostburg State College
Frostburg, Maryland 21532

John Wiley & Sons

New York / Chichester / Brisbane / Toronto / Singapore

Library of Congress Cataloging in Publication Data

Sawyer, Donald T.
Chemistry experiments for instrumental methods.

"This manual derives from a much earlier version by
the late Charles N. Reilley and Donald T. Sawyer"—Pref.
Bibliography
1. Instrumental analysis—Laboratory manuals.
I. Heineman, William R.   II. Beebe, Janice M.
III. Title.

QD79.I5S39   1984   543'.08   83-23297
ISBN 0-471-89303-X

Printed and bound in the United States of America

30 29 28 27 26 25 24 23 22 21

Contents

Preface

One of the problems, as well as one of the virtues, in the preparation and selection of instrumental analysis texts has been and continues to be that there is a wide divergence of opinion as to what the content and emphasis should be. There is more general agreement on what is desirable in the laboratory portion of the course. However, the lack of a challenging, reliable, and really instructive group of experiments has caused many instructors concern. This serious deficiency has led to the development of the present group of laboratory experiments, which are designed to give a basic understanding of the theory as well as to demonstrate the importance of instrumental methods to the modern chemist.

Our primary concern in developing and collecting this material has been to illustrate the fundamental principles of an instrumental method and its general theory. We have endeavored to include experiments that not only illustrate analytical applications for instruments but, whenever possible, also require the use of instruments to determine important physical chemical data. This latter feature helps to ensure that the student will acquire a sound knowledge of the chemical principle involved in the measurement and, in many cases, aids in the selection of the most appropriate conditions for an analysis. A fundamental requirement for each experiment has been that it be well tested and workable, and that "cookbook" procedures be eliminated as far as possible. The purpose of this manual is not to displace other instrumental texts, but to supplement them by providing a set of experiments which will thoroughly test and demonstrate the principles discussed in the lecture section of this course. The manual will serve for many instructors as a foundation upon which lecture material for the course may be built.

This is primarily a laboratory manual; however, each chapter is introduced by a brief resumé of the fundamental theory relative to the experiments in that chapter. Also, each experiment within a chapter is preceded by a more detailed outline of the basic theory and principles upon which it is based; hence, laboratory experiments may be assigned for periods prior to lectures on that subject. Suggestions for preliminary reports to be submitted before performing the experiment are included in many of the experiments to ensure that the student has gained sufficient basic knowledge to undertake the experiment with profit. At the end of each experiment a set of more detailed questions is given which tests the student's understanding and appreciation of the specific instrumental methods involved.

Because efficient use of limited laboratory time is imperative, the experiments have been selected to demonstrate the basic principles of an instrumental method and to give the student a sound grasp of the general applicability of specific instrumental methods to all types of chemical problems. Sample preparation has been minimized as far as possible to provide a maximum amount of time for use of an instrument. An attempt has been made to write the laboratory procedures in a manner which forces the student to think about the basic principles of the method. As an ultimate goal, the student should discover and understand an instrumental technique through laboratory experience. We suggest that a few periods near the end of the course be used for a minor research-type problem, with the students applying their knowledge of the techniques to a project that interests each of them.

The manual is divided into three parts: Part I, Electrochemical Methods, covers potentiometric, conductometric, controlled potential, and controlled-current methods; Part II, Methods Based on Electromagnetic Radiation, covers UV-visible and infrared spectrophotometry, atomic absorption and emission spectroscopy, fluorescence spectroscopy, and NMR spectroscopy; and Part III, Separation Methods, covers chromatographic and electrophoretic methods.

The material is presented in a manner which will permit the instructor to begin with either Part II or Part III, rather than Part I. However, the authors feel that the student is better prepared to begin with electrochemical methods. Most, if not all, students have had a thorough introduction to the electrical properties of solutions and ions in elementary chemistry courses. Concepts of equilibria may be tested and evaluated with electrochemical techniques; this is a logical extension of the principles of oxidation-reduction and electrolytic cells introduced in quantitative analysis courses.

This manual derives from a much earlier version by the late Charles N. Reilley and Donald T. Sawyer. Because each of the present authors has been profoundly influenced by an association with Professor Reilley, we wish to acknowledge his contributions to our understanding of instrumental methods.

The authors wish to acknowledge the ideas, suggestions, and advice of their colleagues and students. We are particularly grateful to Joan Ruth and Karoly Fogassy for their assistance with the manuscript copy and illustrations. We thank the late Charles N. Reilley and the Department of Chemistry at the University of North Carolina at Chapel Hill for their kind hospitality to J.M.B. during a sabbatical leave in 1978. Much of the preliminary planning for this book was accomplished at that time. Thanks also are due to Professor Timothy A. Nieman (University of Illinois, Urbana), Professor Johannes Coetzee (University of Pittsburgh), Professor Geoffrey Coleman (University of Alabama), Professor Frank Meeks (University of Cincinnati), Professor Stewart Karp

(Long Island University, C. W. Post Center) for reading the entire manuscript and offering many valuable suggestions. The editorial efforts of Mary Ann Rosenberg have been creative and constructive and have added significantly to the clarity of the manuscript. Finally, a personal acknowledgment to our spouses (Shirley, Linda, and Palmer) for their patience and understanding of what appeared to be an endeavor without an end.

Donald T. Sawyer
William R. Heineman
Janice M. Beebe

June 1984

Chemistry Experiments for Instrumental Methods

1
Introduction and Bibliography

Instruments are the tools for contemporary experimental chemistry and can be used to equal advantage in research, analysis, and process control. An all-too-common attitude is that *instrumental analysis,* by virtue of its name, is synonymous solely to the work of analysts. Although instruments are vital for the work of the modern analyst, they are equally important tools for all types of chemists, whether their work be routine or of the most fundamental and exploratory nature. Hence, a broader perspective is encouraged. The study of instrumental methods should be directed not only toward the understanding of the elements of a particular instrument and what is critical in the components that make up the instrument, but also toward their use in the solution of chemical problems.

Too often the chemist acquires his first introduction to an instrumental method when faced with a pressing chemical problem. It then becomes necessary to read the instruction manual hurriedly to bring the instrument into immediate operation. Although apparently expedient, this approach is not efficient and may result in data of questionable value. There is an additional and unfortunate tendency for chemists to use only those methods with which they are acquainted, even in those cases where much more convenient and appropriate methods are readily available. Intelligent use of instrumental methods demands that full understanding be gained of the physical principles upon which the instrument is based, and what limitations are brought about by basing an instrument on these principles. Where the instrument may best be applied to take advantage of these principles is a further important consideration. Only through such an appreciation for instrumental methods will it be possible for the well-trained chemist to select the instrumental method which is best suited to the problem at hand.

The above considerations have served as the basis for the development of the experiments in this manual. Each chapter in the manual covers a specific instrumental area. By doing one or more experiments within a chapter, a student should acquire a solid knowledge of the basic principles and should gain an appreciation for where the particular instrumental methods covered may best be applied for maximum effectiveness.

Much more will be gained from experiments and laboratory time if prior to each experiment a prospectus is written. This prospectus should include a brief statement of the object of the experiment, precautions to be observed, and a brief outline of the critical factors in the experimental procedure. Such a prospectus will not only organize one's thoughts with respect to a particular experiment, but will provide an organized outline of what is to be accomplished and, hence, make it possible to use the laboratory time efficiently.

Beyond the experiments which are included in the manual the student is encouraged to design at least one or two experiments of his own. The instructor can provide suggestions, and ideas can be obtained from the literature. These experiments can take the form of minor research projects.

Another type of project can be based on the determination of an element in a particular material, such as an alloy, an ore, or other multicomponent mixtures. First, an analytical method should be developed which is based on classical chemical procedures obtained by reference to the literature. Next a method of analysis based on an instrumental method should be designed or developed, again using the literature as a source of information and ideas. After writing up the two proposed methods in brief but specific terms, they should be submitted to the instructor for approval. The material should then be analyzed by the two methods and comparisons made of the convenience, precision, and accuracy of each.

Many instrumental techniques have not been included in this manual primarily because of their cost, complexity, or the technical skills required for their operation. This should in no way detract from an appreciation of their importance and usefulness to chemists. Such techniques include mass spectrometry; x-ray emission, absorption, diffraction, and fluorescence; electron microscopy; and ultracentrifugation. Some instruments discussed in the manual may not be available in a particular laboratory; for example, a proton NMR spectrometer. It is desirable for the student to have as complete an understanding of instruments that are not available as for the actual instruments considered in the particular course.

The student is sincerely encouraged to spend as much time as possible in outside reading, directing this toward gaining a broad, fundamental appreciation for the usefulness of instrumental methods. Few, if any, future opportunities will present themselves where time and inclination are available to acquire this basic knowledge. It is far better that this introduction occur in an organized fashion and as a part of a formal course. The student who makes the very minimum effort, that is "cookbooks" the experiments, will receive correspondingly minimal return for the time he invests. Only with adequate preparation can students spend their time profitably; this is particularly true for experiments which, at times, can be confusing and complex.

The following bibliography outlines a part of the literature available on instrumental methods and their applications in chemistry.

Bibliography

General

Bauer, H. H., G. D. Christian, and J. E. O'Reilly: "Instrumental Analysis," Allyn & Bacon, Boston, 1978.

Bi-Annual Reviews, *Anal. Chem.*, April issue, even-numbered years.

Ewing, G. W.: "Instrumental Methods of Chemical Analysis," 4th ed., McGraw-Hill, New York, 1975.

Kolthoff, I. M., and P. J. Elving (eds.): "Treatise on Analytical Chemistry," Interscience, New York (a multi-volume series).

Kuwana, T. (ed.): "Physical Methods on Modern Chemical Analysis," Academic Press, New York, 1978 (a multi-volume series).

Mann, C. K., T. J. Vickers, and W. M. Gulick: "Instrumental Analysis," Harper & Row, New York, 1974.

Pecsok, R. L., L. D. Shields, T. Cairns, and I. G. McWilliam: "Modern Methods of Chemical Analysis," 2nd ed., Wiley, New York, 1976.

Skoog, D. A., and D. M. West: "Principles of Instrumental Analysis," 2nd ed., Saunders, Philadelphia, 1980.

Strobel, H. A.: "Chemical Instrumentation," 2nd ed., Addison-Wesley, Reading, Mass., 1973.

Weissberger, A., and B. W. Rossiter (eds.): "Techniques of Chemistry," vol. I: "Physical Methods of Chemistry," Wiley-Interscience, New York, 1971–1977 (six parts).

Willard, H. H., L. L. Merritt, Jr., J. A. Dean, and F. A. Settle, Jr.: "Instrumental Methods of Analysis," 6th ed., Van Nostrand, New York, 1981.

Part I. Electrochemical Methods

General

Adams, R. N.: "Electrochemistry at Solid Electrodes," Dekker, New York, 1969.

Bard, A. J., and L. R. Faulkner: "Electrochemical Methods," Wiley, New York, 1980.

Bockris, J. O'M., and A. K. N. Reddy: "Modern Electrochemistry," vols. I and II, Plenum, New York, 1970.

Delahay, P.: "New Instrumental Methods in Electrochemistry," Interscience, New York, 1954.

Galus, Z.: "Fundamentals of Electrochemical Analysis," Horwood, Halsted Press, London, 1976.

Kissinger, P. T., and W. R. Heineman (eds.): "Laboratory Techniques in Electroanalytical Chemistry," Dekker, New York, 1984.

Kolthoff, I. M., and P. J. Elving (eds.): "Treatise on Analytical Chemistry," part I, vol. 4, sec. D-2, Interscience, New York, 1963.

Lingane, J. J.: "Electroanalytical Chemistry," 2nd ed., Interscience, New York, 1978.

Macdonald, D. D.: "Transient Techniques in Electrochemistry," Plenum, New York, 1977.

Plambeck, J. A.: "Electroanalytical Chemistry," Wiley, New York, 1982.

Sawyer, D. T., and J. L. Roberts, Jr.: "Experimental Electrochemistry for Chemists," Wiley-Interscience, New York, 1974.

Weissberger, A., and B. W. Rossiter (eds.): "Techniques of Chemistry," vol. I: "Physical Methods of Chemistry," part II, Wiley-Interscience, New York, 1971.

Potentiometric Methods

Bates, R. G.: "Determination of pH," Wiley, New York, 1973.

Covington, A. K.: "Ion-Selective Electrode Methodology," vols. I and II, CRC Press, Boca Raton, Fla., 1979.

Durst, R. A. (ed.): "Ion-Selective Electrodes," National Bureau of Standards, Washington, D.C., 1969.

Freiser, H. (ed.): "Ion-Selective Electrodes in Analytical Chemistry," vols. I and II, Plenum, New York, 1978, 1980.

Koryta, J.: "Ion-Selective Electrodes," Cambridge University Press, Cambridge, 1975.

Laitinen, H. A., and W. E. Harris: "Chemical Analysis," 2nd ed., McGraw-Hill, New York, 1975, chaps. 12 and 13.

Ramette, R. W.: "Chemical Equilibrium and Analysis," Addison-Wesley, Reading, Mass., 1981, chaps. 8 and 11.

Conductometric Methods

Britton, H. T. S.: "Conductometric Analysis," Van Nostrand, Princeton, N.J., 1934.

Reilley, C. N.: High Frequency Methods, chap. 15 in "New Instrumental Methods in Electrochemistry," P. Delahay (ed.), Interscience, New York, 1954.

Controlled Potential Methods (Voltammetry)

Adams, R. N.: "Electrochemistry at Solid Electrodes," Dekker, New York, 1969, chap. 5.

Bond, A. M.: "Modern Polarographic Methods in Analytical Chemistry," Dekker, New York, 1980.

Kolthoff, I. M., and J. J. Lingane: "Polarography," 2nd ed., Interscience, New York, 1952.

Meites, L.: "Polarographic Techniques," 2nd ed., Wiley-Interscience, New York, 1965.

Part II. Methods Based on Electromagnetic Radiation

General

Bentley, K. W., and G. W. Kirby (eds.): "Techniques of Chemistry," vol. IV: "Elucidation of Organic Structures by Physical and Chemical Methods," 2nd ed., part I, Wiley-Interscience, New York, 1972.

Elving, P. J., M. M. Bursey, and I. M. Kolthoff (eds.): "Treatise on Analytical Chemistry," 2nd ed., part I, vol. 10, sec. 1, Interscience, New York, 1983.

Elving, P. J., E. J. Meehan, and I. M. Kolthoff (eds.): "Treatise on Analytical Chemistry," 2nd ed., part I, vol. 7, sec. H, Interscience, New York, 1981.

Lambert, J. B., H. F. Shurvell, L. Verbit, R. G. Cooks, and G. H. Stout: "Organic Structural Analysis," Macmillan, New York, 1976.

Mellon, M. G. (ed.): "Analytical Absorption Spectroscopy," Wiley, New York, 1950.

Olsen, E. D.: "Modern Optical Methods of Analysis," McGraw-Hill, New York, 1975.

Pasto, D. J., and C. R. Johnson: "Organic Structure Determination," Prentice-Hall, Englewood Cliffs, N.J., 1969.

Silverstein, R. M., G. C. Bassler, and T. C. Morrill: "Spectrophotometric Identification of Organic Compounds," 4th ed., Wiley, New York, 1981.

Weissberger, A., and B. W. Rossiter (eds.): "Techniques of Chemistry," vol. I: "Physical Methods of Chemistry," part IIIB, Wiley-Interscience, New York, 1972.

Ultraviolet-Visible Absorption Spectroscopy

Jaffe, H. H., and M. Orchin: "Theory and Applications of Ultraviolet Spectroscopy," Wiley, New York, 1962.

Lever, A. B. P.: "Inorganic Electronic Spectroscopy," Elsevier, Amsterdam, 1968.

Infrared Spectroscopy

Bellamy, L. J.: "The Infrared Spectra of Complex Molecules," 3rd ed., Wiley, New York, 1975.

Brame, E. G., and J. Graselli (eds.): "Infrared and Raman Spectroscopy," Dekker, New York, 1977.

Colthup, N. B., L. H. Daly, and S. E. Wiberley: "Introduction to Infrared and Raman Spectroscopy," 2nd ed., Academic Press, New York, 1974.

Atomic Spectroscopy

Dean, J. F., and T. Rains (eds.): "Flame Emission and Atomic Absorption Spectrometry," vols. I and II, Dekker, New York, 1969, 1971.

Fluorescence Spectroscopy

Guilbault, G. G. (ed.): "Practical Fluorescence: Theory, Methods, Techniques," Dekker, New York, 1973.

Pesce, A. J., C. G. Rosen, and T. L. Pasby (eds.): "Fluorescence Spectroscopy: An Introduction for Biology and Medicine," Dekker, New York, 1971.

Udenfriend, S.: "Fluorescence Assay in Biology and Medicine," vols. I and II, Academic Press, New York, 1962, 1969.

Wehry, E. L.: "Modern Fluorescence Spectroscopy," vols. I and II, Plenum, New York, 1976.

White, C. E., and R. F. Argauer: "Fluorescence Analysis. A Practical Approach," Dekker, New York, 1970.

Winefordner, J. D., S. G. Schulman, and T. C. O'Haver: "Luminescence Spectrometry in Analytical Chemistry," Wiley, New York, 1972.

NMR Spectroscopy and ESR Spectroscopy

Akitt, J. W.: "N.M.R. and Chemistry; An Introduction to Nuclear Magnetic Resonance Spectroscopy," Chapman and Hall, London, 1973.

Becker, E. D.: "High Resolution NMR," 2nd ed., Academic Press, New York, 1980.

Emsley, J. W., J. Feeney, and L. H. Sutcliffe: "High Resolution Nuclear Magnetic Resonance Spectroscopy," vols. I and II, Pergamon, Oxford, 1965.

Levy, G. C., and G. L. Nelson: "Carbon-13 Nuclear Magnetic Resonance for Organic Chemists," Wiley-Interscience, New York, 1972.

Strothers, J. B.: "Carbon-13 NMR Spectroscopy," Academic Press, New York, 1972.

Swartz, H. M., J. R. Bolton, and D. C. Borg (eds.): "Biological Applications of Electron Spin Resonance," Wiley-Interscience, New York, 1972.

Yen, T. F. (ed.): "Electron Spin Resonance of Metal Complexes," Plenum, New York, 1969.

Part III. Separation Methods

General

Elving, P. J., E. Grushka, and I. M. Kolthoff: "Treatise on Analytical Chemistry," 2nd ed., part I, vol. 5, sec. G, Interscience, New York, 1982.

Heftmann, E.: "Chromatography," 3rd ed., Van Nostrand Reinhold, New York, 1975.

Helfferich, F., and G. Klein: "Multicomponent Chromatography; Theory of Interference," Dekker, New York, 1970.

Karger, B. L., L. R. Snyder, and C. Horvath: "An Introduction to Separation Science," Wiley, New York, 1973.

Laitinen, H. A., and W. E. Harris: "Chemical Analysis," 2nd ed., McGraw-Hill, New York, 1975.

Gas Chromatography

Grob, R. L. (ed.): "Modern Practice of Gas Chromatography," Wiley-Interscience, New York, 1977.

Littlewood, A. B.: "Gas Chromatography: Principles, Techniques, and Applications," 2nd ed., Academic Press, New York, 1970.

Purnell, H.: "Gas Chromatography," Wiley, New York, 1962.

Liquid Chromatography

Johnson, E. L., and R. Stevenson: "Basic Liquid Chromatography," Varian, Palo Alto, Calif., 1978.

Kissinger, P. T. (ed.): "An Introduction to Detectors for Liquid Chromatography," BAS Press, West Lafayette, Ind., 1981.

Miller, J. M.: "Separation Methods in Chemical Analysis," Wiley, New York, 1975.

Snyder, L. R., and J. J. Kirkland: "Introduction to Modern Liquid Chromatography," 2nd ed., Wiley, New York, 1979.

Yau, W. W., J. J. Kirkland, and D. D. Bly: "Modern Size-Exclusion Liquid Chromatography," Wiley, New York, 1979.

Yost, R. W., L. S. Ettre, and R. D. Conlon: "Practical Liquid Chromatography—An Introduction," Perkin-Elmer, Norwalk, Conn., 1980

I

Electrochemical Methods

2
Potentiometric Methods

Potentiometric methods are based on the measurement of a potential difference between two electrodes immersed in a solution. The electrodes and the solution constitute an *electrochemical cell*. The potential between the two electrodes is usually measured with a pH/mV meter. A schematic representation of typical electrochemical apparatus for potentiometry is shown in Fig. 2-1. The buret is used for a potentiometric titration. One of the two electrodes is termed an *indicator electrode*, which is chosen to respond to a particular species in solution whose activity is to be measured during the experiment. The other electrode is a *reference electrode* whose half-cell potential is invariant.

The potential of an electrochemical cell is given by

$$E_{\text{cell}} = E_{\text{ind}} - E_{\text{ref}} + E_{\text{lj}} \tag{1}$$

where
E_{cell} = potential of the electrochemical cell
E_{ind} = half-cell potential of the indicator electrode (which is considered to be the cathode in this example)
E_{ref} = half-cell potential of the reference electrode
E_{lj} = liquid-junction potential

Liquid-junction potentials develop at the interface between two electrolytes. They are typically found at the junction of the reference electrode and the solution in the cell. The half-cell potentials of most indicator electrodes respond to changes in the activity of the species being determined according to the Nernst equation. For example, in the case of a silver wire immersed in a solution of Ag^+, the electrode reaction is

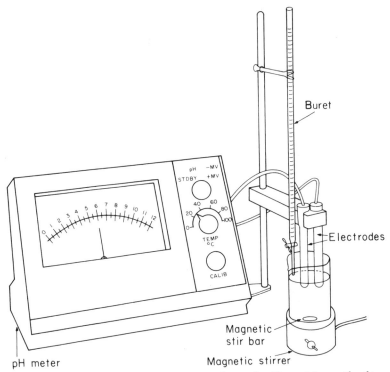

Figure 2-1 / Experimental apparatus for potentiometric methods.

$$\text{Ag}^+ + e \rightleftharpoons \text{Ag} \tag{2}$$

and the Nernst equation for the electrode response is

$$E_{\text{ind}} = E^\circ_{\text{Ag}^+,\text{Ag}} + \frac{RT}{nF} \ln a_{\text{Ag}^+} \tag{3}$$

where

E° = standard reduction potential for the Ag$^+$, Ag couple, V
R = gas constant, 8.314 V · C/K · mol
T = absolute temperature, K
n = number of electrons in the half-reaction, equiv/mol
F = Faraday constant, 96,485 C/equiv
a_{Ag^+} = activity of Ag$^+$

In situations where the measurement of concentration, rather than activity, is of interest, concentration can be substituted into Eq. (3) to give

$$E_{\text{ind}} = E^\circ_{\text{Ag}^+,\text{Ag}} + \frac{RT}{nF} \ln (f_{\text{Ag}^+}[\text{Ag}^+]) \tag{4}$$

where

f_{Ag^+} = activity coefficient for Ag$^+$
$[\text{Ag}^+]$ = molar concentration of Ag$^+$

Substitution of Eq. (4) into Eq. (1) after conversion to base-10 logarithms and correction for 25°C gives the equation for the potential of the electrochemical cell

$$E_{\text{cell}} = E^\circ_{\text{Ag}^+,\text{Ag}} + \frac{0.0591}{n} \log f_{\text{Ag}^+} + \frac{0.0591}{n} \log [\text{Ag}^+] - E_{\text{ref}} + E_{1j} \tag{5}$$

Solution conditions are usually arranged so that $E^\circ_{Ag^+, Ag}$, f_{Ag^+}, E_{ref}, and E_{lj} are constant during the experiment. Thus, a practical version of Eq. (5) is given by

$$E_{cell} = E^* + \frac{0.0591}{n} \log [Ag^+] \tag{6}$$

where E^* is a constant that can be determined by calibration with standard solutions. This linear relationship between E_{cell} and the log of solution concentration (or activity if f is not maintained constant) is the basis of analytical potentiometry.

Reference Electrodes

A reference electrode should have a constant half-cell potential that is independent of the properties of the solution into which it is immersed. The most commonly used reference electrodes are *calomel* and *silver–silver chloride electrodes*.

Of the calomel electrodes, the *saturated calomel electrode* (SCE) is the most popular. A paste of mercurous chloride (calomel) covered with potassium chloride crystals is layered over a sufficient volume of mercury to cover a platinum contact wire; the cell is filled with a solution saturated with Hg_2Cl_2 and KCl. This internal solution is separated from the solution into which the electrode is to be used by a fritted glass disc, asbestos fiber, or a salt bridge, which maintains electrolyte contact between the internal solution and the sample solution, but prevents the internal solution from rapidly leaking out. There are various ways in which these electrodes are constructed commercially. Two types of commonly used SCEs are shown in Fig. 2-2a and b.

The half-cell reaction and the Nernst equation for the SCE are

$$Hg_2Cl_2 + 2e \rightleftharpoons 2Hg + 2Cl^- \tag{7}$$

$$E_{ref} = E^\circ_{Hg_2Cl_2, Hg} - 0.0591 \log (a_{Cl^-})^2 \tag{8}$$
$$= 0.242 \text{ V vs. SHE}$$

The silver–silver chloride reference electrode consists of a silver wire that is coated with AgCl and immersed in a tube, as shown in Fig. 2-2c, filled with a fixed concentration of chloride (usually saturated potassium chloride) plus a few drops

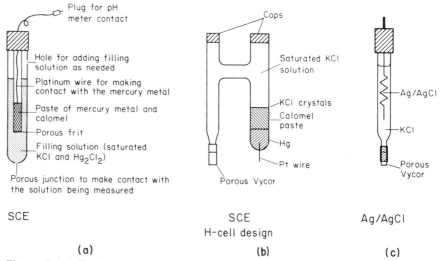

Figure 2-2 / Typical reference electrodes. (*Reprinted in part by permission of Bioanalytical Systems, Inc., West Lafayette, Ind.*)

of 1 M silver nitrate. The half-cell reaction and the Nernst equation for this electrode are

$$AgCl + e \rightleftharpoons Ag + Cl^- \tag{9}$$

$$E_{ref} = E^\circ_{AgCl,Ag} - 0.0591 \log a_{Cl^-} \tag{10}$$
$$= 0.199 \text{ V vs. SHE}$$

This is an excellent reference electrode, easily constructed and long-lasting, which can be used in many electrochemical systems.

Indicator Electrodes

The indicator electrode is of paramount importance in analytical potentiometry. This electrode should interact with the species of interest so that E_{ind} reflects the activity of this species in solution. It is important that the indicator electrode respond selectively to that particular species and not to other compounds in the sample that might constitute an interference.

The practical importance of having indicator electrodes that selectively respond to numerous species of analytical significance has stimulated the development of many types of indicator electrodes

A *redox electrode* responds to the redox potential of a solution as established by one or more redox couples in the solution. An inert metal, such as platinum, responds to many redox couples. For example, a platinum electrode immersed in a solution containing Fe^{2+} and Fe^{3+} gives a potential that is dependent on the ratio of activities of these two species as shown by the following electrode process and Nernst equation:

$$Fe^{3+} + e \rightleftharpoons Fe^{2+} \tag{11}$$

$$E_{ind} = E^\circ_{Fe^{3+},Fe^{2+}} + 0.0591 \log \frac{a_{Fe^{3+}}}{a_{Fe^{2+}}} \tag{12}$$

An electrode of this type is used for the titration of Fe^{2+} in Experiment 2-3.

A *first-order electrode* consists of a metal in contact with a solution containing its ion. An example is a silver wire, which responds to a_{Ag^+} as shown by the electrode process and Nernst equation shown above in Eqs. (2) and (3).

A *second-order electrode* consists of a metal in contact with a solution saturated with one of its sparingly soluble salts. (Saturation can be accomplished by simply coating the salt on the metal itself.) As shown for the Ag/AgCl system below, the Ag electrode

$$\text{Electrode reaction: } \begin{array}{|c|} \hline Cl^- \\ + \\ Ag^+ + \\ \Updownarrow \ K_{sp} \\ AgCl \\ \hline \end{array} e \rightleftharpoons Ag \tag{13}$$

still responds to a_{Ag^+} by means of electron transfer. However, a_{Ag^+} is now controlled by a_{Cl^-} via the precipitation equilibrium constant K_{sp}. Increasing a_{Cl^-} in solution consumes Ag^+ causing a decrease in a_{Ag^+} by the equilibrium process enclosed by the dashed-line box. Consequently, the electrode potential responds to the activity of Cl^- as described by Eq. (10), even though Cl^- undergoes no electron transfer with the Ag wire. Why is a second-order electrode for Cl^- more practical than a first-order electrode would be?

A large number of electrodes with good selectivity for specific ions are based on the measurement of the potential generated across a membrane. Electrodes of this type are referred to as *ion-selective electrodes.* The membrane is usually attached to the end of a tube that contains an internal reference electrode. This membrane electrode and an external reference electrode are then immersed in the solution of interest. Since the potentials of the two reference electrodes are constant, the circuit responds to the membrane potential.

Different membrane materials have proved to give optimal responses for certain ions. For example, a *glass membrane* is unsurpassed for measuring proton activity (pH). This electrode is so important that a special description of the electrode and an explanation of its use are found below. The glass electrode for pH is used extensively in acid-base titrations, as exemplified in Experiments 2-1, 2-4, and 2-6.

Liquid membrane electrodes in which the membrane contains an organic ion-exchange liquid have proved effective for numerous ions such as Ca^{2+}, K^+, and NO_3^-. The calcium electrode is used in Experiment 2-5, which gives a detailed treatment of its operation.

Solid state or *precipitate electrodes* have nonglass, solid state crystals or pellets as the membrane component of the electrode. This approach has proved effective for numerous cations and anions. Perhaps the best example is the excellent electrode for F^-, which is based on a crystal of LaF_3 doped to create crystal defects for improved conductivity.

Gas-sensing electrodes consist of an ion-selective electrode in contact with a thin layer of electrolyte. When a gas, such as ammonia, dissolves in the solution, the resulting pH change is sensed by a glass pH electrode.

Similarly, an enzyme coating on an ion-selective electrode can catalyze the production of the ion to which the electrode is sensitive from a substrate to be measured, e.g., NH_4^+ from urea by the enzyme urease. Enzyme electrodes benefit from the selectivity of enzyme catalysis.

Types of Potentiometric Measurements

In *direct potentiometry* the appropriate indicator electrode and a reference electrode are immersed in the solution to be analyzed; the concentration is then determined by means of the potential measured. Calibration of the electrode by standard solutions is absolutely necessary. A *calibration curve* of E_{cell} vs. log concentration [a plot of Eq. (6)] can be constructed from the potentials measured for a series of standard solutions. The *method of standard additions* involves addition of known volumes of the standard to the sample solution with a potential measurement made after each addition. The advantage of this method is that the standardization is accomplished in the sample matrix.

A *potentiometric titration* involves titrating the sample with a standard solution of titrant; the potential between appropriate indicator and reference electrodes immersed in the sample is monitored during the titration. The end point of the titration is signaled by an abrupt change in the potential. An indicator electrode that responds to a component (reactant or product of sample or titrant) in the titration is used. In this case the amount of species to be determined is calculated from the volume of titrant required to reach the potentiometric end point.

Measurement of pH

The untold millions of times that the measurement of pH is repeated each year is indicative of the importance of knowing solution pH in a variety of situations. The excellent selectivity and wide pH range of the conventional glass electrode makes this measurement a fast, simple operation by direct potentiometry.

The glass electrode consists of a glass (or plastic) tube with a bulb of thin, pH-

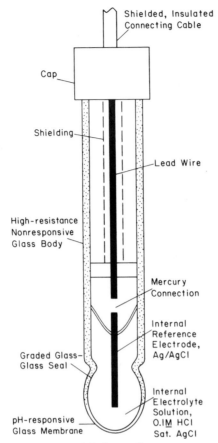

Shielded, Insulated
Connecting Cable

Cap

Shielding

Lead Wire

High-resistance
Nonresponsive
Glass Body

Mercury
Connection

Internal
Reference
Electrode,
Ag/AgCl

Graded Glass-
Glass Seal

Internal
Electrolyte
Solution,
0.1M HCl
Sat. AgCl

pH-responsive
Glass Membrane

**Figure 2-3 / Schematic represen-
tation of a typical glass indicator electrode. (From "Instrumental
Analysis," H. H. Bauer, G. D. Christian, and J. E. O'Reilly, copyright ©
1978 by Allyn and Bacon, Inc., Boston, p. 25. Reprinted by permission.)**

sensitive glass at the end. A typical glass electrode is shown in Fig. 2-3. The bulb contains a silver wire coated with AgCl and is filled with a solution of 0.1 M hydrochloric acid that is saturated with silver chloride. Since the chloride ion concentration remains constant, the silver chloride electrode potential remains constant, as defined by Eq. (10), and serves as an internal reference electrode. An external reference electrode, such as an SCE, is used to complete the circuit for measuring the potential across the pH-sensitive glass membrane. The potential developed at the membrane is a function of the difference in activity of the hydrogen ions on either side of the membrane. The complete cell can be represented schematically as follows:

$$\text{Ag} \mid \text{AgCl (satd.), 0.1 } M \text{ HCl} \mid \underset{\text{membrane}}{\text{Glass}} \mid \underset{\text{solution}}{\text{Sample}} \parallel \text{SCE} \quad (14)$$

Glass electrode

External
reference
electrode

Since a_{H^+} on the inside of the membrane is held constant by the 0.1 M HCl, the cell potential responds to the pH of the sample solution

$$E = E^* + \frac{RT}{nF} \quad \text{pH} \tag{15}$$

or

$$E = E^* + 0.0591 \quad \text{pH} \quad \text{(at 25°C)} \tag{16}$$

A pH meter measures the potential of a glass electrode with respect to the external reference electrode and displays the output on a meter scale calibrated to read directly in pH (or with a digital pH readout). The measured potential is a linear function of pH as described by Eqs. (15) and (16) and shown in Fig. 2-4. Since the meter measures the potential of an electrochemical cell and is unaware of the pH convention, it must be standardized in order for the readout to display the proper pH. Standardization is accomplished by means of one or more buffers of known pH. pH meters have a zero calibration control (also called standardization or asymmetry potential control) whereby the intercept of Eq. (15) can be adjusted as shown in Fig. 2-4. Many meters also have a slope adjustment control that enables the gain of the instrument's amplifier to be adjusted to give the nernstian slope of 59.1 mV per unit of pH at 25°C. A second slope adjustment control enables temperature variations to be compensated for. Thus, the slope adjustment control can be used in conjunction with two buffers to set the nernstian slope at a certain temperature. If the temperature of the samples then changes, the temperature compensator can be set to the new temperature and the slope is then ad-

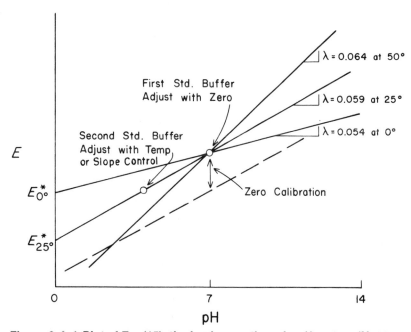

Figure 2-4 / Plot of Eq. (15), the basic equation of a pH meter. (Not to scale.) Zero calibration raises or lowers response line. Temperature or slope control changes slope of response line. (*From "Modern Methods of Chemical Analysis," R. L. Pecsok, L. D. Shields, T. Cairns, and I. G. McWilliam, 2nd ed., copyright © 1976 by John Wiley & Sons, Inc., New York, p. 386. Reprinted by permission.*)

justed to the proper value as shown in Fig. 2-4. Some instruments have only the temperature compensator as a means of adjusting the slope. As shown in Fig. 2-4, the pH electrode system exhibits an isothermal point at which the meter reading is unaffected by varying the slope adjustment and/or temperature compensator. This point occurs in the vicinity of 0 V. For an accurate calibration of the pH meter, two buffers are used. The first buffer is usually pH 7.00 and is used to assign the isothermal point of 0 V to a pH of 7.00. A second buffer (near the other end of the pH range to be measured) is used in conjunction with the slope adjustment to set the nernstian slope of 59.1 mV per pH unit at 25°C.

Practical Measurement of pH

In actual practice, certain hints and precautions will help to ensure the reliable measurement of pH:

1. If accurate measurements in basic solutions are desired, pH measurements should be done in the absence of CO_2, which is present in the air. In any case the readings should be made quickly with minimal exposure to the atmosphere.

2. For solutions more basic than pH 11, glass electrodes of special composition are necessary to avoid interferences due to sodium ions. Electrodes are often supplied with a sodium ion correction monograph.

3. Glass electrodes slowly deteriorate with use. Frequently this condition can be rectified by briefly treating the electrode with a dilute hydrofluoric acid solution. The reliability of a glass electrode should be checked by standardizing with two buffers at two separate pH's, such as pH 4 and pH 9. After standardizing with one buffer, the meter reading should not disagree with that of the second buffer by more than 0.05 pH unit.

4. Buffers must be reliable; a pH measurement is only as reliable as the buffer used to standardize the meter. Table 2-1 lists a group of widely accepted buffers. Buffers for the standardization of pH are commercially available.

5. To obtain stable operation of a new glass electrode, soak it in pH 7 buffer solution for at least 2 h. If the electrode is allowed to become dry, it must be resoaked. Electrodes should be immersed in distilled water or in buffer (pH 4 or 7) when not in use.

6. For pH measurements of very dilute acids or bases (less than 0.01 M), addition of an inert electrolyte (such as KCl) may be necessary to provide sufficient conductance for stable operation of the pH meter.

7. Since the glass electrode requires an electrometer amplifier with extreme current sensitivity for the measurement of the pH (potential), careful shielding of the glass electrode lead is imperative. A break in this shielding will make the electrode extremely unstable.

8. The electrodes used with a pH meter are fragile and easily ruined if scratched or bumped. Although most of the newer electrodes have a plastic shield for protection, it is still possible to damage the tips if handled improperly.

9. A reference electrode usually has a hole for filling with the appropriate electrolyte solution. Normally the hole is open while taking measurements, but kept covered for storage to prevent evaporation of the internal electrolyte. Care must be taken to prevent any of the sample being measured or any wash water from entering this hole.

10. Glass parts of the electrodes should not be touched with anything except soft tissue paper.

11. The electrodes must not be removed from a solution while the meter is in-

Table 2-1 / Standard pH Values Assigned by the National Bureau of Standards†

				PRIMARY STANDARDS			
Temperature °C	KH Tartrate (saturated)	KH$_2$ Citrate (0.05 m)	KH Phthalate (0.05 m)	KH$_2$PO$_4$ (0.025 m) Na$_2$HPO$_4$ (0.025 m)	KH$_2$PO$_4$ (0.008695 m) Na$_2$HPO$_4$ (0.03043 m)	Borax (0.01 m)	NaHCO$_3$ (0.025 m) Na$_2$CO$_3$ (0.025 m)
0		3.863	4.003	6.984	7.534	9.464	10.317
5		3.840	3.999	6.951	7.500	9.395	10.245
10		3.820	3.998	6.923	7.472	9.332	10.179
15		3.802	3.999	6.900	7.448	9.276	10.118
20		3.788	4.002	6.881	7.429	9.225	10.062
25	3.557	3.776	4.008	6.865	7.413	9.180	10.012
30	3.552	3.766	4.015	6.853	7.400	9.139	9.966
35	3.549	3.759	4.024	6.844	7.389	9.102	9.925
38	3.548		4.030	6.840	7.384	9.081	
40	3.547	3.753	4.035	6.838	7.380	9.068	9.889
45	3.547	3.750	4.047	6.834	7.373	9.038	9.856
50	3.549	3.749	4.060	6.833	7.367	9.011	9.828
55	3.554		4.075	6.834		8.985	
60	3.560		4.091	6.836		8.962	
70	3.580		4.126	6.845		8.921	
80	3.609		4.164	6.859		8.885	
90	3.650		4.205	6.877		8.850	

Note: m is molality (mol/kg).
†R. G. Bates, Revised standard values for pH measurements from 0 to 95°C. *J. Res. Nat. Bur. Standards, 66A,* 179 (1962). R. G. Bates, "Determination of pH: Theory and Practice," 2nd ed., Wiley, New York, 1973, chap. 4.

dicating; the meter knob should be set on STANDBY before removing the electrodes.

General Procedure for pH Measurements with the pH Meter

Consult the manufacturer's directions for proper use.

The instrument is allowed to warm up thoroughly in accordance with the manufacturer's directions, 15 to 20 min will usually suffice. All samples and buffers should be allowed to reach the same temperature before any measurements are made.

Standardization Using a Single Buffer

For measurements not requiring maximum precision:

1 Place the clean, dry electrodes in a buffer which has a pH value within 3 pH units of the solution being measured.
2 Set the temperature compensator to the buffer temperature, and the function switch to pH.
3 Rotate the *standardizing* (often called *calibration control* or *asymmetry potential*) knob until the meter is balanced at the known pH of the buffer (or in the case of a digital readout, until the pH value is displayed). Turn to STANDBY.
4 Remove electrodes, rinse with distilled water, and dry gently with clean absorbent tissue.
5 Place the electrodes in the solution to be measured. (If the temperature is different from that of the buffer, rotate the temperature compensator knob to the temperature of the solution.) Set the function switch to pH and take the reading. Reset to STANDBY. Remove electrodes from solution, rinse, and store in distilled water.

Standardization Using Two Buffers

For maximum precision:

1   Use two standard buffers approximately 3 pH units apart, within the region of the solution(s) to be measured. The first buffer is usually pH 7, or at least between pH 6 and pH 8. Place the electrodes in the first buffer, turn the function switch to pH, and adjust the temperature compensator to the buffer temperature.
2   Rotate the standardizing (or calibration control or asymmetry potential) knob until the meter is balanced at the pH of the buffer. Turn to STANDBY. Remove the electrodes from the first buffer, rinse with distilled water, blot, and place in the second buffer. The temperature of the electrodes, standard buffers, solutions, and wash water should be the same, or as close as possible.
3   Turn the function switch to pH. Read the pH without changing the position of the standardizing knob. The reading should be at least within 0.05 pH unit of the actual pH of the second buffer. Use the slope adjustment control to set the pH reading to that of the second buffer.
4   Remove electrodes, rinse with distilled water, and dry gently with clean absorbent tissue.
5   Place the electrodes in the solution to be measured. (If the temperature is different from that of the buffer, rotate the temperature compensator knob to the temperature of the solution.) Set the function switch to pH and take the reading. Reset to STANDBY. Remove electrodes from solution, rinse, and store in distilled water.
6   Ideally, a check should be made at the end of a series of measurements. The standardization should be checked from time to time. Very often, daily checks are sufficient, depending upon the electrodes and the pH meter.

For extremely precise measurements, the recommendations of the American Society for Testing Materials (ASTM) offer guidance. A method which follows these recommendations is given in the textbook by R. G. Bates, "Determination of pH" (Wiley, New York, 1973).

General Procedure for Potentiometric Measurements in Millivolt Readings

The instructions provided by the instrument manufacturer should be followed. Ideally, the temperature of the electrodes, standards, and solutions would be at 25°C. No temperature correction is made for millivolt readings on a pH meter. In general, the meter is zeroed by adjusting the pointer to read zero on the "mV" scale. If the pointer is off scale, the leads should be reversed or the polarity switched. Many meters read directly in millivolts and do not require a zero adjustment. The function switch should always be switched to STANDBY before removing electrodes from solution.

General Instructions for Potentiometric Titrations

Most potentiometric titrations are done in the following manner, with a few modifications for special cases. They can be performed manually or with automatic titration equipment. In either case, the cell potential (or pH) is plotted vs. the volume of titrant.

The electrodes are placed in the sample solution, and the solution is stirred continuously with a magnetic stirrer. The buret is positioned so that reagent is delivered without splashing. The potential or pH reading is recorded before any titrant is added, and then recorded after each addition of reagent. The first few additions of titrant are usually fairly large (1 to 5 mL). The potential should remain constant for 30 s (within 1 or 2 mV or 0.05 pH unit) before taking each reading. Occasionally the magnetic stirrer may cause erratic readings and it might be advisable to turn it off to make a measurement.

The volume of reagent added at each time should be judged by the change in reading after each increment. The size of the additions should be greatly decreased with rapid changes in potential or pH. In the neighborhood of the equivalence point the additions should be reduced to 0.1-mL increments (for convenient location of the end point by the derivative method). The titration is continued beyond the equivalence point by 2 to 3 mL using small increments initially, with increasing volumes being added when there is little additional change in potential or pH. pH titrations are usually carried to a pH of 12 for ordinary neutralization titrations with base. Ordinary potentiometric titrations are carried out until the potential reading levels off and the potentiometric titration curve is complete. The end point is usually taken as the point of maximum (steepest) slope, and can be determined by any of several methods. The most straightforward method is by estimation of the midpoint of the steep portion of the curve when the potential (or pH) is plotted vs. the volume of titrant added. This is not a very accurate method and can be virtually impossible for curves with slopes that are not very steep or symmetrical. The following discussion explains some of the most common procedures for determination of the end point.

Methods of End-Point Determination in Potentiometric Titrations

The addition of standard solution in an amount that is chemically equivalent to the substance in the sample with which it reacts occurs at the *equivalence point* in the titration. The location of the equivalence point can be estimated by observing a physical change that corresponds with the equivalence point and which is referred to as the *end point* of the titration. It is desirable that the observed end point occur at the same volume of titrant as does the actual equivalence point of the titration.

There are many ways to determine the end point of a potentiometric titration. Some commonly used methods are described below.

Method of Bisection

This method can be used for fairly symmetrical curves of p-functions (such as pH) or potential vs. the volume of titrant. There must be a steep slope at the end point, as well as nearly straight lines before and after the end point. This technique is illustrated in Fig. 2-5.

Construction Method

The construction method is very similar to the method of bisection and requires the same type of curve. The method is shown in Fig. 2-6.

Tangential Method Using Parallel Tangents

This method can be used for curves that do not have approximately straight lines before and after the end point and involves the use of a plastic sheet with paired-by-color lines parallel to each other with a center parallel slot wide enough for a pencil point to go through. The plastic template and the method are illustrated in Fig. 2-7.

Circle-Fit Method

A clear, rigid, plastic sheet with various-sized circles, each with a hole through its center, is used for this method. The circles with the same curvature as the curved sections just before and just after the end point are matched. The centers are

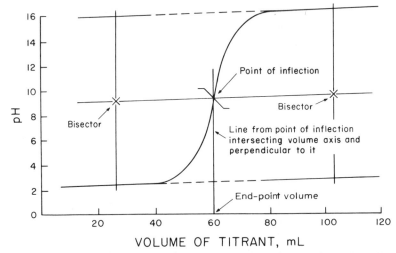

Figure 2-5 / Method of bisection. (*From "Analytical Chemistry," J. G. Dick, copyright © 1973 by McGraw-Hill, New York, fig. 5.15. Reprinted by permission.*)

marked on the graph when lined up; a straight line is drawn between the centers. This line intersects the end point, as shown in Fig. 2-8.

The Derivative Method   A plot of the change in potential with a change in volume $\Delta E/\Delta V$, or change in pH with a change in volume $\Delta pH/\Delta V$ vs. the average volume added produces a first-derivative curve. The slope of the curve increases to a maximum, which corresponds to the maximum slope of the titration curve, then decreases. The

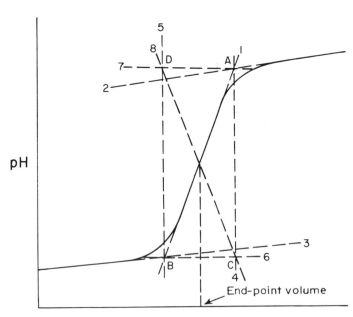

1. Draw a line through the straightest portion of the pH increase.
2. Extend a line (2) from the final portion of the curve to intersect line 1, creating point A.
3. Same as step 2, but creating point B, by using the initial portion of the curve.
4. Through A, draw a line parallel to the pH axis.
5. Through B, draw a line parallel to the pH axis.
6. Through B draw a line parallel to the volume axis creating point C.
7. Through A draw a line parallel to the volume axis creating point D.
8. Connect points D and C with a straight line. The intersection with line 1 is the center of a rectangle which you have constructed around the end point.

Figure 2-6 / Construction method. (*From "Problems and Experiments in Instrumental Analysis," C. E. Meloan and R. W. Kiser, copyright © 1963 by Charles E. Merrill Books, Inc., Columbus, Ohio, p. 128. Reprinted by permission.*)

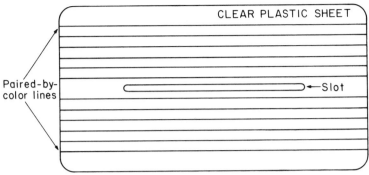

Paired-by-color parallel lines for tangential method.

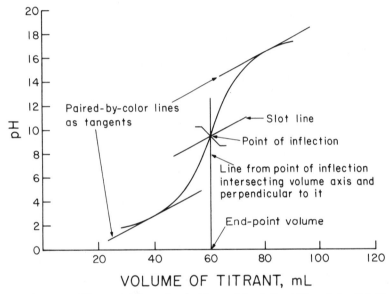

Figure 2-7 / Tangential method. (*From "Analytical Chemistry," J. G. Dick, copyright © 1973 by McGraw-Hill, New York, figs. 5.16 and 5.17. Reprinted by permission.*)

maximum of the curve is the end point of the titration as shown in Fig. 2-9. Sample calculations for this method are shown in Table 2-2.

Subtracting the corresponding data for $\Delta E/\Delta V$ for the first derivative and plotting $\Delta^2 E/\Delta V^2$ vs. the average volume will give the second-derivative curve, the zero point of which gives the end-point volume as shown in Fig. 2-9. Sample calculations are shown in Table 2-2.

Gran's Plot Method

Although this method has been used primarily with ion-selective electrode titrations, it is equally applicable to other potentiometric titrations. The potential vs. volume of titrant added usually results in an *S*-curve on ordinary graph paper. On Gran's Plot† paper, a straight line is obtained which when extrapolated to the horizontal axis, gives the end-point volume. This type of graph paper is a semi-antilog paper which provides for conversion from electrode potential to concentration; a volume correction is made by the skewed horizontal lines. Only a few points are

†Made by Orion Research, Incorporated.

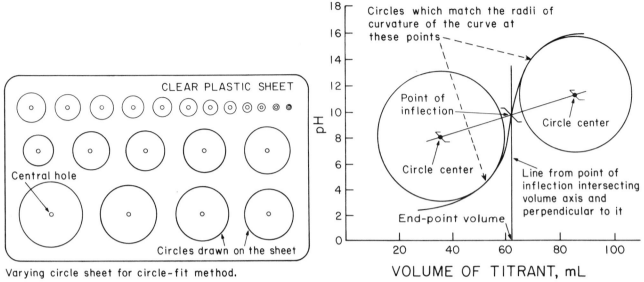

Varying circle sheet for circle-fit method.

Figure 2-8 / Circle-fit method. *(From "Analytical Chemistry," J. G. Dick, copyright © 1973 by McGraw-Hill, New York, figs. 5.18 and 5.19. Reprinted by permission.)*

needed before and after the end point. Usually the line after the end point is extrapolated to obtain the end-point volume.

Titration curves on ordinary graph paper and on Gran's Plot paper are shown in Fig. 2-10. (The paper is available in three forms: no volume correction, 10% volume corrected, and 100% volume corrected.)

Table 2-2 / Sample Potentiometric Titration Data: First- and Second-Derivative† Methods

Volume of Titrant, mL	E, mV	ΔV, first	ΔE	$\dfrac{\Delta E}{\Delta V}$	Average mL, first	$\dfrac{\Delta(\Delta E)}{\Delta V}$	ΔV, second	$\dfrac{\Delta^2 E}{\Delta V^2}$	Average mL, second
3.70	−197								
		0.10	8	80	3.75				
3.80	−189					50	0.10	500	3.80
		0.10	13	130	3.85				
3.90	−176					80	0.10	800	3.90
		0.10	21	210	3.95				
4.00	−155					1440	0.10	14,400	4.00
		0.10	165	1650	4.05				
4.10	+10					1350	0.10	−13,500	4.10
		0.10	30	300	4.15				
4.20	+40					100	0.10	−1000	4.20
		0.10	20	200	4.25				
4.30	+60					100	0.10	−1000	4.30
		0.10	10	100	4.35				
4.40	+70								

$$
† \quad V_{\text{end point}} = V_1 + (V_2 - V_1)\left[\frac{\Delta^2 E_1/\Delta V_1{}^2}{(\Delta^2 E_1/\Delta V_1{}^2) + (|\Delta^2 E_2/\Delta V_2{}^2|)}\right]
$$

$$
= 4.00 + 0.10 \left(\frac{14,400}{14,400 + 13,500}\right)
$$

$$
= 4.05 \text{ mL}
$$

where V_1, V_2 = volume before and after end point, respectively
E_1, E_2 = potential before and after end point, respectively

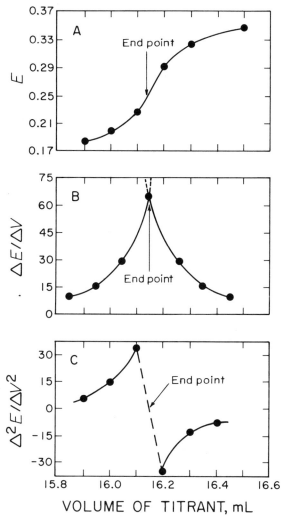

Figure 2-9 / Potentiometric titration curves. Curve A: normal titration curve, showing the region near the end point. Curve B: first-derivative titration curve. Curve C: second-derivative titration curve.

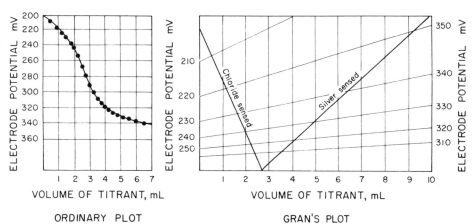

ORDINARY PLOT   GRAN'S PLOT

Figure 2-10 / Comparison of ordinary potentiometric titration curve and Gran's Plot curve.

EXPERIMENT 2-1

pH Titration of H_3PO_4 Mixtures; Calculation of K_1, K_2, and K_3

Purpose

In this experiment the titration of pure H_3PO_4 and H_3PO_4 with HCl or NaH_2PO_4 is followed by measuring the pH of the solution after each addition of NaOH titrant. From this data, K_1, K_2, and K_3 of H_3PO_4 may be calculated. In addition the amount of HCl, H_3PO_4, and NaH_2PO_4 present in the sample may be determined.

References

1 R. G. Bates, "Determination of pH," Wiley, New York, 1973.
2 J. J. Lingane, "Electroanalytical Chemistry," 2nd ed., Interscience, New York, 1958.
3 D. T. Sawyer and J. L. Roberts, Jr., "Electrochemistry for Chemists," Wiley-Interscience, New York, 1974.
4 D. A. Skoog and D. M. West, "Fundamentals of Analytical Chemistry," 4th ed., Saunders, Philadelphia, 1982, chaps. 8, 9, and 16.

Apparatus

pH meter
pH electrode (glass) and reference electrode (SCE or Ag/AgCl)
    or combination pH electrode
Magnetic stirrer
Beakers (2), 300 mL, tall form
Graduated cylinder, 100 mL
Volumetric flasks (2), 100 mL
Buret, 50 mL
Pipet, 25 mL

Chemicals

0.100 M sodium hydroxide [NaOH], standard
Sample 1†: Approximately 0.1 M phosphoric acid [H_3PO_4] diluted to mark in a 100-mL volumetric flask
Sample 2: Approximately 0.1 M H_3PO_4, plus hydrochloric acid [HCl] or sodium phosphate, monobasic [NaH_2PO_4] diluted to mark in a 100-mL volumetric flask
Buffer, for standardization of pH meter: pH 4.0, commercial, or pH 3.57, saturated potassium acid tartrate [$KHC_4H_4O_6$], approximately 0.6 g per 100 mL

Theory

During the course of the titration, addition of OH^- will not significantly increase the pH of the solution until most of the HCl has been neutralized and the H_3PO_4 has been changed into $H_2PO_4^-$.

$$H_3PO_4 + OH^- \rightarrow H_2PO_4^- + H_2O \qquad (1)$$

Further addition of OH^- will increase the pH of the solution, yielding the first break in the titration curve at the equivalence point.

Additional OH^- will then react with the second hydrogen ion, converting $H_2PO_4^-$ into HPO_4^{2-}.

$$H_2PO_4^- + OH^- \rightarrow HPO_4^{2-} + H_2O \qquad (2)$$

---

†Sample 1 is optional; it is included to enable comparison of pure H_3PO_4 with a mixture of H_3PO_4 and another acid.

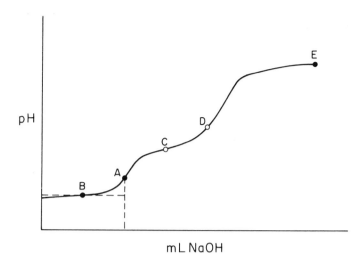

Figure 2-11 / Potentiometric titration curve for titration of H$_3$PO$_4$ with NaOH.

From the pH value at B calculate the H$^+$ concentration actually present in the solution. The difference in H$^+$ concentrations (H$^+_{added}$ − H$^+_{pH}$) is a measure of the H$^+$ used in converting H$_2$PO$_4^-$ to H$_3$PO$_4$.

$$H^+ + H_2PO_4^- \rightarrow H_3PO_4$$

The concentration of H$^+$ used is a measure of the concentration of H$_3$PO$_4$ formed. From the pH, the value of the H$^+$ concentration in the solution can be determined.

The actual concentration of H$_2$PO$_4^-$ in solution at point B is the concentration at point A less the concentration of H$_3$PO$_4$ which is present at point B. Dilution corrections must be made for the H$_2$PO$_4^-$ concentration observed at point A when it is used at point B. With [H$^+$], [H$_3$PO$_4$], and [H$_2$PO$_4^-$] known, K_1 may be determined.

K_2 can be calculated as follows:

$$H_2PO_4^- \rightarrow HPO_4^{2-} + H^+$$
$$K_2 = \frac{[H^+][HPO_4^{2-}]}{[H_2PO_4^-]}$$

Point C in Fig. 2-11 represents a mark halfway to the second end point. Hence,

$$[HPO_4^{2-}] = [H_2PO_4^-]$$
$$K_2 = [H^+]$$

K_2 may be determined from the pH value.
The following describes the calculation of K_3:

$$HPO_4^{2-} \rightarrow PO_4^{3-} + H^+$$
$$K_3 = \frac{[PO_4^{3-}][H^+]}{[HPO_4^{2-}]}$$

Because of the small value of K_3, an indirect method must also be used in calculating its value.

Calculate the amount of NaOH added from the end point D to any selected point E on the top, level part of the curve. Then calculate the concentration of OH^- added.

$$OH^-_{added} = \frac{mL\ NaOH(E - D) \times M}{vol.\ at\ E}$$

The actual measure of the OH^- concentration can be determined from the pH of the solution at E. The difference in the concentrations $(OH^-_{added} - OH^-_{pH})$ is a measure of OH^- used to convert HPO_4^{2-} to PO_4^{3-}, thus their concentrations may be determined.

The OH^- concentration of the solution comes from the pH value.

With the concentrations of H^+, HPO_4^{2-}, and PO_4^{3-} known, K_3 may be determined.

Questions

1 Comment on the statement that a buffer is a *mixture* of a conjugate acid and base.

2 Why is a saturated solution of potassium acid tartrate ($KHC_4H_4O_6$) acceptable as a pH standard? Is it a buffer?

3 How accurate is K_1? K_2? K_3? Why?

4 In the calculation of the K value for a 0.01 M monobasic acid, in what pH *range* may ionization (of HA) and hydrolysis (of A^-) be neglected?

5 Sketch your electrode system and label each part and each chemical. Describe the function of each.

6 What is the difference between pH and pa_H? Which quantity is measured by a pH meter?

EXPERIMENT 2-2

Potentiometric Titration of a Chloride-Iodide Mixture; Calculation of K_{sp} for AgCl and AgI

Purpose

The purpose of this experiment is to determine the amount of chloride and iodide simultaneously in a mixture and to calculate the K_{sp} of AgI and AgCl.

References

1 D. A. Skoog and D. M West, "Fundamentals of Analytical Chemistry," 4th ed., Saunders, Philadelphia, 1982, chaps. 4, 7, and 16.

2 D. A. Skoog and D. M. West, "Principles of Instrumental Analysis," 2nd ed., Saunders, Philadelphia, 1980, chap. 19.

3 H. H. Willard, L. L. Merritt, Jr., J. A. Dean, and F. A. Settle, Jr., "Instrumental Methods of Analysis," 6th ed., Van Nostrand, Princeton, N.J., 1981, chap. 23.

4 H. H. Bauer, G. D. Christian, and J. E. O'Reilly, "Instrumental Analysis," Allyn and Bacon, Boston 1978, chap. 2.

Apparatus (see Fig. 2-12)

Pipet, 20 mL
Beaker, 100 mL

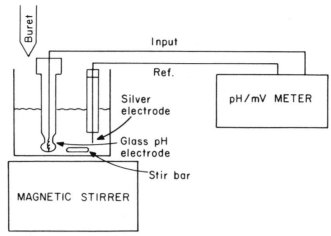

Figure 2-12 / Experimental apparatus for potentiometric titration of Cl⁻ and I⁻.

Magnetic stirrer with stirring bar
Buret, 10 mL
Graduated cylinder, 100 mL
Potentiometer (commercial pH/mV meter or other potentiometer)
Indicator electrode, silver (A piece of silver wire, a platinum electrode electroplated with silver by immersing in a potassium–silver cyanide plating bath for about 30 s, or a commercial silver billet electrode can be used.)
Reference electrode, a glass pH electrode (pH meter must be used and operated on "mV" setting); (Alternatively a Hg/HgSO₄ electrode with a sulfate salt bridge or an SCE or Ag/AgCl electrode with a sulfate salt bridge can be used.)

Chemicals

0.1 *M* silver nitrate [AgNO₃], standardized
Barium nitrate [Ba(NO₃)₂], solid
6 *M* nitric acid [HNO₃]
Unknown: I⁻ and Cl⁻ solution in 100-mL volumetric flask (dilute to mark). Approximately 0.02 *M* in each ion after dilution.

Theory

A metal (M) that is immersed in a solution containing ions of that metal (M^{n+}) can be used as an indicator electrode to monitor the solution concentration of M^{n+}. The electrode potential responds in a nernstian fashion which for the electrode reaction M^{n+} + ne → M (at 25°C) is

$$E = E^* + \frac{0.0591}{n} \log[\text{M}^{n+}]$$

Such an electrode can be used to monitor the precipitation titration of an anion (X⁻) with M^{n+}

$$\text{M}^{n+} + n\text{X}^- \rightarrow \text{MX}_n$$

If their K_{sp}'s are sufficiently different, concentrations of different anions in a mixture can be determined from a potentiometric titration curve.

In such precipitation titrations there is a sudden change in the ion concentration (Ag^+ in this experiment) at each equivalence point because the concentration of X^- determines, through the solubility-product equations, the concentration of Ag^+. To detect the equivalence point an electrode that responds to a change in the Ag^+ concentration during the titration can be used; its potential is measured after each successive addition of the titrating agent. In the precipitation of Cl^- and I^- by Ag^+, with the concentrations used in the experiment, AgI begins to precipitate first, and E changes very slowly until nearly all the I^- in the sample is precipitated. At the first equivalence point the $[Ag^+]$ increases more rapidly until the K_{sp} of AgCl is reached; here again $[Ag^+]$ and E remain fairly constant until all the Cl^- is precipitated, then another sharp increase in E occurs, indicating the second equivalence point.

Usually potentiometric titrations are carried out using SCE or Ag/AgCl reference electrodes. Since these electrodes leak a small amount of chloride, they are unsuitable for a titration in which chloride is the unknown unless a non-halide-containing salt bridge is used to connect the reference electrode with the sample solution. A glass electrode can be used as the reference electrode in this experiment. The potential of the electrode when immersed in a pH buffer is quite constant; the leakage problem associated with the SCE is thus eliminated.

Procedure

Read the operating instructions for the pH meter before turning it on; allow 5 to 10 min for it to warm up. Rinse both electrodes with distilled water. Pipet a 20-mL aliquot of the Cl^-, I^- solution into a 100-mL beaker. Dilute to 75 mL with distilled water. Add a few drops of 6 M HNO_3 and approximately $\frac{1}{2}$ g solid $Ba(NO_3)_2$. Rinse the stirring magnet with distilled water and place it in the solution. Immerse the electrodes in the solution. Zero the potentiometer (noting that the Ag electrode will be negative at the beginning of the titration). Start the stirrer and read the initial potential. Proceed with the titration, using large ($\frac{1}{2}$-mL) increments up to near the first end point, then take smaller (0.1-mL or less) increments as required until the E values become nearly constant past the first end point. Then take $\frac{1}{2}$-mL increments, until near the second end point, and again use 0.1-mL increments. Continue the titration past the end point by a few milliliters. If the meter needle goes to zero before the titration is complete, either rezero the needle at the other end of the scale or reverse the electrode leads and proceed as before, now reading E as positive.

Treatment of Data

Plot E (vertical axis) vs. volume (mL) of $AgNO_3$. Determine the two end points by one or more of the methods for end-point determination in potentiometric titrations (text of Chap. 2).

Determine the number of grams of NaCl and NaI in the solution. Calculate the K_{sp} of AgCl and AgI, using $E = E^* + 0.0591 \log [Ag^+]$. *Note:* The data which have been taken are sufficient for this calculation and do not require that you know the half-cell potential of the reference electrode.

Questions

1 What is the purpose of the $Ba(NO_3)_2$?
2 What percent of AgI remains unprecipitated when the first amount of AgCl starts to precipitate?
3 What effect would the presence of 1 M ammonia have on the above titration? Be specific in your answer. Indicate how you would experimentally evaluate the equilibrium constant for the silver-amine complex.

4 Explain why you do not need to know the half-cell potential for the reference electrode in order to calculate the K_{sp} for AgCl and AgI.
5 Compare the methods that you used to determine the end point.

SUPPLEMENTARY EXPERIMENTS

I. Titration of Chloride and Iodide with Permanganate†

Purpose

The iodide in a chloride-iodide mixture may be determined selectively by an oxidation titration with permanganate ion.

Apparatus (See Fig. 2-12)

Potentiometer (pH meter) with platinum indicator electrode and SCE or Ag/AgCl reference electrode
Buret, 10 mL
Pipet, 20 mL
Beaker, 100 mL
Graduated cylinder, 100 mL
Magnetic stirrer with stirring bar

Chemicals

0.02 M potassium permanganate [$KMnO_4$]
Potassium iodide [KI], solid, reagent grade
3 M sulfuric acid [H_2SO_4]
Unknown: I⁻ and Cl⁻ solution in 100-mL volumetric flask (dilute to mark); approximately 0.02 M in each ion after dilution. (The solution from the first part of the experiment may be use here.)

Procedure

Turn on the pH meter and zero it. The platinum electrode will be at almost the same potential as the SCE initially but will become positive relative to the reference electrode after the titration has been started. Standardize the permanganate solution with KI using appropriate sized samples for 0.02 M $KMnO_4$ in the following manner. Place the KI in the beaker and add 50 mL of water and 5 mL of 3 M H_2SO_4. Titrate potentiometrically, noting the initial potential before adding any $KMnO_4$. During the level portion of the titration curve, ½-mL increments can be added; as the end point is approached the increments should be reduced to 0.1 mL. Continue the titration 1 or 2 mL beyond the end point, then perform a duplicate standardization.

Next, pipet 20 mL of the unknown solution, add 50 mL of water plus 5 mL of 3 M H_2SO_4, and titrate to the end point. Perform a duplicate titration.

Treatment of Data

Plot the titration curves for the standard KI sample and for your unknown.

Report the molarity of the $KMnO_4$ solution and the grams of NaI in the unknown.

Calculate the equilibrium constant for the titration reaction, using (1) potential data from the literature and (2) the data which you obtained in the laboratory for your titration curve.

Questions

1 Which value for the equilibrium constant do you believe more closely approaches the true value for your solution conditions?
2 Why does the chloride ion not interfere with this titration?

†From a similar experiment at Boston College, courtesy of Dr. T. S. Light.

II. Determination of Chloride in Mouthwash†

Purpose

The concentration of chloride in mouthwash is to be determined by potentiometric titration with Ag^+.

Apparatus (See Fig. 2-12)

Beaker, 250 mL
Pipet, 25 mL
Buret, 25 mL
pH/mV meter (potentiometer)
Silver indicator electrode
Glass pH electrode (used as the reference electrode)
Magnetic stirrer and stirring bar

Chemicals

Acetate buffer containing 0.5 M sodium acetate and 0.5 M acetic acid [HOAc], 1 L
0.0400 M sodium chloride [NaCl], primary standard, 250 mL
Silver chloride [AgCl] solution, saturated
0.1 M $AgNO_3$ solution
Lavoris mouthwash

Procedure

Place about 75 mL of acetate buffer into the cell (a 250-mL beaker). Add about 1 mL of saturated AgCl solution. Lower the electrodes into the cell. Turn on the stirring motor. Turn the meter function switch to pH and observe the meter reading. If the reading is not near center scale or not an easily remembered value, adjust the calibration control until the meter reads a convenient value (e.g., 7.00). This should be done each time the cell is refilled.

Pipet 25.00 mL of NaCl standard solution into the cell. Add $AgNO_3$ titrant to the cell using a 25-mL buret while observing the meter. Continue titrating until the meter reads the same value as observed for the saturated AgCl solution. This may be difficult to do as the potential changes rapidly with small additions of titrant near the end point, but minimal error will result if the titration is stopped within ±0.3 of the saturated AgCl value (e.g., 6.7 to 7.3).

Discard the contents of the cell, rinse the electrodes with water, rinse the cell with water, place about 75 mL of buffer into the cell, add about 1 mL of saturated AgCl solution, adjust the calibration control and repeat the titration on second and third aliquots of NaCl solution.

Prepare the cell as for the standards but pipet 25.00 mL of Lavoris mouthwash into the cell instead of standard NaCl. Titrate three such aliquots of Lavoris.

Obtain a titration curve of the chloride in Lavoris by the following procedure. Prepare the cell in the usual manner including the addition of 25.00 mL of Lavoris. Write down the reading of the meter. Add 1.0 mL of $AgNO_3$ titrant and reread the meter. Continue adding 1-mL increments of titrant until the total volume added is within 1 mL of the expected end point. Read the buret carefully at this point. Add $AgNO_3$ *2 drops at a time* and measure the potential after *each 2-drop* addition. Continue until the total volume added is 1 mL past the expected end point. Read the buret carefully at this point. From the total volume added drop-wise and the number of drops added, the volume per drop can be computed. Convert the drops added to volume added—this technique is easier and more accurate than reading the buret after each 2-drop addition. Resume 1-mL additions of $AgNO_3$ followed by potential measurements until the total volume added is 5 mL past the expected end point.

†From a similar experiment at North Carolina State University, courtesy of K. W. Hanck.

note The pH scale is used in this experiment because it is more sensitive than the mV scale—a full scale voltage of 14×100 mV for the mV scale compared to 14×59 mV on the pH scale (temperature control at 25°C). Setting the temperature control to its lowest value decreases the full scale voltage even further. The numbers plotted as potential in the graph are proportional to the real potential—if one wanted to convert them to voltage it could be done. Since we are interested mainly in the shape of the curve this will not be done. It should be emphasized that pH *is not being measured in this experiment even though the pH scale is used!*

Treatment of Data Compute the molarity of the $AgNO_3$ titrant based on the average volume needed to titrate a 25.00-mL aliquot of standard NaCl solution.

Compute the average molarity of chloride in Lavoris.

Plot the measured potential (*y* axis) vs. the volume of $AgNO_3$ added (*x* axis). Locate the potential of the saturated AgCl solution on the graph. Does the point correspond to the steepest point on the titration curve? If not, estimate the difference in volume between the two points.

Questions 1 Does Lavoris contain any substance(s) that might affect the accuracy of this Cl^- determination?
2 Could Cl^- in Lavoris be determined by direct potentiometry with a chloride ion-selective electrode?

EXPERIMENT 2-3
Effect of Complex Formation on the Titration of Fe^{2+}

Purpose A primary objective in this experiment is to show the effect of a complexing agent on the potentiometric oxidation-reduction titration of Fe^{2+} with Ce^{4+}. The use of an inert electrode for this type of titration is illustrated in the determination of the amount of iron in a sample. The derivative method of end-point detection is used for the calculations. Various methods of plotting experimental titration data for the determination of end points are illustrated.

References 1 J. J. Lingane, "Electroanalytical Chemistry," 2nd ed., Interscience, New York, 1958, chap. 7.
2 D. A. Skoog and D. M. West, "Fundamentals of Analytical Chemistry," 4th ed., Saunders, Philadelphia, 1982, chap. 14
3 L. Meites and H. C. Thomas, "Advanced Analytical Chemistry," McGraw-Hill, New York, 1958, pp. 56–64.
4 E. H. Swift, "Introductory Quantitative Analysis," Prentice-Hall, Englewood Cliffs, N.J., 1950, pp. 506–510. An excellent table of formal potentials.
5 W. M. Latimer, "Oxidation Potentials," 2nd ed., Prentice-Hall, Englewood Cliffs, N.J., 1952.
6 G. F. Smith, "Cerate Oxidimetry," The G. Frederick Smith Chemical Company, Columbus, Ohio, 1942.
7 F. R. Duke in "Treatise on Analytical Chemistry," I. M. Kolthoff, P. J. Elving, and E. B. Sandell (eds.), part I, vol. 1, Interscience, New York, 1959, p. 661.

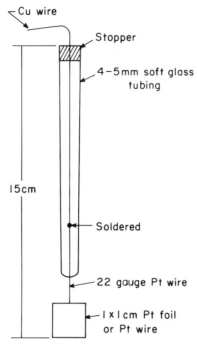

Figure 2-13 / Construction of Pt electrode.

Apparatus

Buret, 50 mL
Beaker, 250 mL
Pipet, 25 mL
Magnetic stirring bar and stirring motor
pH meter or potentiometer
Platinum indicator electrode [This can be prepared by sealing platinum wire into soft glass (see Fig. 2-13).]
SCE or Ag/AgCl reference electrode

Chemicals

Mercury(II) chloride [$HgCl_2$] solution (5%), 50 g $HgCl_2$ per liter.
Tin(II) chloride [$SnCl_2$] solution, 90 g $SnCl_2$ to 200-mL concentrated HCl, then dilute with 400 mL water.

note Keep in contact with metallic tin to avoid disproportionation.

Iron(II) solution, approximately 0.1 M and in acid solution (Sample is issued in 100-mL volumetric flask. Dilute to mark when received.)
0.1 M cerium(IV) solution, 1 M in H_2SO_4, standardized against primary standard iron wire or ferrous ammonium sulfate [Ceric salts and compounds such as $Ce(HSO_4)_4$, $Ce(OH)_4$, $Ce(SO_4)_2(NH_4)_2SO_4 \cdot 2H_2O$, and $Ce(NH_4)_2(NO_3)_6$ are commonly used materials.]

Theory

The ferrous-ferric system serves as a good example of the effect of complex formation on the shape of a potentiometric-titration curve. For this system the Nernst equation takes the form

$$E = E^\circ + \frac{0.0591}{1} \log \frac{a_{Fe^{3+}}}{a_{Fe^{2+}}} \tag{1}$$

The electrode reaction is expressed by the half-reaction

$$Fe^{3+} + e \rightarrow Fe^{2+} \qquad E° = 0.77 \tag{2}$$

Upon addition of H_3PO_4, Fe^{3+} forms a phosphate complex,

$$Fe^{3+} + H_3PO_4 \rightarrow Fe(HPO_4)^+ + 2H^+ \tag{3}$$

which causes the concentration of free Fe^{3+} to become very small. Because Fe^{2+} forms a much weaker complex with phosphoric acid, the decrease in $[Fe^{2+}]$ is much less than for Fe^{3+}. Thus, the effect of H_3PO_4 is to lower the $[Fe^{3+}]$ and to decrease the potential of the indicator electrode as shown by Eq. (1).

Will the addition of H_3PO_4 affect the titration curve, and if so, how?

Procedure

Clean the platinum electrode by heating it in concentrated nitric acid and follow by rinsing with distilled water.

Before any work has been initiated, turn the pH meter on to allow the instrument sufficient time to warm up to operational temperature. All apparatus must be cleaned and rinsed with distilled water.

Place a 25-mL aliquot of iron sample and 10 mL concentrated HCl in a 250-mL beaker and warm (do not boil) over a flame. Slowly, *drop-wise,* add a solution of tin(II) chloride to the beaker until the yellow coloration just disappears, then add 2 or 3 more drops.

$$2Fe^{3+} + Sn^{2+} \rightarrow 2Fe^{2+} + Sn^{4+} \tag{4}$$

The yellow color is due to the presence of iron(III) chloride complex ions. Dilute the solution to approximately 100 mL and add quickly and all at once 10 mL of mercury(II) chloride solution.

$$2Hg^{2+} + 2Cl^- + Sn^{2+} \rightarrow Sn^{4+} + Hg_2Cl_2 \tag{5}$$

This removes excess Sn^{2+} from solution. The solution should now have a milky white precipitate of mercury(I) chloride. If the solution is dark, indicating the presence of free mercury, too large an excess of Sn^{2+} was added. Discard the solution and start over. If no white precipitate appears, an excess of Sn^{2+} was not present, and all of the Fe^{3+} may not be reduced. Discard and start again. If the solution has been prepared properly it is now ready for use.

Immerse the electrodes in the solution, being sure that they are completely covered. Place the beaker on the stirring motor and drop the magnetic stirring bar into the beaker. The speed of the motor should be adjusted until the solution is in gentle agitation.

Zero the pH meter, adjusting the pointer to read zero on the scale (use the "mV" scale). If the pointer does not read on scale, reverse the leads or rezero on the other end of the scale. Some meters read millivolts directly and cannot be zeroed. Follow the manufacturer's instructions. Under no circumstances should the electrodes be removed from the solution without first switching to STANDBY.

Carry out this titration fairly rapidly, since the oxygen in air will slowly reoxidize Fe^{2+} to Fe^{3+}, causing low results to be obtained. Nitrogen or carbon dioxide may be passed over the solution to prevent air oxidation. Drop-wise addition of an $NaHCO_3$ solution from a separatory funnel is recommended as a convenient means of preventing air oxidation. Why?

Read the potential of the solution before adding any of the titrant. The solution should be well stirred at all times. The first few increments may be rather large, 3 to 4 mL. Potential and buret readings are taken after each addition. When the potential begins to change rather rapidly, the size of the increments should be decreased greatly. In the neighborhood of the end point, 0.2 mL on either side, the increments should be reduced to 0.1 mL. After the end point has been passed, the increments may be increased in size.

When repeating a titration for the purpose of checking a previous determination, the reagent may be added rapidly to within 0.5 mL of the end point.

Repeat the above titration with another 25-mL aliquot, but this time add 5 mL of concentrated H_2SO_4 and 15 mL of concentrated H_3PO_4 to the solution just before starting the titration.

Treatment of Data

Plot E vs. volume (mL) of Ce^{4+} for both of the titrations (in the presence and in the absence of H_3PO_4). Calculate $E^{\circ\prime}_{Fe^{3+}, Fe^{2+}}$ and $E^{\circ\prime}_{Ce^{4+}, Ce^{3+}}$ (formal reduction potentials) for both systems, and then calculate the end-point potential by averaging the two $E^{\circ\prime}$ values. Calculate the end-point potential by one of the graphical methods described in the text of the chapter. How well do the two potentials agree?

Plot $\Delta E/\Delta V$ vs. milliliters of Ce^{4+} added (see text of the chapter) and determine the end point. Then plot $\Delta^2 E/\Delta V^2$ against the average volume (mL) of Ce^{4+} and determine the end point.

Calculate the grams of iron present in your 100-mL volumetric flask using the two different methods of end-point detection. How do the different methods compare in terms of giving the same value for the unknown?

Questions

1 Show mathematically that the average of $E^{\circ\prime}_{Fe^{3+}, Fe^{2+}}$ and $E^{\circ\prime}_{Ce^{4+}, Ce^{3+}}$ gives the theoretical equivalence-point potential.

2 Derive a relationship between the potential at the equivalence point for the titration of Fe^{2+} with $Cr_2O_7^{2-}$ and $E^{\circ\prime}_{Cr_2O_7^{2-}, Cr^{3+}}$ and $E^{\circ\prime}_{Fe^{3+}, Fe^{2+}}$.

3 What would be the shape of the titration curve if you neglected to destroy the excess Sn^{2+} ion?

4 Could a standard voltmeter be used in place of the pH meter? Justify your answer.

5 What are the advantages of a derivative plot of the titration curve?

6 Calculate the equivalence-point potentials for each break in the titration of a 25-mL solution containing $0.02\ M\ Cr_2O_7^{2-}$, $0.1\ M\ Cu^{2+}$, and $6\ M\ HCl$ with $0.1\ M\ Cr^{2+}$. For data, see the references listed at the beginning of the experiment.

7 Calculate the equivalence-point potentials for each break in the titration by MnO_4^- of a mixture of V^{3+} and Fe^{2+}. Assume that the solution contains $3\ M\ H_2SO_4$.

8 Of what practical use could the addition of H_3PO_4 to the solution of Fe^{2+} as done in this experiment be?

EXPERIMENT 2-4

Acid-Base Titrations in Nonaqueous Media: Titrations of Nicotine

Purpose

One aim of this experiment is to demonstrate how the solvent medium affects the extent of an acid-base reaction. Nicotine is titrated first with water as the solvent and then with glacial acetic acid as the solvent. A comparison of the two titration curves is then made. A second aim is to extend this type of titration to the determination of nicotine in tobacco.

References

1 J. S. Fritz, "Acid-Base Titrations in Non-Aqueous Solvents," Allyn and Bacon, Boston, 1973.
2 W. Huber, "Titrations in Non-Aqueous Solvents," Academic Press, New York, 1967.
3 J. Kucharsky and L. Safarik, "Titrations in Non-Aqueous Solvents," Elsevier, Amsterdam, 1965.
4 R. H. Cundiff and P. C. Markunas, *Anal. Chem., 27,* 1650 (1955) and *J. Assoc. Offic. Agr. Chemists, 43,* 519 (1960) (Titrations of nicotine and nornicotine).
5 "Titrants, Indicators, Solvents, for Non-Aqueous Titrimetry," Eastman (Distillation Products Industries) Organic Chemical Information booklet I64-1068, Rochester, N.Y.
6 D. A. Skoog and D. M. West, "Fundamentals of Analytical Chemistry," 4th ed., Saunders, Philadelphia, 1982, chap. 11.
7 I. M. Kolthoff and S. Bruckenstein in "Treatise on Analytical Chemistry," I. M. Kolthoff, P. J. Elving, and E. B. Sandell (eds.), part I, vol. 1, Interscience, New York, 1959, p. 475.
8 B. Kratochvil, Annual Reviews, *Anal. Chem., 48,* 355R (April 1976).
9 "Modern Chemical Technology," rev. ed., *A.C.S., 6,* 1072 (1972).

Part I. Titrations with Pure Nicotine: Effect of Aqueous and Nonaqueous Media

Apparatus

pH meter
Glass-pH electrode
Reference electrode (Ag/AgCl)
Magnetic stirrer and stirring bar
Pipet, 10 mL
Buret, 10 mL
Volumetric flask, 50 mL (containing unknown)
Beakers (2), 50 mL
Rubber squeeze bulb or tubing connected to a water aspirator for pipeting

Chemicals

Aqueous Solutions

0.1 *M* nicotine, standard, aqueous, 16.24 g/L (to make unknowns)

safety caution   Nicotine is a toxic substance that absorbs through the skin.

Unknowns: 0.02 to 0.04 *M*, aqueous
0.05 *M* perchloric acid [$HClO_4$], aqueous, standardized with sodium carbonate [Na_2CO_3]

Nonaqueous Solutions

0.1 *M* nicotine, standard, in glacial acetic acid to make unknowns
Glacial acetic acid

0.05 *M* perchloric acid, in 98% glacial acetic acid/2% acetic anhydride standardized with primary standard potassium acid phthalate in glacial acetic acid

safety caution This solution is potentially explosive on long standing if impurities are present.

Methyl violet indicator, 200 mg per 100 mL chlorobenzene
Unknowns: 0.02 to 0.04 *M* in 98% glacial acetic acid/2% acetic anhydride

Theory

A detailed understanding of the effect of solvent on acid-base titrations may be obtained from the references listed at the beginning of the experiment.

A weak base in an aqueous solution gives a poor end point when titrated with a strong acid since water also has some basic characteristics. The water molecules compete with the base molecules for protons and cause a drawn-out end point as illustrated for pyridine in Fig. 2-14A. To correct this condition, a solvent that is a weaker base than water is needed. Likewise, for the titration of weak acids, the desired solvent should be less acidic than water.

Since the strongest acid that can exist in water is the hydronium ion, H_3O^+, strong acids that can completely ionize in water, such as HCl, $HClO_4$ and HNO_3, form hydronium ions with their protons and are "leveled" to the same strength in water and cannot be differentiated. The strongest acid that can exist in glacial acetic acid is the H_2OAc^+ ion. Strong acids such as HCl, $HClO_4$, and HNO_3 do not ionize completely in glacial acetic acid, and their relative strengths can be differentiated. Weak bases which ionize only slightly in water will ionize completely in acetic acid since they are leveled to the strongest base that can exist in acetic acid, that is, the strength of the acetate ion. In other words, these weak bases, such as amines, behave as strong bases in acetic acid even though they behave as weak bases in water. Thus, substances such as pyridine, with a dissociation constant of only about 10^{-9} in water, and nicotine can be titrated in glacial acetic acid since they behave as an appreciably stronger base due to an enhanced tendency to react with the solvent. For example, the equilibrium constant for the reaction of pyridine (C_6H_5N) in glacial acetic acid is considerably larger than for that in water.

$$C_6H_5N + H_2O \rightarrow C_6H_5NH^+ + OH^- \qquad (K_b = 1.5 \times 10^{-9})$$
$$C_6H_5H + HOAc \rightarrow C_6H_5NH^+ + OAc^- \qquad (K_b = 8 \times 10^{-7})$$

Although many solvents have been suggested for the titration of weak bases, glacial acetic acid has been extensively studied because of its desirable qualities of

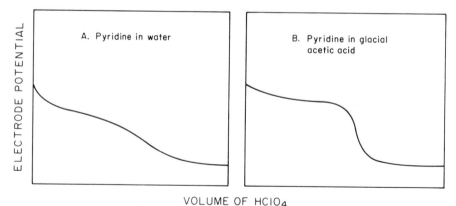

Figure 2-14 / Comparison titration of pyridine in water and glacial acetic acid. (*From "Modern Chemical Technology," rev. ed., copyright © 1972 by The American Chemical Society.*)

stability, availability in pure form, ease of dehydration with acetic anhydride, and ability to dissolve most bases. Acetic anhydride is usually added to the glacial acetic acid to make it anhydrous by reacting with any water present. Since the co-efficient of expansion of acetic acid is greater than that of water (coefficient of expansion of glacial acetic acid = 0.11% per degree Celsius), care must be taken to avoid temperature changes during titrations. It is possible to correct for temperature changes by recording the temperature of the reagent at the time of standardization and at the time of analysis by using the equation:

$$vol_{std} = vol \left[1 + 0.0011 \left(T_{std} - T \right) \right]$$

where
T = temperature when analyzed
T_{std} = temperature of reagent when standardized
vol = measured volume of reagent
vol_{std} = corrected volume

Potentiometric detection of the end point in nonaqueous titrations has the advantage over visual indicators in that it is almost universal. Different visual indicators are required for each type of nonaqueous system, but the glass electrode is applicable to any system with a sufficiently large dielectric constant. In acetic acid, two acid-base indicators have been useful: methyl violet, which changes from violet to green and then to yellow, the end point occurring at the disappearance of the violet color; and crystal violet. However, in very weak acid titrations, the potentiometric end point must be used.

There is a problem with a glass electrode indicating system, however. The glass membrane must be hydrated to function properly. Because of the anhydrous nature of glacial acetic acid, the surface of the glass electrode is slowly dehydrated while in contact with this solvent. If the glass electrode is stored in water and the titration is completed in a reasonable length of time, no difficulties arise from this problem.

A pH meter is usually used as a potentiometer to measure the potential between the glass indicator electrode and a calomel or silver–silver chloride reference electrode during the progress of the titration. Although either pH or millivolt readings can be recorded, it should be remembered that a true pH reading cannot be made in a nonaqueous solution, although relative pH units can be made and subsequently graphed to determine an end point. It is usually preferable to read the millivolt scale.

Perchloric acid is usually used as a titrant, since it is a strong acid in many nonaqueous systems. Potassium hydrogen phthalate is used as a primary standard for perchloric acid in glacial acetic acid, since it behaves as a strong base in glacial acetic acid, converting from the acid salt to the undissociated acid.

Procedures

A. Titration of Aqueous Nicotine

Pipet a 10-mL aliquot of the unknown nicotine into a 50-mL beaker, using a rubber bulb or tubing connected to a water aspirator. Add enough water to cover the electrodes adequately. Stir the solution with a magnetic stirrer throughout the titration. Record the initial potential (in millivolts) and buret reading. Titrate with 0.05 M aqueous perchloric acid using small increments and recording the potential and volume of perchloric acid added. Near the end point, the potential will change more rapidly and the titrant should be added drop by drop. (If derivative curves are to be used to obtain the end point, add increments of 0.1 mL to simplify the calculations.) When repeating a titration for the purpose of checking a previous determination, increments can be added rapidly to within 1.0 mL of the end point of the titration.

B. Titration of
Nonaqueous Nicotine

safety caution Nonaqueous solvents are usually toxic and should not be inhaled. In addition, glacial acetic acid burns the skin. Wear plastic gloves and work in a hood. Keep solutions covered as much as possible with Parafilm or any other suitable product.

note *Care of Glass Electrode for Nonaqueous Experiments* The glass and reference electrodes should be immersed in distilled water at all times when not in use. *Gently* wipe the electrodes dry with lint-free tissue immediately before lowering the electrodes into the solution to be titrated. Titrate as quickly as possible. The electrodes should not remain in the glacial acetic acid for more than 20 min, or the water in the "gel" surface of the glass electrode tip will be leached out, causing the electrode to become nonfunctional. At the end of each nonaqueous titration, rinse the electrodes thoroughly with distilled water, and soak them in distilled water for at least 20 min before proceeding with the next nonaqueous titration.

Pipet 10 mL of nonaqueous unknown into a *dry* 50-mL beaker, using a rubber bulb or aspirator, and add sufficient glacial acetic acid to cover the electrodes. (A 50-mL Erlenmeyer flask may be used if a combination electrode is available.) Add a few drops of nonaqueous methyl violet indicator. Observe the color, and titrate with 0.05 M perchloric acid in glacial acetic acid, recording the potential readings and color changes during the course of the reaction. Repeat the titration only if necessary.

C. Comparison of
HClO$_4$ and HCl as
Titrants (Optional)

Repeat the aqueous and nonaqueous titrations above, using 0.05 M hydrochloric acid solutions in place of the perchloric acid.

Treatment of Data Plot on fine-ruled graph paper (10 × 10 to the cm) the recorded potential (mV) readings vs. the volume (mL) of perchloric acid, placing both the aqueous and nonaqueous plots on the same graph. Determine the end points and calculate the molarity of the base in each case. For the glacial acetic acid titration, indicate the colors of the indicator at the various potential readings observed during the titration. Designate the color that should be used for the end point in the titration of nicotine with perchloric acid.

If you performed the optional titration, plot the titration curves with hydrochloric acid as titrant in the same manner.

Questions

1 Explain the difference in titration curve shapes in the initial stages of the titration (see Ref. 7).
2 Explain the difference in the end points for the two titrations. Write equations for the reactions.
3 Suggest other solvents that would be satisfactory for the titration of nicotine.
4 Suggest several satisfactory solvents and titrants for the titration of phenol.
5 (Optional) Explain which of the two acids—hydrochloric or perchloric—is intrinsically the stronger acid.

Part II. Determination of Nicotine in Tobacco

Apparatus pH meter
glass pH electrode
Reference electrode (Ag/AgCl)

Magnetic stirrer and stirring bar
Buret, 10 mL
Erlenmeyer flask, 250 mL, glass-stoppered
Beaker, 150 mL, or Erlenmeyer flask, 125 mL
Funnel, long stem
Filter paper, Whatman #40, 15 cm (or Whatman #42, depending on sample)
Pipets, 100 mL and 25 mL
Rubber bulb, or tubing connected to a water aspirator

Chemicals

Tobacco, 2.5 to 3.0 g (cigarette, cigar, pipe or chewing tobacco)
Acetic acid (5%), vol/vol
NaOH (36%), wt/vol
Celite or Filter-Cel, 2 to 3 g
0.025 M HClO$_4$ in glacial acetic acid containing 20 mL acetic anhydride in each liter of solution, standardized

safety caution This solution is potentially explosive on long standing if impurities are present.

Methyl violet indicator (0.2%), in chlorobenzene
90% benzene/10% chloroform solution (anhydrous)
Glacial acetic acid

Theory

Tobacco contains a group of so-called "amine alkaloids" that are determined as "nicotine." Included in this group are nicotine, nornicotine, nicotyrine, nornicotyrine, anabasine, antabine, myosime, and N-methyl myosime. Structures of five of these are shown in Fig. 2-15.

Nicotine is an oily, pale yellow liquid, soluble in water and glacial acetic acid. It is extremely toxic; as little as 40 mg has been lethal to humans when taken by mouth. Black Leaf 40, a 40% solution of nicotine sulfate, is commonly used as an agricultural insecticide. It is determined potentiometrically when extracted from tobacco since nornicotine gives a false visual end point (up to 15% error).

Procedure

safety caution Use a hood, avoid getting benzene or chloroform on your skin or inhaling vapors of these solvents! Avoid skin contact with nicotine!

Weigh a 2.5- to 3.0-g sample of tobacco into a 250-mL glass-stoppered Erlenmeyer flask (or an unstoppered Erlenmeyer flask which can be covered with aluminum foil held fast with a rubber band). Wet the tobacco with 15 mL of 5% acetic acid. Pipet 100 mL of 90% benzene/10% chloroform solution into the

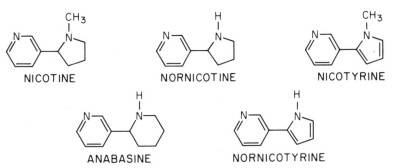

Figure 2-15 / Amine alkaloid constituents of nicotine.

flask (in the hood!), add 10 mL of 36% NaOH and cover the flask tightly. Stir *vigorously* with a magnetic stirrer for at least 20 min or shake for 20 min. Add 4 to 5 g of Celite or Filter-Cel and stir vigorously. (The Celite removes water and improves filterability.) Filter the mixture through Whatman #40 paper using a long-stem funnel. Collect the filtrate in a 250-mL beaker.

safety caution   Do not wash the flask with water or any liquid.

The solution should be nearly colorless; if not, add about 3 g of Celite and filter again. Pipet 25 mL of the filtered solution into a 150-mL beaker. Add 25 to 30 mL of glacial acetic acid and a magnetic stirring bar and position the pH meter electrodes in the solution. After filling a 10-mL buret with 0.025 M $HClO_4$ in glacial acetic acid, add several drops of methyl violet indicator to the solution to be titrated and proceed with the titration. Add 0.5 mL of titrant at a time up to a potential increase of 10 mV, then add 0.1-mL increments around the end point as in the previous experiments. Observe all color changes and continue the titration until very small potential changes (around 4 mV) result when 0.5 mL is added.

Treatment of Data   On fine-ruled graph paper (10 × 10 to the cm), plot the potential (mV) readings vs. the volume (mL) of perchloric acid titrant. Determine the end point and calculate the percent total alkaloids as nicotine. Indicate the colors of the indicator at the various potential readings as observed during the titration.

Questions

1   Did you observe more than one break in the titration curve? Why?
2   What effect would addition of water have on the titration curve? Explain.
3   Did the visual indicator change color at the same place the greatest potential change was recorded? If the answer is no, how much error would result if only a visual indicator were used?

EXPERIMENT 2-5
Ion-Selective Electrodes

Purpose   The determination of a metal ion by direct potentiometry with an ion-selective electrode is illustrated.

References

1   Orion Research, Inc., "Analytical Methods Guide" (1978) and "Newsletters."
2   R. E. Lamb, D. F. S. Natusch, J. E. O'Reilly, and N. Watkins, *J. Chem. Ed., 50,* 432 (1973).
3   B. W. Lloyd, F. L. O'Brien, and W. D. Wilson, *J. Chem. Ed., 53,* 328 (1976).
4   G. J. Moody, R. B. Oke, and J. D. R. Thomas, *Analyst, 95,* 910 (1970).
5   A. Ansaldi and S. I. Epstein, *Anal. Chem., 45,* 595 (1973).
6   G. A. Rechnitz, *Chem. Eng. News, 45,* 146 (1967); *Chem. Eng. News, 37,* 294 (1965).
7   Orion Research, Inc., "Orion Instruction Manual for the Calcium Ion Electrode," Model 93-20 (1975), Cambridge, Mass.
8   R. A. Durst, "Ion-Selective Electrodes," National Bureau of Standards,

Spec. Pub. 314, Nov. 1969. (S D Catalog No. C 13.10: 314, Superintendent of Documents, U.S. Government Printing Office, Washington, D.C. $3.50)

9   J. W. Ross, Jr., *Science, 156,* 1378 (1967) (*calcium electrode*).

10 M. E. Thompson and J. W. Ross, Jr., *Science, 154,* 1643 (1966) (*calcium in sea water*).

11 C. N. Durfor and E. Becker, *J. Am. Water Works Ass'n., 56,* 237 (1964).

12 G. A. Rechnitz, *J. Chem. Ed., 60,* 282 (1983).

13 N. Lakshminarayanaiah, "Membrane Electrodes," Academic Press, New York, 1976.

14 J. Koryta, "Ion-Selective Electrodes," Cambridge University Press, Cambridge, 1975.

15 A. K. Covington, "Ion Selective Electrode Methodology," vols. I and II, CRC Press, Boca Raton, Fla., 1979.

16 W. E. Morf, "The Principles of Ion-Selective Electrodes and of Membrane Transport," Elsevier, New York, 1981.

17 H. Freiser (ed.), "Ion-Selective Electrodes in Analytical Chemistry," vols. I and II, Plenum, New York, 1978.

18 G. J. Moody and J. D. R. Thomas, "Selective Ion Sensitive Electrodes," Merrow, Waterford, 1971.

19 T. S. Light and C. C. Cappuccino, *J. Chem. Ed., 52,* 247 (1975).

Apparatus

pH meter with expanded scale (or specific-ion meter)
Reference electrode (SCE or Ag/AgCl)
Magnetic stirrer with asbestos pad
Beakers (5), 150 mL, polyethylene (or plastic round wide-mouth bottles)
Pipets, 1 mL, 2 mL, and 10 mL
Volumetric flasks (5), 100 mL
Thermometer
Calcium ion–selective electrode or the following apparatus for construction of the electrode:

   Glass tubing, 5-cm pieces, i.d. 5 to 8 mm
   Polyethylene bottle cap, i.d. 50 mm
   Filter paper, 7.5 cm
   Tygon tubing, i.d. to fit o.d. of glass tubing above, 1-cm pieces
   Silver wire, 20-22 gauge, 8-in lengths, coated with AgCl
   Coaxial cable lead wire with adapter to pH meter
   Stoppers for glass tubing above

Chemicals

0.100 *M* calcium chloride [$CaCl_2$], standard solution
Unknowns: made up from 0.1 *M* $CaCl_2$ standard solution
Ionic Strength Adjuster (ISA): 4 *M* sodium chloride [$NaCl$], 29.8 g diluted to 100 mL
1 *M* potassium hydroxide [KOH] or 2 *M* hydrochloric acid [HCl] for pH adjustment of samples
Samples to be analyzed (soil, milk, water, wine, serum)

note $CaCl_2$ solutions must be in 5.5–11 pH range. Adjust with pH 10 NH_3 buffer.

The following reagents are needed for construction of the calcium electrode:

Polyvinylchloride (PVC), chromatographic grade, 0.4 mg for eight electrodes (Polysciences, Ind. or Breon 113, B.P. Chemicals, U.K. Ltd.). Polyvinylchloride tubing can also be used, see Ref. 3.
Tetrahydrofuran, 14 mL for eight electrodes

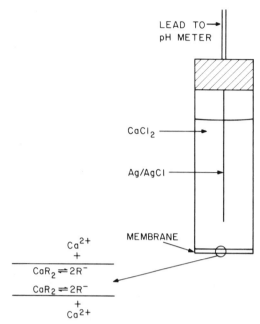

Figure 2-16 / Schematic diagram of calcium ion–selective electrode.

Calcium ion liquid ion exchanger, 1.0 g for eight electrodes (Orion Cat. #92-20-01, special order). Liquid ion exchanger can be prepared with didecylphosphoric acid and dioctylphenylphosphonate. See Ref. 4. 1.0 M CaCl$_2$

Preparation of Calcium Ion–Selective Electrode (Ref. 3)

If a commercial calcium ion–selective electrode is unavailable, a good, inexpensive electrode can be prepared by the following procedure.

Membrane Preparation

Dissolve 0.4 g of polyvinylchloride (PVC) and 1.0 g of calcium ion liquid ion exchanger in 14 mL of tetrahydrofuran, stirring approximately 15 min until completely dissolved. Pour into a 50-mm-i.d. polyethylene bottle cap and allow to evaporate in a hood for approximately 3 days. A flexible, yellow, transparent membrane results. Approximately eight discs 12 mm in diameter can be cut out with a cork borer.

Electrode Assembly

For each electrode, cut a piece of glass tubing (i.d. 5 to 8 mm) about 5 cm long and fire-polish both ends. Place a membrane (2 to 4 mm larger than the o.d. of the glass tube) over the end of the glass tubing. Slip a piece of Tygon tubing (i.d. slightly larger than the o.d. of the glass tube) over the end to hold the membrane in place. The membrane is delicate; care must be taken to prevent stretching or tearing it. (Alternatively, a broken glass electrode may be used. Cement a membrane by dipping the end of a piece of Tygon tubing into tetrahydrofuran to soften and attaching the membrane to this end and allowing to dry; then slipping the tubing over the end of the inverted, filled electrode.) Fill the glass tube with 0.1 M CaCl$_2$ solution. Insert a silver wire, previously coated with AgCl, into the open end of the tube to form the internal electrode. Use a stopper to hold the wire in place. Connect the completed electrode to the pH meter by a coaxial cable with the external sheath grounded. See Fig. 2-16 for a diagram of the complete assembly. Soak the electrode in 1 M CaCl$_2$ for 24 h before use and store in 1 M CaCl$_2$ when not in use.

Theory

The calcium ion-selective electrode is a liquid–liquid membrane type electrode. The membrane, which is saturated with an organic ion exchanger, separates the solution being measured from the electrode filling solution, as shown in Fig. 2-16. The membrane is saturated with a liquid ion exchange solution such as a didecylphosphate complex of calcium, CaR$_2$, dissolved in dioctylphenylphosphonate where R$^-$ is

$$C_{10}H_{21}O - \overset{\displaystyle \overset{O}{\|}}{\underset{\displaystyle \underset{OC_{10}H_{21}}{|}}{P}} - O^-$$

The dioctylphenylphosphonate solvent and the didecylphosphate are relatively insoluble in water. This minimizes dissolution of the ion exchange solution out of the membrane when the electrode is immersed in aqueous samples. The potentiometric response of the electrode is due to the selective extraction of Ca^{2+} into the membrane by R$^-$ by the following equilibrium at the membrane/solution interface:

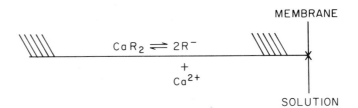

R$^-$ is constrained in the membrane by the hydrophobicity of the didecyl groups. Boundary potentials exist at both the internal and the external interfaces. The inner boundary potential is kept constant by filling the electrode with a fixed concentration of Ca^{2+} (0.1 M CaCl$_2$). The outer boundary potential then varies depending on the activity of Ca^{2+} in the sample solution which shifts the equilibrium at the interface. The potential across the membrane is measured by a millivolt meter in contact with a reference electrode on either side of the membrane. The internal reference is the Ag/AgCl wire with a constant concentration of Cl$^-$ (0.1 M CaCl$_2$); the external reference electrode is usually a conventional SCE or Ag/AgCl reference.

The potential across the membrane responds to calcium ion activity, $a_{Ca^{2+}}$, in the sample solution according to the Nernst equation

$$E = E_{constant} + \frac{0.0591}{2} \log a_{Ca^{2+}} \tag{1}$$

The calcium ion activity is related to the calcium ion concentration by the ionic activity coefficient, $f_{Ca^{2+}}$

$$a_{Ca^{2+}} = f_{Ca^{2+}} [Ca^{2+}] \tag{2}$$

Since activity coefficients depend greatly on total ionic strength, the background ionic strength is usually made high and constant by means of an ionic strength adjuster (ISA). As a result of this stabilization of the ionic strength to a relatively constant value, the activity coefficient becomes nearly constant and the potential is then proportional to concentration of calcium as in the following equation:

$$E = E^* + \frac{0.0591}{2} \log [Ca^{2+}] \tag{3}$$

where E^* includes contributions from $f_{Ca^{2+}}$, E_{1j}, and the two reference electrodes. (See text of the chapter.)

The calcium electrode also responds to metal ions that complex with R^- and thus can compete in the surface equilibrium to generate a boundary potential. Such ions constitute an interference if present in sufficiently large concentration. The magnitude of the interference is related to the formation constant of MR_2 relative to that of CaR_2. The response of the electrode to another interfering cation can be included in the Nernst equation which is shown below for a general case.

$$E = E_{constant} + \frac{0.0591}{n} \log (a_A + K_{AB} a_B^{n/z}) \tag{4}$$

where a_A = activity of species A with charge n
                     a_B = activity of interferent B with charge z
                     K_{AB} = the selectivity constant of the electrode for A over B

The selectivity constant is a measure of the extent of the interference posed by a particular ion that might be present in the sample. Selectivity constants for typical ions are shown in Table 2-3 for the calcium electrode.

The pH range for satisfactory measurements with the calcium electrode is from 5.5 to 11. Below 5.5, hydrogen ions interfere; above 11, $Ca(OH)^+$ is formed. Sulfate concentrations must be less than $5 \times 10^{-4}\ M$ to prevent formation of $CaSO_4$. In the presence of carbonates and bicarbonates, the pH must be less than 7 to avoid precipitation of $CaCO_3$ or formation of $CaHCO_3^+$ and the combined carbonate and bicarbonate concentration must be less than $3 \times 10^{-3}\ M$. Since the calcium electrode responds only to *free* calcium ions, any ions bound or complexed to other species will not be measured.

In practice, a calibration curve is prepared for the response of a particular cell to a series of standard calcium solutions. The measured potentials are plotted vs. the logarithms of the concentrations or activities of the calcium ion. The result is a straight line with a slope of 29.58 mV if a nernstian response is obtained. A typical plot is represented by Fig. 2-17. Alternatively, the method of standard additions can be used for the analysis of samples.

Table 2-3 / Selectivity Constants for Calcium Ion–Selective Electrode

Ion	K	Ion	K
H^+	10^7	Ni^{2+}	0.08
Zn^{2+}	3.2	Sr^{2+}	0.02
Fe^{2+}	0.8	Mg^{2+}	0.01
Pb^{2+}	0.6	Ba^{2+}	0.01
Cu^{2+}	0.3	Na^+	0.0016

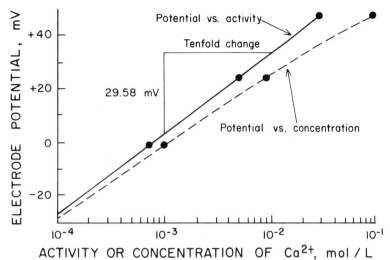

Figure 2-17 / Calibration curve for Ca²⁺. (From "Orion Instruction Manual for the Calcium Ion Electrode," copyright © 1975 by Orion Research Incorporated. Reprinted by permission. "ORION" is a registered trademark of Orion Research Incorporated.)

The electrode response curve will not only shift but will change slope with changes in temperature. A 1°C change in temperature will result in approximately 4% error for measurements at the 10^{-3} M level. Ideally, standards and samples should be kept at a constant temperature, preferably at 25°C.

Procedures

A. Potential Response of Electrodes

Prepare 200 mL of each of the following calcium solutions by serial dilution of the 0.100 M CaCl₂ standard: 1×10^{-2}, 1×10^{-3}, 1×10^{-4}, 1×10^{-5}. Pipet exactly 100 mL of each solution into plastic beakers or bottles and adjust the ionic strength to 0.1 M with 10 mmol NaCl. Measure the potential for each solution. Be sure to rinse the electrodes with distilled water and blot dry between measurements. Measurements should be recorded while the solution is being stirred by a magnetic stirring bar and after the potential has stabilized. (Sometimes the potential will drift. If the potential is stable for a short period prior to the drifting, this reading should be used. If it just continually drifts, readings should be taken as soon as possible. Be consistent whatever method you use, and mention the drift in the experiment write-up.) The asbestos pad should be placed between the stirring motor and the 150-mL beaker containing the sample in order to prevent heating of the sample. Record the temperature of each solution after the potential is measured.

Treatment of Data

Obtain a nernstian plot [Eq. (3)] for the electrode by plotting the electrode potential (mV) vs. log [Ca²⁺]. Calculate the slope of the nernstian plot and compare it with the expected slope.

Determine the range over which the calcium electrode responds linearly to log [Ca²⁺]. Estimate the detection limit of the electrode.

(*Optional*) According to the Debye-Hückel theory, the expression for the activity coefficient of an ion, f_M, is given by

$$\log f_M = \frac{-AZ^2 I^{1/2}}{1 + Bb_M I^{1/2}}$$

where A and B are constants for a given solvent at a specified temperature, Z is the integral charge of the ion (i.e., $+1$, $+2$, -3), I is the ionic strength, and b_M is the so-called "effective diameter" of the ion in solution. For water at 25°C, $A = 0.5085$, $B = 0.3281 \times 10^8$, $b_{H^+} = 9 \times 10^{-8}$, $b_{Na^+} = 4 \times 10^{-8}$, $b_{K^+} = 3 \times 10^{-8}$, $b_{NH_4^+} = 2.5 \times 10^{-8}$, $b_{Mg^{2+}} = 8 \times 10^{-8}$, $b_{Ca^{2+}} = 6 \times 10^{-8}$, and $b_{Ba^{2+}} = 5 \times 10^{-8}$.

Calculate the ionic strength of each solution. Then calculate the activity coefficient of Ca^{2+} ($f_{Ca^{2+}}$) for each solution using the Debye-Hückel equation. Using $f_{Ca^{2+}}$, calculate $a_{Ca^{2+}}$ for each solution [Eq. (2)]. Obtain a nernstian plot [Eq. (1)] of E vs. $\log a_{Ca^{2+}}$. Is the electrode response more closely related to activity or concentration?

B. Determination of Ca^{2+} in an Unknown

Obtain 250 mL of an unknown. Add 25 mmol of NaCl to adjust the ionic strength to 0.1 M.

Calibration Curve Method

Transfer exactly 100 mL of the unknown solution into a 150-mL plastic beaker and measure the potential. Do not discard; use for the method of standard additions.

Method of Standard Additions

Add 10.00 mL of the Ca^{2+} standard which is about 10 times more concentrated than the anticipated concentration of the diluted unknown as determined from the calibration curve to the unknown solution in the 150-mL beaker. Mix and record the new potential.

Known Addition Using Gran's Plots (optional)

The Gran's Plot known addition method allows measurements to be made with greater precision than can be obtained by direct measurement or from known addition tables. A series of small incremental additions of the species sensed by the electrode are made to the sample. The readings are plotted on Gran's Plot paper and a straight line is drawn through the points and is extrapolated to the horizontal axis. The original sample concentration is proportional to the distance from the intercept to the origin. For example, 100 mL of 5×10^{-5} M Ca^{2+} are taken exactly, and five 1-mL increments of 2×10^{-3} M Ca^{2+} standard are added, taking potential readings after each addition. Extrapolation to zero on the horizontal axis should give a reading of 5×10^{-5} M Ca^{2+}. For your unknown, repeat this procedure using a 100-mL sample of the unknown and adding 1-mL increments of a Ca^{2+} standard that is 100 times as concentrated as the unknown. If the unknown is too concentrated for the standard, it can be diluted first.

Treatment of Data

Calibration Curve Method

Determine the concentration of Ca^{2+} from the calibration curve obtained in Procedure A.

Method of Standard Additions

Calculate the concentration of Ca^{2+} in the unknown in the 150-mL beaker from the equation for standard additions:

$$C_x = \frac{C_s V_s}{(V_x + V_s) \times 10^{-n\Delta E/0.0591} - V_x}$$

where $\quad C_x =$ concentration of unknown
$C_s =$ concentration of standard
$V_x =$ volume of unknown
$V_s =$ volume of standard
$\Delta E = (E$ after standard addition$) - (E$ of unknown$)$

Report the concentration of Ca^{2+} in the sample.

Known Addition Using Gran's Plots (optional)

Plot the data obtained with the Gran's Plot procedure and plot it on Gran's paper for a divalent ion with 10% volume correction. Report the concentration of Ca^{2+} in the sample.

Compare the results obtained by the three methods for Ca^{2+} concentration in the unknown sample.

C. Determination of Ca²⁺ in Real Samples

Measure the concentration of Ca^{2+} in one of the samples listed below. Prepare the sample as indicated. Measure the concentration of Ca^{2+} by following the procedures outlined in Procedure B for the calibration curve method, the method of standard additions, and the Gran's Plot method (optional).

Clays: Mix dry powder with 100 mL (volumetric) of water and let stand. Adjust to 0.1 M with NaCl, filter, and measure.
Milk or skim milk: Adjust ionic strength to 0.1 M with NaCl and measure directly.
Water: Adjust ionic strength to 0.1 M with NaCl and measure directly.
Wine: Dilute sample 1:9 and stabilize ionic strength at 0.1 M with NaCl. Measure directly.

Treatment of Data

Calculate the concentration of Ca^{2+} in the real samples by the various methods as outlined under Treatment of Data in Procedure B.

Compare the results obtained by the various methods. Do the results differ significantly (you must establish what a significant difference is)? If so, explain why they differ and explain which results you have the most confidence in.

Evaluate the possibility of interferences from other ions in the sample(s) that you analyzed.

Questions

1 How do you account for the fact that the slopes of the calibration curves for sodium and calcium are positive while that for fluoride is negative?
2 Compute the single ion activity coefficients for Na^+, K^+, NH_4^+, Mg^{2+}, Ca^{2+}, and Ba^{2+}, assuming $I = 0.1$. Use the Debye-Hückel equation on p. 47.
3 The potential response of the calcium electrode is given by Eq. (4). Compute the values for the terms for the cations appearing within the logarithmic portion of the equation for the cell potential, assuming that $C_{Ca^{2+}} = 0.65 \times 10^{-3}$ M, $C_{Mg^{2+}} = 0.26 \times 10^{-3} M$, $C_{Na^+} = 0.52 \times 10^{-3} M$, $C_{K^+} = 0.03 \times 10^{-3} M$, and pH $= 7.0$ using the activity coefficients computed in Question 2 and the selectivity constants in Table 2-3. On the basis of these results, decide which ions you would expect to cause significant interference when analyzing tap water for calcium.
4 What factor(s) would you expect to determine the detection limit of a specific ion electrode?
5 Describe in detail an experiment by which the activity coefficient, f_{MX}, of an electrolyte such as NaF can be determined. In particular, show what rela-

tionship between f_{MX} and experimentally determined quantities such as molality, electrical potential, etc. enables f_{MX} to be computed.

SUPPLEMENTARY EXPERIMENTS

I. Sodium Electrode

The sodium electrode is based on a selective glass membrane. Selectivity constants for some of the common interferences are $Li^+ = 0.1$, $K^+ = 0.005$, $NH_4^+ = 0.001$, $Ag^+ = 250$, and $H^+ = 10$.

Chemicals

Sodium chloride [NaCl]
Tetramethylammonium hydroxide $[(CH_3)_4NOH]$ or calcium hydroxide $[Ca(OH)_2]$

Procedure

A. Preparation of Standards

Make a master standard solution of 10^{-2} M concentration by dissolving 1.0 mmol of NaCl in 100 mL of water (use a volumetric flask). From the master standard prepare 1×10^{-3}, 5×10^{-4}, 2×10^{-4}, and 5×10^{-5} M solutions by pipetting the appropriate amount of the master standard into 100-mL volumetric flasks and diluting to the mark.

Adjust each standard to the same ionic strength (0.1 M) by adding 10 mmol of tetramethylammonium hydroxide or $Ca(OH)_2$ to each.

B. Suggested Unknowns

Water: Adjust ionic strength with tetramethylammonium chloride to 0.1 M and measure.

Wine: Dilute 9:1 with 1 M $Ca(OH)_2$ to give pH > 7.0, ionic strength 0.1 M

C. Method

Measure the potential of each solution using an expanded scale pH meter. Do not use glass beakers for any measurements. Allow 20 s for the readings to stabilize. Why can glass beakers *not* be used?

Treatment of Data

Plot potential (mV) vs. concentration of the standard on suitable graph paper. Is any part of the plot linear? What is its slope, and how is this related to the Nernst equation? What is the Na^+ concentration in your unknown solution?

II. Fluoride Electrode

The fluoride electrode is based on a crystal of LaF_3 that is doped with Eu^{2+} to improve conductivity. The selectivity of this electrode is excellent, OH^- being the main interference (selectivity constant = 0.1).

Chemicals

Sodium fluoride [NaF]
Sodium chloride [NaCl]
Acetic acid [HOAc]

Procedure

A. Preparation of Standards

Make a master standard solution of 10^{-2} M concentration by dissolving 1.0 mmol of NaF in 100 mL of water (use a volumetric flask). From the master standard prepare 1×10^{-3}, 5×10^{-4}, 2×10^{-4}, 1×10^{-4}, and 5×10^{-5} M solutions by pipetting the appropriate amount of the master standard into 100-mL volumetric flasks, and diluting to the mark. Adjust each standard to the same ionic strength (0.1 M) by adding 10 mmol of NaCl to each.

B. Suggested Unknowns

Carbonated beverages: Decarbonate by heating and aerating, and add NaCl to adjust ionic strength to 0.1 M. Measure directly.

Tap water: Adjust ionic strength to 0.1 M with NaCl and measure directly.

Saliva (after using fluoride toothpaste): Acidify to pH 4.7 with acetic acid and measure directly.

Toothpaste (Ref. 19): Weigh 200 mg of toothpaste in a small polyethylene weigh boat and transfer quantitatively to a 250-mL beaker using 50 mL TISAB (total ionic strength adjustment buffer) and a rubber policeman. Boil the mixture for 2 min. Cool, transfer quantitatively to a 100-mL volumetric flask, and dilute to the mark with distilled water. TISAB is available commercially or can be prepared in the laboratory by mixing 57 mL glacial acetic acid, 58 g NaCl, and 4 g CDTA (1,2-cyclohexylene dinitrilo tetraacetic acid) in approximately 500 mL H_2O, adjust pH to 5.0–5.5 with 5 M NaOH and dilute to a total volume of 1 L.

C. Method

Measure the potential of each solution using an expanded scale pH meter. Do not use glass beakers for any measurements. Why? Allow at least 20 s for the potential readings to stabilize.

Treatment of Data

Plot potential (mV) vs. concentration of the standards on suitable graph paper. Is any part of the plot linear? What is its slope and how is this related to the Nernst equation? What is the F^- concentration of your unknown?

III. Nitrate Electrode

The nitrate electrode is of the liquid ion exchanger type. The essential ingredient is a substituted 1,10-phenanthroline complex of nickel(II) which functions as the ion exchanger with nitrate. The electrode is not particularly selective as evidenced by the following selectivity constants: $I^- = 20$, $Br^- = 0.1$, $NO_2^- = 0.04$, $Cl^- = 0.004$, $CO_3^{2-} = 0.0002$, $ClO_4^- = 1000$, $F^- = 0.00006$, $SO_4^{2-} = 0.00003$.

Chemicals

Sodium nitrate [$NaNO_3$]
Sodium sulfate [Na_2SO_4]

Procedure

A. Preparation of Standards

Make a master standard solution of 10^{-2} M concentration by dissolving 1.0 mmol of $NaNO_3$ in 100 mL of water (use a volumetric flask). From the master standard prepare 1×10^{-3}, 5×10^{-4}, 2×10^{-4}, 1×10^{-4}, and 5×10^{-5} M solutions by pipetting the appropriate amount of the master standard solution into 100-mL volumetric flasks and diluting to the mark. Adjust each standard to the same ionic strength (0.1 M) by adding 10 mmol of Na_2SO_4 to each.

B. Suggested Unknowns

Potatoes: Peel and dice a 250-g sample, blend with 250-mL water. Bring ionic strength to 0.1 M with Na_2SO_4 and measure directly in slurry.

Soil: Disperse air-dried soil in water and filter. Adjust ionic strength to 0.1 M with Na_2SO_4 and measure directly.

Water: If water has a low level of chlorine no pretreatment is needed. Otherwise adjust ionic strength to 0.1 M with Na_2SO_4.

C. Method

Measure the potential of each solution using an expanded scale pH meter. Do not use glass beakers for any measurements. Allow at least 20 s for the readings to stabilize.

Treatment of Data   Plot potential (mV) vs. concentration of the standards on suitable graph paper. Is any part of the plot linear? What is its slope, and how well does it agree with that predicted by the Nernst equation? What is the NO_3^- concentration of your unknown?

EXPERIMENT 2-6

Determination of the Copper(II) Ethylenediamine Stability Constants by pH Titration

---

Purpose   The purpose of this experiment is to determine the stability constants for the copper(II) ethylenediamine complexes by pH titration.

References
1 G. A. Carlson, J. P. McReynolds, and F. H. Verhoek, *J. Am. Chem. Soc.*, *67*, 1333 (1945).
2 J. Bjerrum, "Metal Ammine Formation in Aqueous Solution," P, Haase and Son, Copenhagen, 1941.
3 J. Bjerrum and E. J. Nielsen, *Acta. Chem. Scand., 2*, 297 (1948).

Apparatus   pH meter, electrodes, and standard buffer solution
Buret, 50 mL
Pipet, 25 mL
Beaker, 150 mL
Magnetic stirrer and stirring bar

Chemicals   Stock solution containing 0.30 M ethylenediamine dihydrochloride [en · 2HCl], 1.0 M potassium nitrate [KNO_3], and 0.10 M cupric chloride [$CuCl_2$]
Stock solution containing 0.30 M ethylenediamine dihydrochloride, 1.0 M KNO_3, and 0.10 M barium chloride [$BaCl_2$]
Sodium hydroxide [$NaOH$] solution (about 3.00 M), standard

Copper(II) forms a stable complex with two molecules of ethylenediamine (en) per metal ion. In this experiment copper(II) is in solution with an excess of ethylenediamine dihydrochloride (en · 2HCl) and the solution is titrated with strong base in order to liberate the free amine. Ionic strength is maintained constant with 1.0 M KNO$_3$

The copper(II) ethylenediamine complex is so stable that Cu^{2+} competes with protons for the amine. The protons liberated from the amine by the metal ions lower the pH of the solution. By following the course of the titration with a pH meter and employing the proper calculations information can be obtained to calculate the stability constants for the metal-amine complex.

A qualitative idea of how to interpret the experimental curve of pH vs. volume of titrant added can be obtained by referral to Fig. 2-18.

Curve A is simply the titration curve for the titration of ethylenediamine dihydrochloride with a strong base. This titration is done by substituting BaCl$_2$, which does not form a stable complex with ethylenediamine, for the copper(II) salt.

If copper(II) formed only a stable 1:1 complex, Cu(en)$^{2+}$, with the amine, a curve similar to that of curve B would be expected. The curve is initially shifted to lower pH because metal liberates two protons from the amine in forming the complex. The second equivalent of amine is titrated by the strong base so the curve is similar to that of the uncomplexed amine during the addition of the second equivalent of base.

If the metal ion formed a 1:1 complex with the amine at low pH and then a 1:2 complex, Cu(en)$_2^{2+}$, at higher pH, the curve would resemble curve C. The exact shape of the curve will depend upon the stability of the two complexes.

If the metal replaces the protons from two ligand molecules at low pH and

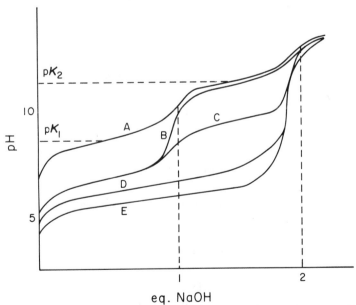

eq. NaOH

Figure 2-18 / A. Titration of en·2HCl with NaOH. B. Same as (A) except a metal ion that forms a stable 1:1 complex, M(en)$^{2+}$. C. A metal ion present that forms a 1:1 complex, M(en)$^{2+}$, and then a 1:2 complex, M(en)$_2^{2+}$, at higher pH. D. A metal ion present that forms a weak 1:2 complex. E. A metal ion present that forms a stable 1:2 complex.

forms a stable 1:2 complex then the titration curve will resemble the curve found by titrating a strong acid with a strong base. This curve is shown by curve E.

Curve D is the type of curve expected if the equilibrium constants for the metal-amine complex and the constants for protonation of the amine are of the same order of magnitude. In this case, the results of competition between the metal ion and protons for the amine are evident.

The following derivation for the more general case can be found in Refs. 1 and 2. Only the case of formation of $Cu(en)^{2+}$ and $Cu(en)_2^{2+}$ will be treated, and it will be assumed that no higher complex or reaction of a single amine molecule with more than one metal ion will occur. The justification for the assumptions can be found in Ref. 3.

The reactions that occur and the pertinent acid dissociation constants for enH_2^{2+} and stepwise formation constants for the metal complexes are

Titration of the amine salt (en · 2HCl) *and acid dissociation constants*

$$enH_2^{2+} + OH^- \rightleftharpoons enH^+ + H_2O \tag{1}$$

$$K_{enH_2^{2+}} = \frac{[enH^+][H^+]}{[enH_2^{2+}]}$$

$$enH^+ + OH^- \rightleftharpoons en + H_2O \tag{2}$$

$$K_{enH^+} = \frac{[en][H^+]}{[enH^+]}$$

Reaction of the metal ion with the amine

$$Cu^{2+} + en \rightleftharpoons Cu(en)^{2+} \tag{3}$$

$$Cu(en)^{2+} + en \rightleftharpoons Cu(en)_2^{2+} \tag{4}$$

Formation constants for the complexes

$$K_1 = \frac{[Cu(en)^{2+}]}{[Cu^{2+}][en]} \tag{5}$$

$$K_2 = \frac{[Cu(en)_2^{2+}]}{[Cu(en)^{2+}][en]} \tag{6}$$

A formation constant (β_2) for the overall complexation reaction may be written as follows:

$$Cu^{2+} + 2en \rightleftharpoons Cu(en)_2^{2+}$$

$$\beta_2 = K_1 K_2 = \frac{[Cu(en)_2^{2+}]}{[Cu^{2+}][en]^2}$$

Terms used in the derivation

C_A Total concentration of all en in all forms
C_M Total concentration of all metal ion in all forms

[en]   Total concentration of *free* ligand

C_H Total concentration of acid in the solution

$\bar{n}$ The ratio of complex-bound ligand to the total concentration of metal ion

α The fraction of en not complex-bound which exists as the free amine

$\bar{n}_A$ Mean number of hydrogen ions bound to not-complex-bound amine

Total amine concentration

$$C_A = [\text{en}] + [\text{enH}^+] + [\text{enH}_2^{2+}] + \bar{n}C_M \tag{7}$$

We may define α and $\bar{n}_A$ in terms of the equilibrium expressions from Eqs. (1) and (2) and of the hydrogen ion concentration

$$\alpha = \frac{[\text{en}]}{[\text{en}] + [\text{enH}^+] + [\text{enH}_2^{2+}]} = \frac{K_{\text{enH}+}K_{\text{enH}_2 2+}}{K_{\text{enH}+}\,K_{\text{enH}_2 2+} + K_{\text{enH}_2 2+}[\text{H}^+] + [\text{H}^+]^2} \tag{8}$$

$$\bar{n}_A = \frac{[\text{enH}^+] + 2[\text{enH}_2^{2+}]}{[\text{en}] + [\text{enH}^+] + [\text{enH}_2^{2+}]} = \frac{K_{\text{enH}_2 2+}[\text{H}^+] + 2[\text{H}^-]^2}{K_{\text{enH}+}\,K_{\text{enH}_2 2+} + K_{\text{enH}_2 2+}[\text{H}^-] + [\text{H}^+]^2} \tag{9}$$

The average number of ligand molecules per metal ion

$$\bar{n} = \frac{[\text{Cu(en)}^{2+}] + 2[\text{Cu(en)}_2^{2+}]}{[\text{Cu}^{2+}] + [\text{Cu(en)}^{2+}] + [\text{Cu(en)}_2^{2+}]} \tag{10}$$

Substituting the expressions from (5) and (6) into Eq. (10), we get $\bar{n}$ in terms of the formation constants for the complex and of the concentration of the free amine in solution.

$$\bar{n} = \frac{K_1[\text{en}] + 2K_1 K_2[\text{en}]^2}{1 + K_1[\text{en}] + K_1 K_2[\text{en}]^2} \tag{11}$$

From Eqs. (7) and (9)

$$\bar{n}_A = \frac{[\text{enH}^+] + 2[\text{enH}_2^{2+}]}{C_A - \bar{n}C_M} \tag{12}$$

Total concentration of acid in solution

$$C_H = [\text{H}^+] + [\text{enH}^+] + 2[\text{enH}_2^{2+}] \tag{13}$$

Equations (12) and (13) can now be solved for $\bar{n}$ in terms of quantities that are known or can be measured.

$$\bar{n} = \frac{C_A - \{(C_H - [\text{H}^+])/\bar{n}_A\}}{C_M} \cong \frac{C_A - (C_H/\bar{n}_A)}{C_M}† \tag{14}$$

[H$^+$] can be neglected compared to C_H under the conditions used. C_A and C_M are known while C_H is merely the difference between the concentration of the acid

---

†Note that Eq. (14) need not be corrected for volume changes since the dilution factor would be the same for each term and the corrections cancel.

present originally as the amine salt and that neutralized by addition of the base. $\bar{n}_A$ can be calculated if K_{enH^+} and $K_{enH^{2+}}$ and the experimentally determined $[H^+]$ are known. The pK_a's for the amine differ by about 2.5 so the break in the titration curve of the free amine is not as sharp as desirable for accurate measurement and the pK_a values listed in one of the references may have to be used.

From Eqs. (8), (9), and (13), [en] can be found:

$$[en] = \frac{\alpha}{\bar{n}_A} (C_H - [H^+]) \cong \frac{\alpha}{\bar{n}_A} (C_H) \tag{15}$$

Each experimental value of $\bar{n}$ and [en] will give an equation of the form (11) which can be solved for K_1, K_2, and $\beta_{12} = K_1 K_2$. These equations converge quite rapidly, and caution must be used in selection of the experimental points. Calculations can be minimized by plotting a formation curve of $\bar{n}$ vs. p[en] $= -\log$ [en] and solving for the constants by the procedure used in Ref. 1.

The formation constants can be expressed in terms of the overall formation constant and a parameter x, so that

$$K_1 = 2x\sqrt{\beta_2} \quad \text{and} \quad K_2 = \frac{\sqrt{\beta_2}}{2x} \tag{16}$$

Substitution into Eq. (11) gives

$$\bar{n} = \frac{2x\sqrt{\beta_2}[en] + 2\beta_2[en]^2}{1 + 2x\sqrt{\beta_2}[en] + \beta_2[en]^2} \tag{17}$$

When $\bar{n} = 1$ as found on the formation curve,

$$\beta_2 = \frac{1}{[en]_{\bar{n}=1}^2} \tag{18}$$

Differentiating Eq. (17) logarithmically

$$\frac{d\bar{n}}{d\ln[en]} = \frac{2x\sqrt{\beta_2}[en] + 4\beta_2[en]^2 + 2x\beta_2^{3/2}[en]^3}{(1 + 2x\beta_2^{1/2}[en] + \beta_2[en]^2)^2} \tag{19}$$

Substitution of Eq. (18) into Eq. (19) gives

$$\left(\frac{d\bar{n}}{d\ln[en]}\right)_{\bar{n}=1} = -0.4343 \left(\frac{d\bar{n}}{dp[en]}\right)_{\bar{n}=1} = \frac{1}{1+x} \tag{20}$$

Since $\left(\dfrac{d\bar{n}}{dp[en]}\right)_{\bar{n}=1}$ is merely the slope of the formation curve, the slope can be measured, and x can be calculated from Eq. (20). From Eq. (16) and the values of x and β_2 from Eqs. (20) and (18), K_1 and K_2, the formation constants for the metal-amine complex, can be calculated.

Procedure

Standardize the pH meter with the standard buffer solution provided.

Pipet 75 mL of the stock solution containing $CuCl_2$ into a 150-mL beaker. Insert the electrodes of the pH meter into the beaker and begin stirring the solution with the magnetic stirrer. Titrate the solution to about pH 13 with the standard base and record the pH with each addition of titrant.

Discard the solution in the beaker and repeat the titration using the stock solution containing $BaCl_2$ in place of the $CuCl_2$. This gives the titration curve for the free amine under the conditions used in the experiment.

Treatment of Data

On the same piece of graph paper plot pH vs. milliliters of titrant added for the two titrations.

Measure K_{enH^+} and $K_{enH_2^{2+}}$ for the amine in the absence of copper(II).

Plot a formation curve for the copper ethylenediamine system using the values of en and $\bar{n}$ calculated from Eqs. (14) and (15). Calculate several points near $\bar{n} = 1$ so an accurate value of the slope can be obtained at this point, but plot only enough points to sketch the curve.

A convenient table for recording data:

Volume Titrant, mL	pH	[H$^+$]	C_H	α	$\bar{n}_A$	[en]	p[en]	$\bar{n}$
		Eqs.		(8)	(9)	(15)		(14)

Measure the slope of the formation curve at $\bar{n} = 1$ and calculate the formation constants (K_1 and K_2) and the overall formation constant (β_2) for the complexes.

Questions

1 Draw curves analogous to those in Fig. 2-18 for the tridentate ligand triethylenetetramine ("trien").

2 Would this approach for determining the stoichiometry and formation constants of a metal complex be adaptable to use with a ligand that contained carboxylic acid groups in place of amine groups.

3 Show the effect of varying β_2 on curve B in Fig. 2-18.

3
Conductometric Methods

Solutions that contain many mobile ions conduct electric current well, and solutions that contain few or relatively immobile ions conduct electric current poorly. In this manner the conductance of solutions provides the analytical chemist with another method by which changes in the composition of solutions can be detected and thus with another method of end-point detection.

The conductance of a solution varies with the *number, size,* and *charge* of the ions and also with some characteristics of the solvent, such as viscosity. Thus ions of different species would be expected to contribute differently to the conductivity of a given solution, so that if one ionic species were replaced by another ionic species of different size or charge through a chemical reaction, a noticeable change in the conductivity of the solution would result.

More specifically, the conductance L of a solution can be represented by the expression

$$L = B \ \Sigma \ C_i \lambda_i Z_i$$

where B is a constant characteristic of the geometry and size of the conductance cell, C is the molar concentration of the individual ions in the solution, λ is the equivalent ionic conductance of the individual ions, and Z is the ionic charge for the individual ions.

Table 3-1 lists some equivalent ionic conductances, which illustrate the relative values for the various ions and how each value is related to the charge and size of the ion.

The analyst makes use of the conductance of ionic solutions by devising a sys-

**Table 3-1 / Equivalent Ionic
Conductances at 25°C**

Cation	λ_+°	Anion	λ_-°
H_3O^+	349.8	OH^-	198
Li^+	38.7	Cl^-	76.3
Na^+	50.1	Br^-	78.4
K^+	73.5	I^-	76.8
NH_4^+	73.4	NO_3^-	71.4
Ag^+	61.9	ClO_4^-	68.0
$\frac{1}{2}Mg^{2+}$	53.1	$C_2H_3O_2^-$	40.9
$\frac{1}{2}Ca^{2+}$	59.5	$\frac{1}{2}SO_4^{2-}$	79.8
$\frac{1}{2}Ba^{2+}$	63.6	$\frac{1}{2}CO_3^{2-}$	70
$\frac{1}{2}Pb^{2+}$	73	$\frac{1}{2}C_2O_4^{2-}$	24
$\frac{1}{3}Fe^{3+}$	68	$\frac{1}{4}Fe(CN)_6^{4-}$	110.5
$\frac{1}{3}La^{3+}$	69.6		

From "Principles of Instrumental Analysis,"
2nd ed., by Douglas A. Skoog and Donald
M. West, copyright © 1980 by Saunders
College/Holt, Rinehart and Winston. Re-
printed by permission of Holt, Rinehart and
Winston, CBS College Publishing.

tem so that replacement of the ionic species to be determined by another species
of significantly different conductance occurs. In this way, by following the
changes in the conductance of the solution as the replacing species is added, determination of when the replacement is complete is possible. This point will be taken
as the end point of the titration. Thus, in the titration of HCl by NaOH, the addition of the latter decreases the hydrogen ion concentration by the formation of
water.

$$[H^+ + Cl^-] + [Na^+ + OH^-] \rightarrow H_2O + [Na^+ + Cl^-]$$

Figure 3-1a shows the titration curve, and Fig. 3-1b indicates how each of the ions
contributes to the conductance. The shape of the titration curve can be predicted
by summing the ionic conductances of the various species at any point during the
course of the titration; the resulting summation gives the titration curve.

Although the hydrogen ions are replaced by sodium ions, the conductance of
the solution is decreased because the sodium ion has a lower mobility than the
hydrogen ion and thus conducts less current. After the equivalence point is

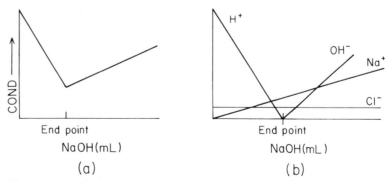

**Figure 3-1 / Conductometric titration of HCl by NaOH. (a) Titration
curve. (b) Relative contributions of each ion to conductance during
titration.**

passed, the addition of NaOH results in an increase in conductance because of the addition of the fast-moving hydroxyl ions. The end point of the titration is taken as the intersection of the two straight lines in Fig. 3-1a.

A conductance titration of any two substances is possible if a reaction occurs in which one ion is substituted for a second ion of different mobility either before or after the equivalence point. Thus not only may acids be titrated with bases but salts of weak acids may be titrated with strong acids. Precipitation reactions may also be followed conductometrically.

The conductance of a solution is equal to the reciprocal of the resistance of the solution. For a column of solution A square centimeters in cross-sectional area between two electrodes d centimeters apart, the resistance R is given by

$$R = \rho \frac{d}{A} \tag{1}$$

where ρ is the *specific resistance* in ohm-centimeters.

Therefore the *conductance* is given by

$$L = \frac{1}{R} = \kappa \frac{A}{d} \tag{2}$$

where κ is the *specific conductance*. The quantity d/A is called the *cell constant* and is specific for a given conductance cell. It can be measured by determining the conductance for a cell containing an exact known concentration of a solution, usually one of potassium chloride. (Why is KCl used? Look at the mobilities of K^+ and Cl^-.) The *equivalent conductance* is equal to the conductance of a solution containing 1 gram-equivalent of solute between electrodes separated by a distance of 1 cm. If C is the concentration of the solution in gram-equivalents per liter, the volume of solution in cubic centimeters per equivalent is equal to $1000/C$ and the equivalent conductance Λ is described by the following equation:

$$\Lambda = \frac{1000\kappa}{C} \tag{3}$$

Substituting for κ gives

$$\Lambda = \frac{1000Ld}{CA} \quad \text{and} \quad L = \frac{A\Lambda C}{1000d} \tag{4}$$

If κ is known for a solution of concentration C, Λ can be determined from Eq. (3). Λ increases with decreasing concentration due to decreasing interionic attraction and repulsion forces. At low concentrations, *for strong electrolytes*, the increase in Λ with dilution is linear and

$$\Lambda = \Lambda° - B\sqrt{C} \tag{5}$$

where $\Lambda°$ is the equivalent conductance at infinite dilution, and B is a constant. $\Lambda°$ can be obtained by extrapolation to zero concentration of the line resulting from plotting Eq. (5). For weak electrolytes, an increase in the degree of dissociation occurs with increasing dilution, and $\Lambda°$ may be obtained by the relationship

$$\Lambda° = \lambda_+° + \lambda_-° \tag{6}$$

where λ_+° and λ_-° are the equivalent ionic conductances for the cation and anion, at infinite dilution, and

$$\Lambda = \lambda_+ + \lambda_- \tag{7}$$

where λ_+ and λ_- are the equivalent ionic conductances at a given concentration.

Conductance Measurements

Although direct current can be used to measure electrolytic resistance, special techniques are required to circumvent the changes in ionic concentrations, and hence resistance, which result from the electrode reactions. Therefore, measurement with alternating current in the frequency range from about 60 to 10,000 hertz (Hz) is the usual practice. Most of the circuits commonly used are of the Wheatstone-bridge type indicated in Fig. 3-2. R_1, R_2, and R_3 are precision variable resistors, and R_C is the cell resistance. A pure sine-wave alternating voltage is applied to points A and C, and an ac null detector ND is connected across points B and D. The "magic-eye" tube is often used as the null detector. A variable condenser is placed across R_3 to balance the cell capacitance.

When the bridge is balanced, points B and D will be at the same potential, and if there is no inductance or capacitance in any part of the circuit, then

$$i_{ABC}R_1 = i_{ADC}R_C \tag{8}$$

$$i_{ABC}R_2 = i_{ADC}R_3 \tag{9}$$

from which

$$R_C = \frac{R_1 R_3}{R_2} \tag{10}$$

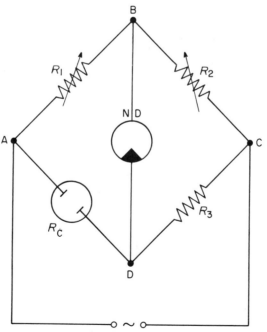

Figure 3-2 / Wheatstone bridge.

Most conductivity bridges read directly in resistance and/or conductance. Inverting the bridge equation (10)

$$\frac{1}{R_C} = L_C = \frac{R_2}{R_1 R_3} \tag{11}$$

shows that the conductance of the cell L_C varies directly with the resistance of the variable resister R_2, and the various ranges are selected by varying R_1 in decade steps.

Preparation of Electrodes

The electrodes used in conductometric titrations consist of two platinum sheets held parallel in a fixed position. The distance between the electrodes is chosen for the particular solution to be measured. The electrode area is large and the plates are spaced closer together for solutions of low conductance, whereas the area of the electrode is small and the plates farther apart for solutions of high conductance. The electrodes must be coated with platinum black to reduce the polarization effect to a minimum by having a maximum surface area exposed to the solution. Platinizing is accomplished by passing a direct current between the electrodes immersed in a solution containing 3 g of chloroplatinic acid and 0.03 g of lead acetate per 100 mL of water. The direction of the current is reversed every $\frac{1}{2}$ min. When a black velvety deposit is obtained, the electrodes should be washed in water and the occluded gases and liquid removed by electrolyzing in dilute sulfuric acid for 30 min with reversal of current every minute. The electrodes are again washed and allowed to stand in low conductivity water for a few hours before using.

Figure 3-3 shows a convenient form for conductivity electrodes. The glass bell protects the platinum electrodes from damage and also provides a rigid support for them.

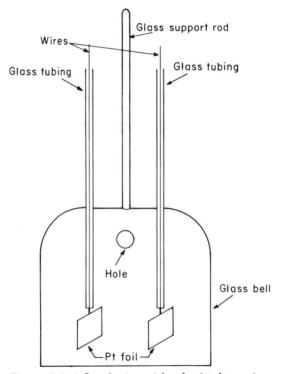

Figure 3-3 / Conductometric electrode system.

Temperature Effects

Most electrolytes have a temperature coefficient for ionic mobility in the order of $2\frac{1}{2}\%/°C$ at room temperature, the conductivity increasing with an increase in temperature. To correct the measured or specific resistance to 25°C, the following formula may be used:

$$R_{25} = R_t (1 + 0.025 \, \Delta t)$$

where R_{25} = measured or specific resistance at 25°C
            R_t = measured or specific resistance at lower or higher temperature
            Δt = difference in temperature between 25° and the temperature of test measurement, taken as positive when the test temperature is higher than 25°, as negative when lower than 25°C

Dilution Effects

In actual practice the dilution of the sample solution by the titrant in a conductometric titration causes the observed conductance to be smaller than that which would be found if the volume remained unchanged during the titration. In order to correct for this dilution, the observed conductance at each point should be multiplied by the ratio $(V + v)/V$, where V is the volume of the original solution and v is the volume of titrant added.

This correction can be minimized by using a titrant which is much more concentrated than the sample solution. If the ratio of molarities is at least 20:1 the correction can be neglected altogether for most systems.

EXPERIMENT 3-1

Conductometric Titrations of HCl and HOAc with NaOH; Determination of K_a for Acetic Acid

Purpose

In this experiment, the relationship between the nature of the substance (weak acid or strong acid) titrated and the order of titration (weak acid titrated with strong base, or strong base titrated with weak acid) is illustrated. From the conductance data obtained in the course of the titration, it is also possible to calculate the dissociation constant of the weak acid.

References

1   H. T. S. Britton, "Conductometric Analysis," Van Nostrand, Princeton, N.J., 1934.
2   C. N. Reilley, High Frequency Methods, chap. 15 in "New Instrumental Methods in Electrochemistry," P. Delahay (ed.), Interscience, New York, 1954.
3   J. W. Loveland in "Treatise on Analytical Chemistry," I. M. Kolthoff, P. J. Elving, and E. B. Sandell (eds.), part I, vol. 4, Interscience, New York, 1963, p. 2569.

Apparatus

Conductance meter with "magic-eye" null detector (procedure is written for this instrument), meter readout or digital readout
Dip-type conductance cell
Pipet, 25 mL

Volumetric flasks (3), 100 mL containing unknown HCl, acetic acid, and
    NaOH samples, approximately 0.05 M when diluted to the mark
Graduated cylinder, 100 mL
Wash bottle
Beaker, 400 mL (or 250-mL tall form)
Magnetic stirrer with stirring bar
Burets (2), 10 mL

Chemicals

0.2 M sodium hydroxide [NaOH] in 10-mL buret, accurately standardized
0.2 M acetic acid [HOAc] in 10-mL buret, accurately standardized

Theory

The general theory of conductance titrations is reviewed in the chapter text.

The dissociation constant of acetic acid can be calculated from the data recorded during the titrations of acetic acid with NaOH and of HCl with NaOH. All that is necessary is a knowledge of (1) the concentration of acetic acid at the beginning of the titration (of acetic acid with NaOH), which is readily obtained from the titration data; (2) the actual conductance of the solution of the weak acid L_{HOAc}, which is actually measured in the experiment; and (3) the conductance of the solution of the weak acid if it were 100% ionized $L_{HOAc(100\%)}$. This last quantity is readily calculated from the various titration curves, knowing that

$$L_{HOAc(100\%)} = L_{NaOAc} + L_{HCl} - L_{NaCl}$$

where all conductances have been corrected to the concentration C of the HOAc used to measure L_{HOAc}.

The degree of ionization α of HOAc is obtained from the conductances.

$$\alpha = \frac{L_{HOAc}}{L_{HOAc(100\%)}}$$

The dissociation constant can then be calculated by the following expression

$$K_a = \frac{[H^+][OAc^-]}{[HOAc]} = C \frac{\alpha^2}{1 - \alpha}$$

where C is the molar concentration of acetic acid used for the conductance measurements.

Procedure

Dilute each unknown sample to the mark and shake well. Pipet a 25-mL aliquot of the HCl sample into a 400-mL beaker (or 250-mL tall-form beaker) and dilute with 150 mL of distilled water (measured in a graduated cylinder). Place the beaker containing your sample on the magnetic stirrer. Lower the conductivity cell into the beaker until it is covered but not so low that the magnetic stirrer will hit against the cell. Slowly increase the stirring speed but do not allow the speed to be so great as to cause cavitation (tornado structure obtained at fast speeds).

Turn on the conductance meter and allow it a minute or so to warm up. With the range switch in the proper place (see instructor) the instrument is adjusted until the eye is as wide open as possible. Do not turn too rapidly. If the reading is erratic, run the dial back and forth several times. Make sure that the lead connections are tight and that the wires make good connection with the conductivity cell.

The conductance is the value of the dial reading multiplied by the value of the range switch.† Take the initial reading and then at 0.5-mL intervals until the buret reads 10 mL. No particular care need be taken in obtaining conductance values near the end point; this is one of the advantages of the conductometric method. Repeat this titration if duplicate checks are desired. Be sure to rinse the cell and beaker *thoroughly* between runs.

In the same manner titrate 25-mL aliquots of the acetic acid unknown sample. In this case obtain readings in 0.2-mL increments at the beginning of the titration up to 2 mL. Continue the titration in 0.5-mL increments until 10 mL of the 0.2 *M* NaOH is added. Repeat if checks are desired.

In the same manner as the first titration, titrate a 25-mL aliquot of the NaOH unknown sample with 0.2 *M* acetic acid, using 0.5-mL increments.

Rinse the conductivity cell with distilled water and store in distilled water. Be sure that your instrument and stirrer are turned off before you leave.

Treatment of Data

Plot the conductance values (corrected for dilution) vs. volume (mL) of titrant added and draw straight lines through the points, giving little weight to the points in the immediate neighborhood of the end point, i.e., where the slope of the line changes abruptly. The end point is determined by extending the straight lines until they meet. Calculate the grams of HCl, NaOH, and acetic acid in your volumetric flasks. Calculate the dissociation constant for acetic acid.

Questions

1 Explain the basis for the difference in titration curves when a weak acid (e.g., acetic acid) or a strong acid (e.g., HCl) is titrated with a strong base.

2 Explain the reasons underlying the different titration curve shapes obtained when HOAc is titrated with NaOH and when NaOH is titrated with HOAc.

3 If the straight-line portion of the HOAc (with NaOH) titration prior to the end point were extrapolated back to 0.0-mL titrant, what value of conductance should be obtained? Why? What value did you obtain? How can you explain this value? A proper answer to this question may explain a disagreement between your value for the dissociation constant and that reported in the literature.

4 In conductometric titrations it is essential that the volume change be as small as possible. For this reason the titrant is very much more concentrated than the solution being titrated. Calculate the percentage error in conductance at the end point caused by neglect of this dilution correction.

5 Draw the titration curve that you would expect for the conductometric titration of a mixture of HCl and HOAc with ammonia.

SUPPLEMENTARY EXPERIMENTS

I. Titration of Mixed Acids with Strong Base‡

Purpose

A mixture of two acids in the presence of a salt is titrated with strong base and the end points are determined conductometrically.

---

†On many instruments the readings on the dial are spaced such that resistance readings are more accurate than conductance readings. Resistance readings can readily be converted to conductance by taking their reciprocals.
‡From a similar experiment developed by Dr. R. S. Juvet, Jr.

Apparatus

Conductance meter
Dip-type conductance cell
Pipet, 25 mL
Volumetric flask, 100 mL
Graduated cylinder, 25 mL
Beaker, 150 mL (or 180-mL tall form)
Magnetic stirrer with stirring bar
Buret, 10 mL

Chemicals

0.1 M NaOH, standard
0.025 M hydrochloric acid [HCl]
0.025 M boric acid [H_3BO_3]
0.025 M ammonium chloride [NH_4Cl]
Mannitol, 0.25 g
Unknown: containing HCl, H_3BO_3, and NH_4Cl (in 100-mL volumetric flask). 0.25 g of mannitol should be added to the flask, then the solution should be diluted to the mark. The final concentration for each acidic component is approximately 0.01 M.

Theory

Mixtures of some acids are conveniently titrated using conductometric end-point detection. The principles are the same as reviewed in the earlier parts of this chapter, but the extension to mixtures is useful and interesting. A mixture of HCl, H_3BO_3, and NH_4Cl is to be titrated with NaOH. Mannitol is added to increase the acidity of H_3BO_3.

Procedure

Place 10 mL of 0.025 M HCl, 15 mL of 0.025 M H_3BO_3, 0.25 g mannitol, 7.5 mL of 0.025 M NH_4Cl, and 20 mL of water into the 150-mL beaker (or 180-mL tall-form beaker) using the graduated cylinder to measure the volumes. Place the beaker on the magnetic stirrer, add the stirring bar, and lower the conductance cell into the beaker until it is covered. Adjust the speed of the stirrer so as not to cause cavitation.

Turn on the conductance meter and allow it to warm up for about 1 min. Set the range switch as directed by the instructor and adjust the eye until it is as wide open as possible. The conductance is the dial reading times the value of the range switch.

Take the initial reading before adding titrant and then titrate the sample with 0.1 M NaOH taking readings at 0.5-mL increments. Continue titrating until a total of 10 mL has been added. Rinse and clean the electrodes and beaker.

Pipet 25 mL of the unknown solution into the beaker, add 25 mL of water, and titrate as in the preceding paragraph. Repeat the titration on a second aliquot.

Treatment of Data

Plot the conductance vs. volume (mL) of 0.1 M NaOH for the known mixture of acids. Dilution corrections should be made for the conductance values before plotting. From the known amounts of each acid, write appropriate titration reactions for each straight-line portion of the graph and indicate the species to which each end point corresponds.

Plot the dilution-corrected conductance vs. volume (mL) of standard 0.1 M NaOH for the unknown-sample titrations. From the end points calculate the grams of HCl, H_3BO_3, and NH_4Cl, respectively, contained in the 100-mL volumetric flask.

Questions

1 What is the function of the mannitol? Explain, writing reactions to justify your answer.
2 Could a mixture of HCl, HOAc, and HNO₃ be titrated conductometrically so that the quantity of each acid could be evaluated? Indicate what solvents and titrants would be necessary for a satisfactory titration.

II. Titration of a Salt of a Weak Acid with a Strong Acid

Purpose

The salt of a weak acid is titrated with a strong acid and the end point is determined conductometrically.

Apparatus

Conductance meter
Dip-type conductance cell
Pipet, 25 mL
Volumetric flask, 100 mL
Graduated cylinder, 25 mL
Beaker, 150 mL (or 180-mL tall form)
Magnetic stirrer with stirring bar
Buret, 10 mL

Chemicals

0.200 M HCl, standard
Unknown: approximately 0.01 M sodium phosphate [Na_3PO_4]

Theory

The titration of the salt of a weak acid with a strong acid can be monitored conductometrically since the weak acid product is only partially dissociated.

Procedure

Titrate a 25-mL aliquot of the unknown to which 150 mL of distilled water has been added, in the same manner as in the previous experiments. Be sure you reach all end points!

Treatment of Data

Plot conductance vs. volume (mL) of 0.2 M HCl for the titration of the unknown Na_3PO_4. Dilution corrections should not be necessary. Why? Calculate the number of grams of Na_3PO_4 in your sample.

Questions

1 Write an equation for each step of the titration curve.
2 Explain the shape of the titration curve.

III. A Titration Involving a Neutralization and a Precipitation

Purpose

A mixture of H_2SO_4 and K_2SO_4 is titrated with $Ba(OH)_2$, which forms the precipitate $BaSO_4$.

Apparatus

Conductance meter
Dip-type conductance cell
Pipet, 25 mL
Volumetric flask, 100 mL
Graduated cylinder, 25 mL
Beaker, 150 mL (or 180-mL tall form)
Magnetic stirrer with stirring bar
Buret, 10 mL

Chemicals	0.075 M barium hydroxide [Ba(OH)$_2$], standard Unknown: approximately 0.01 M in sulfuric acid [H$_2$SO$_4$] and potassium sulfate [K$_2$SO$_4$]
Theory	The conductance method is well suited for the determination of end points in precipitation titrations as well as neutralization titrations. The method of analysis of the curve is similar to the neutralization curves. The solubility of the precipitate formed will have an effect on the titration curve; however, this effect can be tolerated whereas it would be unfit for gravimetric analysis. The effect of appreciable solubility of the precipitate is to round off the intersection of the arms of the graph.
Procedure	Titrate a 10-mL aliquot of the unknown mixture plus 150 mL of distilled water with the standard Ba(OH)$_2$. With the addition of Ba^{2+}, the insoluble BaSO$_4$ is formed in the solution. With the formation of this precipitate, a longer period of time is required for the solution to equilibrate so that the conductance can be measured.
Treatment of Data	Plot the conductance of the mixture vs. the volume (mL) of Ba(OH)$_2$. Indicate which species is titrated first. Why? From the two end points calculate the grams of H$_2$SO$_4$ and the grams of K$_2$SO$_4$ in the sample.
Question	1 What effect would the addition of some alcohol have on the precipitation titration? Explain.

IV. Analysis of Aspirin Tablets

Purpose	This experiment demonstrates a simple, yet accurate method of determining the amount of acetylsalicylic acid (aspirin) in commercial aspirin tablets.
Apparatus	Conductance meter and cell Magnetic stirrer and stirring bar Volumetric flask, 250 mL Pipet, 100 mL Electrolytic beaker, 250-mL tall form Buret, 10 mL
Chemicals	0.100 M sodium hydroxide [NaOH], standard Ethanol, 30 mL Aspirin tablets
Theory	Acetylsalicylic acid (commonly known as aspirin) is a weak acid. The amount present in an aspirin tablet can be readily determined by a conductometric titration.
Procedure	Weigh one aspirin tablet and transfer it to a 250-mL volumetric flask. Add approximately 15 mL of distilled water and swirl until the tablet breaks up and becomes partially dispersed. Add 30 mL of ethanol and swirl again until the tablet is finely dispersed. Dilute to 250 mL with distilled water and mix well. Pipet a 100-mL aliquot into a 250-mL tall-form electrolytic beaker, drop in a magnetic stirring bar, and titrate with 0.100 M NaOH standard, adding 0.5-mL increments until at least 10 mL have been added.

Treatment of Data

Plot conductance vs. volume (mL) of sodium hydroxide added on graph paper ruled 10 by 10 to the centimeter.

Calculate the grams of acetylsalicylic acid in the *original tablet* (MW of acetylsalicylic acid = 180, formula is $CH_3OCO \cdot C_6H_4 \cdot COOH$). Calculate the number of grains in the original tablet (1 g = 15.4 grains).

Calculate the percent by weight of aspirin in the original tablet.

Explain the shape of the titration curve.

Questions

1 What effect does undissolved material from the aspirin tablet have on solution conductance?

2 Describe how this procedure could be automated for the repetitive analysis of aspirin tablets.

3 Discuss sampling problems that might be involved in monitoring quality control for the production of aspirin tablets.

EXPERIMENT 3-2

Conductometric Measurement of the Solubility of a Relatively Insoluble Compound: Silver Acetate

---

Purpose

The objective of this experiment is to determine the solubility of silver acetate by conductometric measurements. By determining the cell constant and determining the equivalent conductance of solutions of silver nitrate, sodium acetate, and sodium nitrate, the solubility can be calculated.

References

1 H. T. S. Britton, "Conductometric Analysis," Van Nostrand, Princeton, N.J., 1934.

2 C. N. Reilley, High Frequency Methods, chap. 15 in "New Instrumental Methods in Electrochemistry," P. Delahay (ed.), Interscience, New York, 1954.

3 J. W. Loveland in "Treatise on Analytical Chemistry," I. M. Kolthoff, P. J. Elving, and E. B. Sandell (eds.), part I, vol. 4, Interscience, New York, 1963, p. 2569.

Apparatus

Conductance meter
Dip-type conductance cell
Pipets, 1, 3, 5, 10, 30, 50, and 100 mL
Beaker, 800 mL
Beaker, 250-mL tall form
Erlenmeyer flasks, stoppered (2), 2000 mL
Magnetic stirrer with stirring bar
Thermometer (to tenths of °C)

Chemicals

0.100 N silver nitrate [$AgNO_3$] standard, 110 mL
0.100 N sodium acetate [NaOAc] standard, 110 mL
0.100 N sodium nitrate [$NaNO_3$] standard, 110 mL
Silver acetate [AgOAc] solutions, saturated at 25 and 35°C (see procedure for preparation)
0.01 N potassium chloride [KCl], standard, 0.7459 g/L

**Table 3-2 / Specific Conductances
of 0.01 N KCl at Various Temperatures**

Temperature, °C	Specific Conductance (κ), mhos/cm
18	0.001225
19	0.001251
20	0.001278
21	0.001305
22	0.001332
23	0.001359
24	0.001386
25	0.001413
26	0.001441
27	0.001468
28	0.001496
29	0.001524

Theory

A discussion of the theory involving conductance of solutions is given in the text of the chapter. The cell constant can be determined by measuring the conductance of a standardized solution of potassium chloride (1 N, 0.1 N, or 0.01 N) and obtaining the specific conductance of this solution from literature values. The specific conductance of 0.01 N KCl is given in Table 3-2 for a number of temperatures.

$$\text{Cell constant} = \frac{d}{A} = \frac{\kappa(\text{table})}{L(\text{measured})}$$

The equivalent conductance of silver acetate can be obtained from the individual conductances of three solutions (silver nitrate, sodium nitrate, and sodium acetate), since

$$\Lambda(\text{AgOAc}) = \Lambda(\text{AgNO}_3) + \Lambda(\text{NaOAc}) - \Lambda(\text{NaNO}_3) \tag{1}$$

From a plot of equivalent conductance times normality vs. normality, the normality of the saturated silver acetate solutions can be obtained.

By plotting the equivalent conductance of each salt vs. the square root of the normality, and extrapolating to zero, equivalent conductance at infinite dilution can be obtained. Such a plot should be a straight line, in accordance with

$$\Lambda = \Lambda^\circ - B\sqrt{N} \tag{2}$$

The straight lines can be represented by the following equations:

$$\Lambda(\text{AgNO}_3) = \Lambda^\circ(\text{AgNO}_3) - a\sqrt{N} \tag{3}$$

$$\Lambda(\text{NaOAc}) = \Lambda^\circ(\text{NaOAc}) - b\sqrt{N} \tag{4}$$

$$\Lambda(\text{NaNO}_3) = \Lambda^\circ(\text{NaNO}_3) - c\sqrt{N} \tag{5}$$

It is possible to find N by solving the following equation:

$$1000\frac{d}{A}\frac{1}{RN} = \Lambda^\circ(\text{AgNO}_3) + \Lambda^\circ(\text{NaOAc}) - \Lambda^\circ(\text{NaNO}_3) - (a + b - c)\sqrt{N} \tag{6}$$

where R = the measured resistance of the N normal silver acetate solution. The normality calculated by this equation can then be compared with the normality obtained graphically.

Procedure

Place exactly 500 mL of water in a clean dry 800-mL beaker and measure its conductance. Add 1.00 mL of 0.100 N silver nitrate, stir thoroughly, and measure the conductance again. Then add successively 1.00, 3.00, 5.00, 10.0, 30.0, and 50.0 mL of silver nitrate, stirring thoroughly and measuring the conductance of the mixture after each addition. (The total volume of silver nitrate added will be 100.0 mL.)

Repeat the measurements as before, using first 0.100 N sodium nitrate and then 0.100 N sodium acetate in place of the silver nitrate.

In each of two 2000-mL glass-stoppered Erlenmeyer flasks, place approximately 25 g of air-dried silver acetate and 1900 mL of distilled water. Warm one flask to 35°C and hold it at that temperature 1 h with frequent shaking; it may be necessary to add more of the silver acetate to make sure an excess is present. The other flask should be kept at 25°C and treated in the same manner. Let them stand at room temperature (25°C) until equilibrium is attained. Dilute exactly 100 mL of each of the saturated silver acetate solutions with exactly 400 mL of water, and measure the conductances of these solutions.

Measure the conductance of the standard 0.01 N potassium chloride solution secured from the instructor. Rinse the cell thoroughly and store it in a beaker of distilled water.

Treatment of Data

Calculate the cell constant, using Table 3-2 for the specific conductance. (If a potassium chloride solution of any other normality is used, consult a handbook for the specific conductance.)

Calculate the normality (molarity times ionic charge, Z) and the square root of the normality for each dilution of each solution. (These actually need be calculated for only one of the solutions since all will be alike provided all the initial normalities were the same—0.100 N.)

Calculate the equivalent conductance for each of the salt solutions used. For each normality calculate the equivalent conductance for silver acetate, using Eq. (1). Be sure to correct for the conductance of water, by subtracting this value from each of the conductance values of the salts. Multiply each value for the equivalent conductance of silver acetate by the normality ($\Lambda N = \kappa$).

Plot κ vs. normality. Read the normality of each of the saturated silver acetate solutions from the graph after multiplying Λ times N for each to obtain κ.

Plot the equivalent conductance vs. the square root of the normality of the salt for each of the salts on the same sheet of graph paper. Extrapolate each line to zero normality (infinite dilution) to obtain $\Lambda°$ for each salt. Compare these values with those secured from Table 3-1. Account for any discrepancies observed.

Using Eqs. (3), (4), and (5) solve for a, b, and c. Calculate the normality of each saturated silver acetate solution by using Eq. (6). Compare these values with those obtained from the κ vs. normality graph.

Questions

1 What are the main sources of error in this experiment?
2 How well do these values of $\Lambda°$ obtained in the experiment compare with the equivalent ionic conductances in Table 3-1?

4
Controlled Potential Methods (Voltammetry)

Electrochemical techniques in which a potential is imposed upon an electrochemical cell and the resulting current is measured are generally categorized as *voltammetric* methods. A variety of such methods have been developed. They differ in the type of potential waveform impressed on the cell, the type of electrode used, and the state of the solution in the cell (quiescent or flowing). Some features of the most commonly used types of voltammetry are summarized in Table 4-1.

Voltammetry has proved to be very useful for analyzing dilute solutions, both quantitatively and qualitatively, for inorganic, organic and biological components, measuring thermodynamic parameters for metal-ion complexes and oxidation-reduction systems, and studying the kinetics of chemical reactions. The experiments in this section are designed to illustrate the principles and applications of several voltammetric techniques.

Principles of Voltammetry

Voltammetric techniques are based on controlling the electrode potential and measuring the resulting current. Electrode and solution phenomena related to potential control and current are dealt with in the following sections.

Potential Control and the Nernst Equation

A key feature in understanding voltammetric methods is the relationship between the potential applied to an electrode and the concentration of redox species at the electrode surface. Consider an electrode at equilibrium with the solution in which it is immersed. The electrode will exhibit a potential, invariant with time and related thermodynamically to the composition of the solution. Assume that the solution contains species O which is capable of being reduced to R at the electrode by the following reversible electrochemical reaction:

$$O + ne \rightleftharpoons R \tag{1}$$

Table 4-1 / Controlled Potential Methods

Name of Technique	Potential Excitation Signal		Mass Transfer	Measurement	Analytical Relation to Bulk Concentration	Typical Display
Polarography (dc or normal)	Slow linear scan (or constant E)		Diffusion	i vs. E	$i_d \propto C$	
AC polarography	Slow linear scan + low amplitude sine wave		Diffusion	i_{ac} vs. E	$i_p \propto C$	
Pulse polarography	Square voltage pulses of increasing amplitude		Diffusion	i vs. E	$i_d \propto C$	
Differential pulse polarography	Square voltage pulses of constant amplitude + linear ramp		Diffusion	Δi vs. E	$i_p \propto C$	
Single sweep voltammetry	Linear scan E		Diffusion	i vs. E	$i_p \propto C$	
Cyclic voltammetry	Triangular scan E		Diffusion	i vs. E	$i_p \propto C$	
Chronoamperometry	Step E		Diffusion	i vs. t	$i_t \propto C$	
Chronocoulometry	Step E		Diffusion	Q vs. t	$Q \propto C$	
Hydrodynamic voltammetry	Linear scan E (or constant E)		Convection/diffusion	i vs. E	$i_\ell \propto C$	
Controlled potential coulometry	Constant E		Convection/diffusion	Q vs. t	$Q = \int_0^t i\, dt$ $= nFVC$	
Controlled potential electrogravimetry	Constant E		Convection/diffusion	Weight of deposit	Weight $\propto VC$	None
Amperometric titration (one or two polarized electrodes)	Constant E + titrant addition		Convection/diffusion	i vs. volume	Volume of titrant $\propto VC$	
Stripping voltammetry	Constant E followed by linear scan or differential pulse scan		Convection/diffusion	i vs. E	$i_p \propto C$	

If by some external means the potential of the electrode is forced to assume a different value, current in the electrode circuit will change the composition of the electrode and/or solution sufficiently to exhibit this new potential.

The fundamental equation that relates the potential E, which is applied to the electrode, and the concentrations of species O and R at the electrode surface is the Nernst equation

$$E = E^{\circ\prime} + \frac{0.0591}{n} \log \frac{C_O^s}{C_R^s} \qquad (2)$$

where
E = potential applied to electrode
$E^{\circ\prime}$ = formal reduction potential of the couple vs. reference electrode
n = number of electrons in reaction (1)
C_O^s = surface concentration of species O
C_R^s = surface concentration of species R

Table 4-2 illustrates how the ratio C_O^s/C_R^s changes as E is changed. This variation of C_O^s/C_R^s as a function of E is the basis of all voltammetric methods. The Nernst equation describes this relationship for reversible systems, i.e., systems for which the electrode reaction in Eq. (1) is rapid in both directions. This means that the surface concentration ratio instantly responds to any change in E. As shown in Table 4-2, the surface concentration can be altered to essentially all O ($C_O^s/C_R^s > 1000$) by making E sufficiently positive of $E^{\circ\prime}$ or essentially all R ($C_O^s/C_R^s < \frac{1}{1000}$) by making E sufficiently negative of $E^{\circ\prime}$. Making E equal to $E^{\circ\prime}$ forces the surface concentrations of O and R to be equal ($C_O^s/C_R^s = 1$).

Mass Transfer in Solution

As a result of changes in surface concentration of O and R at the electrode surface due to an applied E, a concentration imbalance exists between the solution at the electrode surface and the solution at a distance from the electrode. *Mass transfer* (the movement of material from one location in solution to another) mechanisms such as diffusion and convection will act to remove such a concentration gradient by movement of material from a high concentration to a low concentration. The concentration gradients in the solution adjacent to an electrode are illustrated by means of a concentration-distance profile. Such a profile for a planar electrode immersed in a stirred solution of species O (1 mM) is illustrated in Fig. 4-1. The vertical axis represents concentration and the horizontal axis represents distance from the electrode into solution. The interface, or boundary, between electrode and solution is indicated by the vertical line at distance = 0. The dashed

Table 4-2 / Relationship of E to Surface Concentrations†

E, mV	C_O^s/C_R^s
236	10,000/1
177	1,000/1
118	100/1
59	10/1
0	1/1
− 59	1/10
−118	1/100
−177	1/1,000
−236	1/10,000

†For a reversible system, $n = 1$, $E^{\circ\prime} = 0$ V.

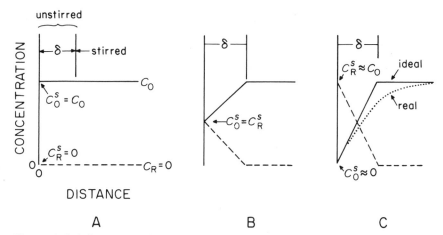

Figure 4-1 / Concentration-distance profiles for voltammetry in stirred solution. (A) Open circuit or E positive of $E^{\circ\prime}$ so $C_O^s/C_R^s > 1000$, (B) $E = E^{\circ\prime}$, (C) E negative of $E^{\circ\prime}$ so $C_O^s/C_R^s < 1000$.

horizontal line is the concentration of R in solution, $C_R = 0$. The continuous horizontal line is the concentration of O in solution, $C_O = 1$ mM. Figure 4-1A illustrates the homogeneous concentration of O and R throughout the solution and up to the electrode surface.

Some understanding of liquid flow along a solid-solution interface is useful in understanding concentration-distance profiles for an electrode immersed in a stirred solution. Three regions of solution flow can be identified.

1 Turbulent flow comprises the solution bulk.
2 As the electrode surface is approached, a transition to laminar flow occurs. This is a nonturbulent flow in which adjacent layers slide by each other parallel to the electrode surface.
3 The rate of this laminar flow decreases near the electrode due to frictional forces until a thin layer of stagnant solution is present immediately adjacent to the electrode surface. It is convenient, although not entirely correct, to consider this thin layer of stagnant solution as having a discrete thickness δ, called the Nernst diffusion layer.

Figure 4-1 illustrates the Nernst diffusion layer in terms of concentration-distance profiles for a solution containing species O. As pointed out above, the concentration of redox species at the electrode-solution interface is determined by the Nernst equation. Figure 4-1A illustrates the concentration-distance profile for O under the condition that its surface concentration has not been perturbed. Either the cell is at open circuit (no potential applied) or a potential which is sufficiently positive of $E_{O,R}^{\circ\prime}$ not to measurably alter the surface concentrations of the O,R couple has been applied, i.e., $C_O^s/C_R^s > 1000$.

The profiles in Fig. 4-1B represent the situation when a potential is applied that requires equal concentrations of O and R at the electrode surface to satisfy the Nernst equation, i.e., $E = E_{O,R}^{\circ\prime}$ (assuming equal diffusion coefficients for O and R). To fulfill this requirement, the electrode electrolyzes O to R at the rate required to maintain equal concentrations of O and R at the surface. If this potential is maintained, a continuous electrolysis of O to R is necessary to maintain these surface concentrations because R diffuses away from the interface across the stagnant layer and is then swept away by the laminar flow. Two regions of this profile can be considered.

1 At distances greater than δ, concentrations are maintained homogeneous by the stirring action. So long as the electrode area is small (microelectrode) relative to the solution volume and the experiment is not prolonged, the bulk concentrations will not be altered appreciably by the electrolytic conversion of O to R at the surface.
2 The removal of O at the electrode surface by its conversion to R sets up a concentration gradient across the stagnant solution layer. The species O diffuses from the stirred region across this layer to the electrode surface where it is electrolyzed to R, which then diffuses back across the stagnant layer to the bulk solution. Thus, the electrolysis process is controlled by a combination of (a) mass transfer of O to the edge of the stagnant layer by the laminar flow and (b) subsequent diffusion of O across the stagnant layer under the influence of the concentration gradient caused by electrolysis of O to R at the electrode surface to satisfy the Nernst equation.

Figure 4-1C shows the profiles that result when the applied potential is sufficiently negative that the concentration of O at the electrode surface is effectively zero, i.e., $C_O^s/C_R^s < 0.001$. In this case, essentially all of O at the electrode surface must be electrolyzed to R in order to satisfy the Nernst equation. Consequently, O is converted to R as rapidly as it can diffuse to the electrode surface. Since this is a limiting condition, application of even more negative potentials causes only negligible change in the profiles.

Although the transition between stagnant and flowing solution is considered to be abrupt in this example, the transition is in reality gradual. Consequently, the profiles are rounded as shown by the dotted line in Fig. 4-1C. However, the hypothetical situation of an abrupt transition is a useful approximation in mathematical treatments.

Current

The current at an electrode is related to the *flux* (rate of mass transfer) of material to the electrode as described by the following equation:

$$i_t = nFAD_O \left(\frac{\partial C_O}{\partial x} \right)_{x=0,t} \tag{3}$$

where i_t = current at time t, A
n = number of electrons transferred per molecule
F = Faraday's constant, 96,485 C/equiv
A = electrode area, cm^2
C_O = concentration of O, mol/cm^3
D_O = diffusion coefficient of O, cm^2/s
t = time, s
x = distance from the electrode, cm

Thus, the current is directly proportional to the slope of the concentration-distance profile at the electrode surface, i.e., to $\left(\frac{\partial C_O}{\partial x} \right)_{x=0,t}$. This equation can be expressed in terms of the Nernst diffusion layer concept by simply approximating ∂x by δ and ∂C by $C_O - C_O^s$ for the case in Fig. 4-1

$$i = nFAD_O \frac{C_O - C_O^s}{\delta} \tag{4}$$

Examination of the profiles in Fig. 4-1 shows that this is a valid substitution for the slope of the profile at the electrode surface for the ideal case of an abrupt transition at $x = \delta$.

Current which results from reduction at the electrode is termed cathodic current; current from an oxidation is anodic current.

Hydrodynamic Voltammogram

A *voltammogram* (current-potential curve) is typically obtained by scanning the potential of an electrode from positive to negative, or vice versa, and recording the current.

A typical voltammogram that would be obtained for a solution containing a 1 mM concentration of O and no R is shown in Fig. 4-2. The shape of this curve can be understood by considering the slopes of the concentration-distance profiles that are depicted for several representative potentials. The current is determined by the slope of the profile for O. As the slope increases due to decreasing C_O^s, the current increases. A limiting cathodic current $i_{\ell c}$ is reached when the surface concentration C_O^s becomes effectively zero. Substitution of $C_O^s = 0$ into Eq. (4) gives the equation for the limiting current

$$i_{\ell c} = \frac{nFAD_O C_O}{\delta} \tag{5}$$

Voltammograms such as the one shown in Fig. 4-2 give useful information for both quantitative and qualitative analysis. As formulated in Eq. (5), $i_{\ell c}$ is directly proportional to the concentration of O. Thus, the magnitude of a limiting (or peak) current on a voltammogram is frequently used to measure concentration. Since the value of the half-wave potential $E_{1/2}$ (the potential at which current equals $i_{\ell c}/2$) is related to $E^{\circ\prime}$ of the redox couple involved in the electrode

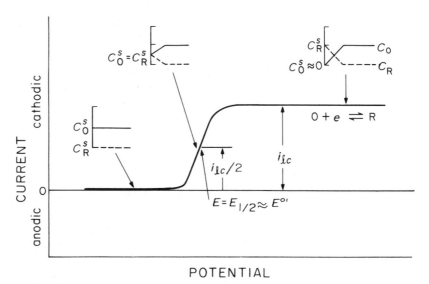

Figure 4-2 / Hydrodynamic voltammogram with representative concentration-distance profiles for a solution containing 1 mM of O and no R.

reaction, $E_{1/2}$ can be used for qualitative identification of species O, inasmuch as O has a unique value of $E_{1/2}$.

The above description is specific for hydrodynamic voltammetry in which the solution flows past a stationary electrode. In some techniques the solution is quiescent, and in others the electrode is not stationary. However, the concepts of changing surface concentrations by potential control and the resulting mass transfer (by convection and/or diffusion) in solution are fundamental to all voltammetric techniques. [What would the voltammogram in Fig. 4-2 look like if the experiment were performed in a nonstirred solution (see Experiment 4-1)?]

Instrumentation

The main instrument for voltammetry is a potentiostat, which applies a potential to the electrochemical cell, and a current to voltage converter, which measures the resulting current. The current is typically displayed on a recorder as a function of the applied potential. Many types of voltammetry require that the potentiostat be capable of scanning from one potential to another potential.

Modern potentiostats utilize a three-electrode configuration as shown in Fig. 4-3. The potentiostat applies the desired potential between a *working electrode* and a *reference electrode*. The working electrode is the electrode at which the electrolysis of interest takes place. The current required to sustain the electrolysis at the working electrode is provided by the *auxiliary electrode*. This arrangement prevents the reference electrode from being subjected to large currents that could change its potential. Some instrumentation is based on a two-electrode system. Here the auxiliary electrode is absent and the reference electrode is subjected to the entire cell current.

A typical *electrochemical cell* is illustrated in Fig. 4-4. Such a cell usually consists of a glass container with a cap having holes for introducing electrodes and nitrogen. Provision is made for oxygen removal from solution by bubbling with nitrogen gas. The cell is then maintained oxygen-free by passing nitrogen over the solution. The nitrogen gas itself may be deoxygenated by procedures such as passing through a hot copper furnace or through solutions of strong reductants such as vanadous or chromous ion. The reference electrode is typically a saturated calomel electrode (SCE) or a Ag/AgCl electrode, which is often isolated from the solution by a tube of electrolyte called a *salt bridge*. The salt bridge prevents contamination of the sample by Cl^- and impurities in the SCE that slowly leak out at the junction. Reference electrodes are discussed in Chap. 2. The auxiliary electrode is typically a coiled platinum wire that is often placed directly into the solution. Since the limiting (or peak) current in any type of voltammetry is temperature-dependent, the cell should be thermostated for the most exacting work. Cells that require as little as 1 to 2 mL of solution for analysis are available.

A large variety of working electrodes have been used with voltammetry. The voltammetric techniques termed *polarography* utilize the *dropping mercury elec-*

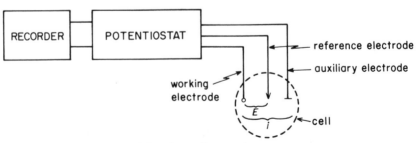

Figure 4-3 / Instrumentation for voltammetry.

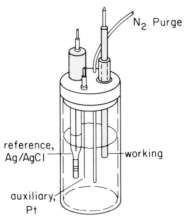

Figure 4-4 / Electrochemical cell for voltammetry. (From Bioanalytical Systems, Inc., West Lafayette, Indiana. Reprinted by permission.)

trode (DME). This electrode consists of repetitive mercury drops continuously extruding from the end of a capillary. (See Experiment 4-4 for a detailed discussion of this electrode.) The *hanging mercury drop electrode* (HMDE) is commonly used in voltammetry. Here a drop of mercury is suspended at the end of a capillary. A thin coat of mercury can be deposited on a substrate such as graphite to form a *mercury film electrode* (MFE). A significant advantage of mercury is its good negative potential range. Solid electrodes such as platinum, gold, glassy carbon, and wax impregnated graphite are commonly used in voltammetry. Such electrodes have a better positive potential range than mercury. Solid electrodes can be rapidly rotated as in the case of the *rotating disk electrode* (RDE). More detailed descriptions of the applicable electrodes are presented in the subsequent experiments on voltammetry.

EXPERIMENT 4-1
Cyclic Voltammetry†

Purpose

Cyclic voltammetry is used to determine the $E^{\circ\prime}$ and n values of the $Fe^{III}(CN)_6^{3-}/Fe^{II}(CN)_6^{4-}$ couple. The effects of sweep rate, concentration of electroactive species, supporting electrolyte, electrode material, and irreversibility are evaluated.

References

1 P. T. Kissinger and W. R. Heineman, *J. Chem. Ed., 60,* 702 (1983).
2 J. J. Van Benschoten, J. Y. Lewis, W. R. Heineman, D. A. Roston, and P. T. Kissinger, *J. Chem. Ed., 60,* 772 (1983).
3 W. R. Heineman and P. T. Kissinger in "Laboratory Techniques in Electroanalytical Chemistry," P. T. Kissinger and W. R. Heineman (eds.), Dekker, New York, 1984, chap. 3.

†Reprinted in part with permission from Refs. 1 and 2. Copyright 1983, Division of Chemical Education, American Chemical Society.

4 A. J. Bard and L. R. Faulkner, "Electrochemical Methods," Wiley, New York, 1980, chap. 6.

5 D. T. Sawyer and J. L. Roberts, Jr., "Electrochemistry for Chemists," Wiley-Interscience, New York, 1974, chap. 7.

6 R. N. Adams, "Electrochemistry at Solid Electrodes," Dekker, New York, 1969, chap. 5.

7 E. Gileadi, E. Kirowa-Eisner, and J. Penciner, "Interfacial Electrochemistry," Addison-Wesley, New York, 1975, chap. III-9.

8 I. M. Kolthoff and W. J. Tomsicek, *J. Phys. Chem., 39,* 945 (1935).

9 W. R. Heineman and P. T. Kissinger, *Amer. Lab., 11,* 29 (November 1982).

10 D. H. Evans, K. M. O'Connell, R. A. Peterson, and M. J. Kelly, *J. Chem. Ed., 60,* 290 (1983).

11 J. T. Maloy, *J. Chem. Ed., 60,* 285 (1983).

Apparatus

Instrument for cyclic voltammetry (such as Bioanalytical Systems CV-1B, CV27, or Electrochemical Analyzer; Princeton Applied Research 173/175; IBM EC/225)

x-y Recorder (oscilloscope can also be used)
Electrochemical cell
Platinum working electrode
Platinum auxiliary electrode
SCE or Ag/AgCl reference electrode
Volumetric flasks, 25 mL and 100 mL
Fine alumina or diamond powder (paste)

Chemicals

10 mM potassium ferricyanide [$K_3Fe(CN)_6$] in 1.0 M potassium nitrate [KNO_3] stock solution
1.0 M KNO_3
4 mM $K_3Fe(CN)_6$ in 1 M sodium sulfate [Na_2SO_4]
Unknown: $K_3Fe(CN)_6$ in 1.0 M KNO_3

Theory

Cyclic voltammetry (CV) is perhaps the most versatile electroanalytical technique for the study of electroactive species. Its versatility combined with ease of measurement has resulted in extensive use of CV in the fields of electrochemistry, inorganic chemistry, organic chemistry, and biochemistry. CV is often the first experiment performed in an electrochemical study of an inorganic or organic compound, a biological material, or an electrode surface. The effectiveness of CV results from its capability for rapidly observing redox behavior over a wide potential range. The resulting voltammogram is analogous to a conventional spectrum in that it conveys information as a function of an energy scan.

CV consists of cycling the potential of an electrode, which is immersed in an unstirred solution, and measuring the resulting current. The potential of this working electrode is controlled vs. a reference electrode such as an SCE or Ag/AgCl electrode. The controlling potential that is applied across these two electrodes can be considered an *excitation signal*. The excitation signal for CV is a linear potential scan with a triangular waveform as shown in Fig. 4-5. This triangular potential excitation signal sweeps the potential of the electrode between two values, sometimes called the *switching potentials*. The excitation signal in Fig. 4-5 causes the potential to first scan negatively from +0.80 to −0.20 V vs. SCE at which point the scan direction is reversed, causing a positive scan back to the original potential of +0.80 V. The scan rate, as reflected by the slope, is 50 mV/s. A second cycle is indicated by the dashed line. Single or multiple cycles can be

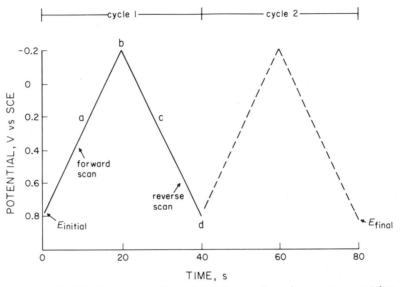

Figure 4-5 / Typical excitation signal for cyclic voltammetry—a triangular potential waveform with switching potentials at 0.8 and −0.2 V vs. SCE. [*Reprinted with permission from P. T. Kissinger and W. R. Heineman, J. Chem. Ed., 60, 702 (1983). Copyright © 1983, Division of Chemical Education, American Chemical Society.*]

used. Modern instrumentation enables switching potentials and scan rates to be easily varied.

A cyclic voltammogram is obtained by measuring the current at the working electrode during the potential scan. The current can be considered the *response signal* to the potential excitation signal. The voltammogram is a display of current (vertical axis) versus potential (horizontal axis). Because the potential varies linearly with time, the horizontal axis can also be thought of as a time axis. This is helpful in understanding the fundamentals of the technique.

A typical cyclic voltammogram is shown in Fig. 4-6 for a platinum working electrode in a solution containing 6.0 mM $K_3Fe(CN)_6$ as the electroactive species in 1.0 M KNO_3 in water as the supporting electrolyte. The potential excitation signal used to obtain this voltammogram is that shown in Fig. 4-5 but with a negative switching potential of −0.15 V. Thus, the vertical axis in Fig. 4-5 is now the horizontal axis for Fig. 4-6. The *initial potential* E_i of 0.80 V applied at *a* is chosen to avoid any electrolysis of $Fe^{III}(CN)_6^{3-}$ when the electrode is switched on. The potential is then scanned *negatively,* forward scan, as indicated by the arrow. When the potential is sufficiently negative to reduce $Fe^{III}(CN)_6^{3-}$, *cathodic current* is indicated at *b* due to the electrode process

$$Fe^{III}(CN)_6^{3-} + e \rightarrow Fe^{II}(CN)_6^{4-} \tag{1}$$

The electrode is now a sufficiently strong reductant to reduce $Fe^{III}(CN)_6^{3-}$. The cathodic current increases rapidly ($b \rightarrow d$) until the concentration of $Fe^{III}(CN)_6^{3-}$ at the electrode surface approaches zero, and the current peaks at *d*. The current then decays ($d \rightarrow g$) as the solution surrounding the electrode is depleted of $Fe^{III}(CN)_6^{3-}$ due to its electrolytic conversion to $Fe^{II}(CN)_6^{4-}$. The scan direction is switched to positive at −0.15 V (*f*) for the *reverse scan*. The potential is still sufficiently negative to reduce $Fe^{III}(CN)_6^{3-}$, so cathodic current continues even though the potential is now scanning in the positive direction. When the electrode

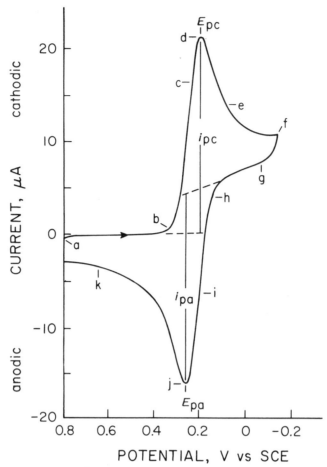

Figure 4-6 / Cyclic voltammogram of 6 mM K$_3$Fe(CN)$_6$ in 1 M KNO$_3$. Scan initiated at 0.8 V vs. SCE in negative direction at 50 mV/s. Platinum electrode area = 2.54 mm^2. [*Reprinted with permission from P. T. Kissinger and W. R. Heineman, J. Chem. Ed., 60, 702 (1983). Copyright © 1983, Division of Chemical Education, American Chemical Society.*]

becomes a sufficiently strong oxidant, Fe^{II}(CN)$_6^{4-}$, which has been accumulating adjacent to the electrode, can now be oxidized by the electrode process

$$\text{Fe}^{II}(\text{CN})_6^{4-} \rightarrow \text{Fe}^{III}(\text{CN})_6^{3-} + e \tag{2}$$

This causes *anodic current* ($i \rightarrow k$). The anodic current rapidly increases until the surface concentration of Fe^{II}(CN)$_6^{4-}$ approaches zero and the current peaks (j). The current then decays ($j \rightarrow k$) as the solution surrounding the electrode is depleted of Fe^{II}(CN)$_6^{4-}$. The first cycle is completed when the potential reaches +0.80 V. Now that the cyclic voltammogram is obtained, it is apparent that any potential positive of approximately +0.4 V would be suitable as an initial potential in that reduction of Fe^{III}(CN)$_6^{3-}$ would not occur when the potential is applied. This procedure avoids inadvertent electrolysis as a result of applying the initial potential.

Simply stated, in the forward scan Fe^{II}(CN)$_6^{4-}$ is electrochemically generated from Fe^{III}(CN)$_6^{3-}$ as indicated by the cathodic current. In the reverse scan this

$Fe^{II}(CN)_6^{4-}$ is oxidized back to $Fe^{III}(CN)_6^{3-}$ as indicated by the anodic current. Thus, CV is capable of rapidly generating a new oxidation state during the forward scan and then probing its fate on the reverse scan. (This very important aspect of the technique is illustrated by Experiment 4-2.)

A more detailed understanding of the cyclic voltammogram waveform can be gained by considering the Nernst equation and the changes in concentration that occur in solution adjacent to the electrode during electrolysis. (See Refs. 1, 3, 4, and 11.)

The important parameters of a cyclic voltammogram are the magnitudes of the *anodic peak current* (i_{pa}), *cathodic peak current* (i_{pc}), *anodic peak potential* (E_{pa}), and *cathodic peak potential* (E_{pc}). These parameters are labeled in Fig. 4-6. One method for measuring i_p involves extrapolation of a base-line current as shown in the figure. The establishment of a correct base line is essential for the accurate measurement of peak currents. This is not always easy, particularly for more complicated systems.

A redox couple in which both species rapidly exchange electrons with the working electrode is termed an *electrochemically reversible* couple. The formal reduction potential $E^{o\prime}$ for a reversible couple is centered between E_{pa} and E_{pc}:

$$E^{o\prime} = \frac{E_{pa} + E_{pc}}{2} \tag{3}$$

The number (n) of electrons transferred in the electrode reaction for a reversible couple can be determined from the separation between the peak potentials:

$$\Delta E_p = E_{pa} - E_{pc} \approx \frac{0.059}{n} \tag{4}$$

Thus, a one-electron process such as the reduction of $Fe^{III}(CN)_6^{3-}$ to $Fe^{II}(CN)_6^{4-}$ exhibits a ΔE_p of approximately 0.059 V. Slow electron transfer at the electrode surface, "irreversibility," causes this peak separation to increase.

The peak current for a reversible system is described by the Randles-Sevcik equation for the forward sweep of the first cycle

$$i_p = 2.69 \times 10^5 n^{3/2} A D^{1/2} C v^{1/2} \tag{5}$$

where
i_p = peak current, A
n = electron stoichiometry
A = electrode area, cm²
D = diffusion coefficient, cm²/s
C = concentration, mol/cm³
v = scan rate, V/s

Accordingly, i_p increases with $v^{1/2}$ and is directly proportional to concentration. The relationship to concentration is particularly important in analytical applications and in studies of electrode mechanisms. The values of i_{pa} and i_{pc} should be close for a simple reversible (fast) couple. That is,

$$\frac{i_{pa}}{i_{pc}} \approx 1 \tag{6}$$

However, the ratio of peak currents can be significantly influenced by chemical reactions coupled to the electrode process, as is demonstrated in Experiment 4-2.

Procedure

Pretreatment of the platinum working electrode surface may be required. Simply polishing the surface with powdered alumina and rinsing thoroughly with distilled water should suffice. The electrode can then be sonicated in an ultrasonic bath if available.

The cell is assembled and filled with 1 M KNO_3 so that the ends of the electrodes are immersed. The cell is deoxygenated by purging with N_2 for approximately 10 min. Following this, N_2 is directed over the solution to prevent oxygen from re-entering the cell during the remainder of the experiment.

While the cell is being deoxygenated, the scan parameters can be set. The working electrode should be disconnected or switched off during this procedure. The initial potential is set at 0.80 V, and the scan limits at 0.80 V and −0.12 V using the recorder (or a digital voltmeter) as a monitor. All scans are initiated in the negative direction with a scan rate of 20 mV/s. These settings are to be used unless otherwise specified.

When deoxygenation is complete, the working electrode is switched on. After allowing the current to attain a constant value (in about 10 s), the potential scan is initiated and a background CV of the supporting electrolyte solution is obtained.

After turning off the working electrode, the cell is cleaned and refilled with 4 mM $K_3Fe(CN)_6$ in 1 M KNO_3. Following the same procedure as above, a CV of the $Fe^{III}(CN)_6^{3-}/Fe^{II}(CN)_6^{4-}$ couple is obtained.

The effect of the scan rate (v) on the voltammograms is observed by using this same solution and recording CV's at the following rates: 20, 50, 75, 100, 125, 150, 175, and 200 mV/s. Between each scan, initial conditions at the electrode surface are restored by moving the working electrode gently up and down without actually removing it from solution or by activating a stirring bar. Care should be taken that no bubbles remain on the electrodes. Allow a minute or two after stirring for the solution to come to rest before obtaining a CV.

Concentration likewise affects the magnitude of the peak current. This is seen by obtaining CV's on 2, 6, 8, and 10 mM $K_3Fe(CN)_6$ using a scan rate of 20 mV/s. A voltammogram of the unknown $K_3Fe(CN)_6$ solution should be obtained as well.

The effect of the supporting electrolyte on the appearance of the CV is demonstrated by recording voltammograms of (1) 4 mM ferricyanide in 1 M KNO_3 and (2) 4 mM ferricyanide in 1 M Na_2SO_4.

Treatment of Data

Determine $E^{\circ\prime}$ and n for the $Fe^{III}(CN)_6^{3-}/Fe^{II}(CN)_6^{4-}$ couple in 1.0 M KNO_3 from one of the cyclic voltammograms on Pt. Compare your value with one reported in the literature. (See Ref. 8.)

Determine the effect of scan rate on peak height by calculating i_{pc} and i_{pa} for the various scan rates used in the scan rate experiment. Plot i_{pc} and i_{pa} vs. $v^{1/2}$.

Determine the effect of scan rate on ΔE_p by plotting ΔE_p vs. v. Explain what causes ΔE_p to increase.

Determine the effect of concentration by plotting i_{pa} and i_{pc} vs. $[Fe^{III}(CN)_6^{3-}]$.

Discuss the effect of supporting electrolyte on the shape of the voltammogram, $E^{\circ\prime}$, and reversibility.

Questions

1 Sketch the concentration-distance profiles for $Fe^{III}(CN)_6^{3-}$ and $Fe^{II}(CN)_6^{4-}$ that would be expected at points a through k on the CV in Fig. 4-6.

2 Using the profiles from Question 1 and Eq. (3) in the text of the chapter, explain why the current increases rapidly, then peaks and decays during the forward scan in Fig. 4-6.

3 What would the reverse scan look like if a stirring bar were switched on at point f during the CV in Fig. 4-6? (*Hint:* See Fig. 4-2.)

4 Explain why larger peak currents are obtained for faster scan rates. (*Hint:* What is the effect of a faster scan rate on the concentration distance profiles?)

5 Sketch the voltammogram that would be obtained if $Fe^{II}(CN)_6^{4-}$ reacted extremely rapidly to give another Fe^{II}-containing species that is not electroactive within the potential range of 0.8 to -0.2 V vs. SCE.

6 Comment on the thermodynamic validity of an $E^{\circ\prime}$ obtained from a cyclic voltammogram for which ΔE_p is substantially greater than $0.059/n$ volts.

EXPERIMENT 4-2
Study of Electrode Mechanism by Cyclic Voltammetry†

Purpose

Cyclic voltammetry is used to study the electrode mechanism of acetaminophen oxidation, which involves coupled chemical reactions.

References

1 J. J. Van Benschoten, J. Y. Lewis, W. R. Heineman, D. A. Roston, and P. T. Kissinger, *J. Chem. Ed., 60,* 772 (1983).

2 D. J. Miner, J. R. Rice, R. M. Riggin, and P. T. Kissinger, *Anal. Chem., 53,* 2258 (1981).

3 C. R. Preddy, D. J. Miner, D. A. Meinsma, and P. T. Kissinger, *Current Separations,* 1984.

4 R. S. Nicholson, *Anal. Chem., 37,* 1351 (1965).

5 R. S. Nicholson and I. Shain, *Anal. Chem., 36,* 705 (1964).

6 R. S. Nicholson and I. Shain, *Anal. Chem., 37,* 178 (1965).

7 M. L. Olmstead, R. G. Hamilton, and R. S. Nicholson, *Anal. Chem., 41,* 260 (1969).

8 D. H. Evans, *Acct. Chem. Res., 10,* 313 (1977).

9 M. D. Hawley in "Laboratory Techniques in Electroanalytical Chemistry," P. T. Kissinger and W. R. Heineman (eds.), Dekker, New York, 1984, chap. 17.

Apparatus

Instrument for cyclic voltammetry (such as Bioanalytical Systems, CV-1B, CV27 or Electrochemical Analyzer; Princeton Applied Research 173/175; IBM EC/225)

x-y Recorder (oscilloscope can also be used)

Electrochemical cell

Platinum auxiliary electrode

SCE or Ag/AgCl reference electrode

Carbon paste working electrode (such as the MF 2010 from Bioanalytical System, Inc., West Lafayette, Ind.; instructions for preparing the electrode are available from the manufacturer)

Chemicals

McIlvaine buffers with 0.5 *M* ionic strength:
  pH 2.2, 500 mL
  pH 6, 200 mL
1.8 *M* sulfuric acid [H_2SO_4], 200 mL

---

†Reprinted in part with permission from Ref. 1. Copyright 1983, Division of Chemical Education, American Chemical Society.

Stock solution of 0.070 M acetaminophen in 0.05 M perchloric acid [$HClO_4$] (store in refrigerator)

Tylenol tablet

Theory

There are inorganic ions, metal complexes, and a few organic compounds that undergo electron transfer reactions without the making or breaking of covalent bonds. The vast majority of electrochemical reactions involve an electron transfer step that leads to a species that rapidly reacts with components of the medium via so-called *coupled chemical reactions*. One of the most useful aspects of cyclic voltammetry (CV) is its application to the qualitative diagnosis of these homogeneous chemical reactions that are coupled to the electrode surface reaction. CV provides the capability for generating a species during the forward scan and then probing its fate with the reverse scan and subsequent cycles, all in a matter of seconds or less. In addition, the time scale of the experiment is adjustable over several orders of magnitude by changing the potential scan rate, enabling some assessment of the rates of various reactions.

Acetaminophen (*N*-acetyl-*p*-aminophenol, APAP), the active ingredient in Tylenol, is commonly used as an aspirin substitute. However, unlike aspirin, it is known to cause liver and kidney damage when administered in large amounts. It is suspected that a metabolite of APAP is the actual hepatotoxic agent, thus APAP and its metabolites have been extensively investigated (Ref. 2).

Voltammetric studies in aqueous solution have revealed chemical as well as electrochemical steps (Ref. 3). The APAP system therefore is useful in demonstrating the mechanistic information that can be obtained from CV's.

The oxidation mechanism of APAP is as follows:

APAP is electrochemically oxidized in a pH-dependent, two-electron, two-proton process to *N*-acetyl-*p*-quinoneimine (NAPQI) (step 1). The occurrence of followup chemical reactions involving NAPQI is pH-dependent. By varying the pH of the media and the scan rate of the cyclic voltammetry experiment, chemical reactions involving NAPQI can be "mapped-out."

At pH values ≥ 6, NAPQI exists in the stable unprotonated form (B). Cyclic voltammograms recorded for APAP at pH 6 are shown in Fig. 4-7. Reasonably well-defined anodic and cathodic waves are evident. The anodic current represents step 1 in the mechanism detailed above while the cathodic current represents the reverse of this step. The similarity in appearance of the pH 6 cyclic voltammograms observed with 40- and 250-mV/s scan rates indicates that the involved species are stable in the time domain of the cyclic voltammetry experiment. The large separation between the anodic and cathodic peak currents in the pH 6 cyclic voltammograms is a manifestation of sluggish heterogeneous electron transfer kinetics.

Under more acidic conditions, NAPQI is immediately protonated (step 2), yielding a less stable but electrochemically active species (C) which rapidly yields (step 3) a hydrated form (D) that is electrochemically inactive at the examined po-

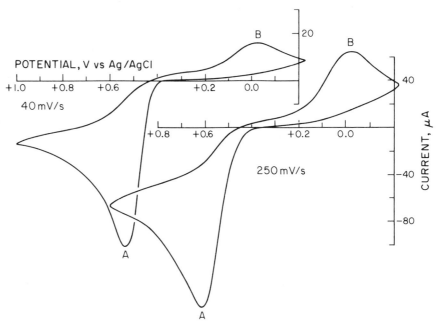

Figure 4-7 / Cyclic voltammograms of 3.6 m*M* APAP in pH 6 McIlvaine buffer. Carbon paste electrode. [*Reprinted with permission from J. J. Van Benschoten, J. Y. Lewis, W. R. Heineman, D. A. Roston, and P. T. Kissinger, J. Chem. Ed., 60, 772 (1983). Copyright © 1983, Division of Chemical Education, American Chemical Society.*]

tentials. Cyclic voltammograms shown in Fig. 4-8 are consistent with this mechanism. The pH of the media is 2. A small cathodic wave due to the reduction of protonated NAPQI (C) is evident when the scan rate of 250 mV/s is employed. This wave is even more pronounced when faster scan rates are employed; however, faster scan rates require the use of an oscilloscope to record the voltammogram. With a slower scan rate of 40 mV/s, a cathodic wave for the reduction of protonated NAPQI is not observed. All of the protonated NAPQI (C) is converted to the inactive hydrated form (D) before sufficiently negative potentials are reached during the reverse scan of the cyclic voltammetry experiment.

Hydrated NAPQI (D) converts (step 4) to benzoquinone; however, the medium has to be extremely acidic for the rate of the process to be significant enough that reduction of benzoquinone is observed during the cyclic voltammetry experiment. The medium for the cyclic voltammograms detailed in Fig. 4-9 is 1.8 *M* H_2SO_4. A poorly defined cathodic wave for the reduction of benzoquinone (E) is observed when a scan rate of 250 mV/s is employed. The reduction wave is broad because the formation of benzoquinone (E) from hydrated NAPQI (D) occurs during the reverse scan. When the scan rate is 40 mV/s, the increased length of time required to reach negative enough potentials during the reverse scan allows for the accumulation of benzoquinone (E). Consequently, a well-defined reduction wave is observed for benzoquinone (E) when the slower scan rate is employed. The second scan in the positive direction yields an anodic wave, in addition to that of APAP, which corresponds to the oxidation of hydroquinone, the reduction product of benzoquinone.

Procedure

Prepare the carbon paste electrode according to the manufacturer's instructions. The electrode surface should be polished to a shiny finish. Care must be taken not to scratch the carbon paste surface once it has been polished.

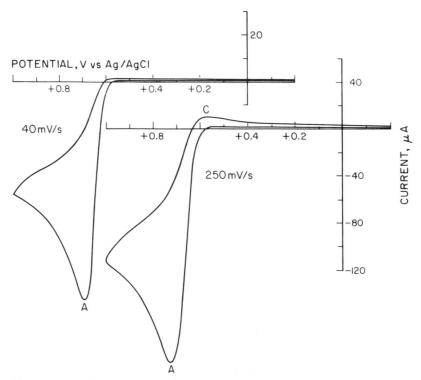

Figure 4-8 / Cyclic voltammograms of 3.6 m*M* APAP in pH 2 McIlvaine buffer. Carbon paste electrode. [*Reprinted with permission from J. J. Van Benschoten, J. Y. Lewis, W. R. Heineman, D. A. Roston, and P. T. Kissinger, J. Chem. Ed., 60, 772 (1983). Copyright © 1983, Division of Chemical Education, American Chemical Society.*]

Prepare a 3 m*M* APAP solution in the pH 2.2 buffer. (The concentrations of all APAP solutions should be accurately known.) Set the scan limits of the potentiostat at 1.0 V and −0.2 V vs. Ag/AgCl. Initiate cyclic voltammograms at 0.0 V with a positive scan. Record cyclic voltammograms at scan rates of 40 mV/s and 250 mV/s. (If an oscilloscope is available, record voltammograms at a few faster scan rates.) Stir the solution briefly and then allow 2 min for the solution to quiet between the recording of each voltammogram.

Repeat the above procedure for the following two solutions: 3 m*M* APAP in pH 6 buffer and 3 m*M* APAP in 1.8 *M* H_2SO_4.

Drop an accurately weighed Tylenol tablet into a 250-mL volumetric flask, add some pH 2.2 buffer and shake until the tablet dissolves, then dilute to volume with pH 2.2 buffer. Dilute a 5.00-mL aliquot of this solution to 50.00 mL using a pipet and a volumetric flask. Prepare four standard solutions of APAP (in addition to the 3 m*M* solution that has already been prepared) that span the concentration range of 0.10 to 5.0 m*M* by appropriate dilution of the APAP stock solution in pH 2.2 buffer. Record cyclic voltammograms of the five standard solutions and the diluted Tylenol solution. These voltammograms should be recorded under identical conditions (such as scan rate).

Treatment of Data   Write the electrode reaction that is occurring for each peak of the cyclic voltammograms obtained for the three supporting electrolytes.

Construct a calibration curve by plotting peak current vs. concentraion of APAP for the standard solutions of APAP. Determine the concentration of

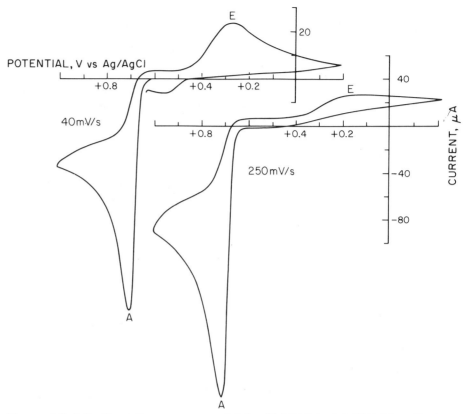

Figure 4-9 / Cyclic voltammograms of 3.6 m*M* APAP in 1.8 *M* H₂SO₄. Carbon paste electrode. [*Reprinted with permission from J. J. Van Benschoten, J. Y. Lewis, W. R. Heineman, D. A. Roston, and P. T. Kissinger, J. Chem. Ed., 60, 772 (1983). Copyright © 1983, Division of Chemical Education, American Chemical Society.*]

APAP in the diluted Tylenol solution. Calculate the percent weight of APAP in the Tylenol tablet. Compare your experimental result with the value on the label of the Tylenol bottle.

Questions

1 An electrode mechanism in which the electrogenerated species reacts chemically is termed an EC mechanism and can be described by the following equations:

Electrode reaction, E: $O + ne \rightleftharpoons R$

Chemical reaction, C: $R \xrightarrow{k}$ product

Draw cyclic voltammograms for the following cases. (Assume the electrode reaction to be reversible.)

   a The rate constant k is zero.
   b The rate constant k is so large that the chemical reaction is essentially instantaneous relative to the scan rate.
   c k has an intermediate value between those in **a** and **b**.

2 What effect would lowering the temperatures be expected to have on the voltammograms in Fig. 4-8?

3 Explain why faster scan rates are necessary to study mechanisms involving faster chemical reactions.

4 What problems can you anticipate encountering for very fast scan rates (>100 V/s)?

EXPERIMENT 4-3
Cyclic Voltammetry of O_2 and SO_2 in Aprotic Media

Purpose

The electrochemical techniques of potential sweep voltammetry and cyclic voltammetry are utilized to determine the following properties for O_2 and SO_2 in aprotic media: (a) the formal reduction potential ($E^{\circ\prime}$), (b) the electron stoichiometry of the electrode reaction, (c) the heterogeneous electron transfer kinetics and the electrode mechanisms, (d) the solution thermodynamics, (e) the effects of the electrode material and its preconditioning upon electron transfer kinetics, (f) the effects of solvents, supporting electrolytes, and solution acidity upon oxidation-reduction reactions, (g) adsorption processes in heterogeneous catalysis, and (h) the pre- and post-chemical reactions that are associated with the electron transfer reactions.

References

1 D. T. Sawyer and J. L. Roberts, Jr., "Experimental Electrochemistry for Chemists," Wiley-Interscience, New York, 1974, chap. 7.
2 H. H. Bauer, G. D. Christian, and J. E. O'Reilly, "Instrumental Analysis," Allyn and Bacon, New York, 1978, chap. 3.
3 E. Gileadi, E. Kirowa-Eisner, and J. Penciner, "Interfacial Electrochemistry," Addison-Wesley, New York, 1975.
4 Z. Galus, "Fundamentals of Electrochemical Analysis," Ellis Harwood, Ltd., New York, 1976.
5 P. Delahay, "New Instrumental Methods in Electrochemistry," Interscience, New York, 1954.
6 R. S. Nicholson and I. Shain, *Anal. Chem., 36,* 705 (1964).
7 W. H. Reinmuth, *Anal. Chem., 32,* 1891 (1960).
8 A. Y. Gokhstein and Y. P. Gokhstein, *Doklayd, Akad. Nauk. SSSR., 131,* 601 (1960).
9 R. S. Nicholson, *Anal. Chem., 37,* 1351 (1965).
10 H. Matsuda and Y. Ayabe, *Z. Elektrochem., 59,* 495 (1955).
11 J. M. Saveant and E. Vianello, "Advances in Polarography," I. S. Langmuir (ed.), vol. I, Pergamon Press, New York, 1960.
12 J. M. Saveant and E. Vianello, *Compt. Rend., 256,* 2597 (1963).
13 J. M. Saveant and E. Vianello, *Electrochim. Acta, 8,* 905 (1963).

Apparatus

Instrument for cyclic voltammetry (such as Bioanalytical Systems CV-1B, CV27, or Electrochemical Analyzer; Princeton Applied Research 173/175; IBM EC/225)
Oscilloscope (optional)
x-y Recorder
Platinum working electrode
Platinum auxiliary electrode
Ag/AgCl reference electrode
Electrochemical cell
Alumina polishing powder, 600 mesh
Pipet, 50 mL
Pipet, 200 μL, with tips
Glass frit, medium porosity
Magnetic stirrer and stirring bar
Luggin capillary

Chemicals

Tetraethylammonium perchlorate [TEAP]
Dimethylsulfoxide [DMSO]

safety cautions

TEAP is potentially explosive.

Avoid skin contact with DMSO, which absorbs rapidly through skin thereby transporting solutes into the body.

Cylinders of oxygen gas and nitrogen gas
~2.5 M SO_2–DMSO solution, standardized

Theory

The technique of cyclic voltammetry is described in Experiment 4-1.

The fundamental equations of linear-potential sweep and cyclic voltammetry have been developed by Delahay, Nicholson, Shain, and others. The equations derived apply only if there are no concentration gradients in the experiment before it is started. For a simple cathodic charge transfer under reversible conditions $(O + ne \rightleftharpoons R)$, the peak current i_{pc} is given by the relation

$$i_{pc} = 0.4463 nFA(Da)^{1/2}C \qquad (1)$$

with

$$a = \frac{nFv}{RT} = \frac{nv}{0.026} \qquad \text{at } 25°C$$

where
n = number of electrons involved in the reduction
F = faraday
A = area of the working electrode, cm^2
v = scan rate, V/s
C = concentration of the bulk species, mol/cm^3
D = diffusion coefficient of the electroactive species, cm^2/s

In terms of the adjustable parameters the peak current is given by the Randles-Sevcik equation

$$i_{pc} = 2.69 \times 10^5 n^{3/2} A D^{1/2} C v^{1/2} \qquad (2)$$

The peak potential E_p for a reversible process is related to the polarographic half-wave potential $E_{1/2}$ (see Experiment 4-4) by the expression

$$E_{pc} = E_{1/2} - 1.11 \frac{RT}{nF} = E_{1/2} - \frac{0.0285}{n} \qquad \text{at } 25°C \qquad (3)$$

$$E_{pa} = E_{1/2} + 1.11 \frac{RT}{nF} = E_{1/2} + \frac{0.0285}{n} \qquad \text{at } 25°C \qquad (4)$$

The polarographic half-wave potential is related to the standard electrode potential $E°$ by the equation

$$E_{1/2} = E° - \frac{RT}{nF} \ln \frac{f_{red}}{f_{ox}} \left(\frac{D_{ox}}{D_{red}}\right)^{1/2} = E° - \frac{RT}{nF} \ln \left(\frac{D_{ox}}{D_{red}}\right)^{1/2} \qquad (5)$$

assuming that the activity coefficients (f_{ox}, f_{red}) are equal for the oxidized and reduced species involved in the electrochemical reaction.

By taking the difference between the anodic and cathodic peak potentials of Eqs. (4) and (3) a good criterion for determining electrode-process reversibility is obtained.

$$E_{pa} - E_{pc} = 2.22 \frac{RT}{nF} = \frac{0.0595}{n} \qquad \text{at } 25°C \qquad (6)$$

This also provides a rapid and convenient means for establishing the number of electrons in a reversible electrochemical reaction. Because of the dynamic nature of the electrode process, the difference between the anodic and cathodic peak potentials depends on the degree of the reversibility of the process. If the process is fast and the scan rate slow, the process is reversible. If the process is slow and the scan rate fast, the process becomes irreversible and the difference between the anodic and cathodic peak potentials is greater than $0.0595/n$.

Irreversible processes obey a distinctly different expression for the peak current.

$$i_{pc} = 2.99 \times 10^5 n (\alpha N_a)^{1/2} A D^{1/2} C v^{1/2} \qquad (7)$$

where α is the transfer coefficient (which indicates the symmetry of the potential energy function for the transition state) and N_a is the number of electrons in the rate-determining step for the process.

The peak potential for the irreversible process is given by the expression derived by Matsuda and Ayabe (Ref. 10).

$$E_{pc} = E° - \frac{RT}{\alpha N_a F} \left(0.780 + \ln \sqrt{\frac{D_{ox} \alpha N_a F v}{RT}} - \ln k_s \right) \qquad (8)$$

The peak potential expressed in terms of the Tafel slope ($b = 2.303 RT / \alpha N_a F$) and the polarographic half-wave potential is

$$E_p = E_{1/2} - b \left(0.52 - \tfrac{1}{2} \log \frac{b}{D} - \log k_s + \tfrac{1}{2} \log v \right) \qquad (9)$$

For this first-order charge-transfer reaction the heterogeneous rate constant k_s and the Tafel slope can be calculated from a plot of E_{pc} vs. $\log v$. In the reversible portion of the graph the slope will be zero. In the irreversible portion of the graph the slope will be equal to the Tafel slope divided by 2. The transformation from a reversible system to an irreversible system occurs at a characteristic scan rate v_c. At this scan rate the values of E_p calculated from Eqs. (3) and (9) are equal and k_s may be calculated once the value of the diffusion coefficient is known.

Several other methods are available for determining the product αN_a and k_s. Using the half-peak potential $E_{p/2}$ as the reference point, Nicholson and Shain (Ref. 6) showed that

$$E_p - E_{p/2} = -1.857 \left(\frac{RT}{\alpha N_a F} \right) \qquad (10)$$

From Eqs. (8) and (10) both the peak potential and the half-peak potential are functions of the scan rate. By scanning the voltammogram at different scan rates, αN_a can be evaluated by the expression

$$(E_p)_2 - (E_p)_1 = (E_{p/2})_2 - (E_{p/2})_1 = \frac{RT}{\alpha N_a F} \ln\sqrt{\frac{v_1}{v_2}} \tag{11}$$

Gokhstein (Ref. 8) derived the relationship between peak current and peak potential for an irreversible reaction and showed that

$$i_{pc} = 0.227 n F A C k_s \exp\left[\frac{-\alpha N_a F}{RT}(E_p - E^\circ)\right] \tag{12}$$

A plot of $\ln i_p$ vs. $(E_p - E^\circ)$ or $(E_{p/2} - E^\circ)$ for different scan rates (which should be varied by several orders of magnitude) yields a straight line with a slope proportional to αN_a and an intercept proportional to k_s.

Still another approach to obtaining kinetic information was described by Reinmuth (Ref. 7) who showed that for an irreversible reaction the current at the foot of the voltammogram is independent of the scan rate. For this condition the instantaneous current is measured at less than one-tenth that of the peak current. It is related to the kinetic parameters by the expression

$$i = n F A C k_s \exp\left[\frac{-\alpha N_a F}{RT}(E - E_i)\right]$$

where E is the potential at the measured instantaneous current and E_i is the potential at which the scan was initiated.

In the preceding discussions, all of the kinetic parameters were derived using the linear potential-sweep method. The equations derived apply only if there are no concentration gradients in the solution just before the scan is initiated. Cycling the potential several times creates complex concentration gradients in the solution near the electrode surface and this boundary layer problem has not been solved. Cyclic voltammetry is well suited for determining reversibility [see Eq. (6)], identifying the steps in the overall reaction, and determining new species which appear in solution as a result of combined electrochemical and chemical steps. Cyclic voltammetry provides information about chemical reactions occurring before and after electron transfer (see Experiment 4-2). This information is obtained from a comparison of peak currents between oxidation and reduction peaks which are coupled and from peak current data as a function of scan rate. For a more detailed discussion, consult the original treatments by Nicholson, Shain, and Saveant and the comprehensive review by Galus (Refs. 4, 6, 9, 11, 12, and 13).

Procedures

A. Cyclic Voltammetry of O₂

Polish the platinum working electrode surface with alumina polishing paste until a mirror bright finish is obtained. Rinse the electrode thoroughly with distilled water, then allow it to dry. (For very accurate work the electrode should be electrochemically treated. See Refs. 1 and 3 for further details.) Determine the amount of tetraethylammonium perchlorate (TEAP) needed to make 50 mL of 0.1 M solution. Add the TEAP to the cell container and assemble the electrochemical cell. Add 50 mL of dimethylsulfoxide (DMSO) to the cell. Deoxygenate the solution with nitrogen for at least 10 min; continue to pass nitrogen over the cell during the experiment. Determine the background current from a voltammogram recorded with a range of -2.0 V to $+1.0$ V vs. SCE with a scan rate of 0.1 V/s. E_i should be approximately -0.050 V vs. SCE.

Bubble oxygen through the DMSO solution until it is saturated. Continue to pass oxygen over the solution during the experiment. The solubility of oxygen in DMSO is 2.1×10^{-3} M at 25°C. Record voltammograms at the following scan rates: 20, 60, 80, 100, and 200 mV/s. If an oscilloscope is available, use it in place of the recorder and record additional voltammograms with scan rates between 1 and 10 V/s. The cell should be bubbled with oxygen for 20 s followed by a 2-min rest period between each run. (Proceed to Procedure B of this experiment.)

Treatment of Data

Calculate the diffusion coefficient at a slow scan rate from Eq. (2). At fast scan rates calculate the value of αN_a using Eqs. (10) and (11).

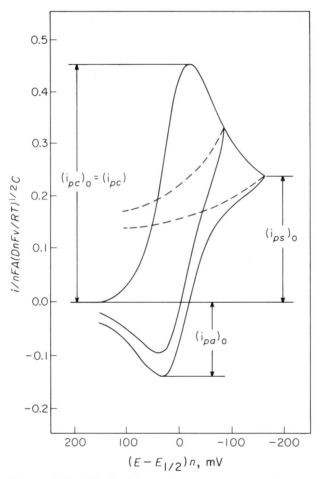

Figure 4-10 / Method for measurement of peak currents and peak-current ratios of cyclic voltammograms

$$\frac{i_{\mathrm{pa}}}{i_{\mathrm{pc}}} = \frac{(i_{\mathrm{pa}})_0}{(i_{\mathrm{pc}})_0} + \frac{0.485(i_{\mathrm{ps}})_0}{(i_{\mathrm{pc}})_0} + 0.086$$

(Adapted from "Experimental Electrochemistry for Chemists," D. T. Sawyer and J. L. Roberts, Jr., copyright © 1974 by Wiley-Interscience, New York, p. 340. Reprinted with permission.)

Calculate the diffusion coefficient for high scan rate from Eq. (7). Does this answer agree with the value you obtained at a slow scan rate? Discuss the results you obtained.

Plot E_{pc} vs. log v; obtain the Tafel slope and calculate the heterogeneous rate constant using Eq. (9).

Plot ΔE_p vs. log v (ΔE_p is obtained by measuring the distance between the cathodic and anodic peaks). Discuss the effect of the scan rate on the shape of the curve and the reversibility of the reaction.

Plot i_{pc} vs. log v; determine the slope. Is there a break in the plot? Is so, why?

B. Cyclic Voltammetry of SO$_2$

Deoxygenate the solution with nitrogen until a voltammogram shows that no oxygen is present. Add enough of the standardized SO$_2$–DMSO solution to make the solution 1 mM in SO$_2$. Record voltammograms at the following scan rates: 20, 40, 80, 100, 200, and 500 mV/s. Add an additional aliquot of the standardized SO$_2$–DMSO solution so the solution in the cell is 5 mM. Record voltammograms at the same scan rates used above. Continue to "blanket" the solution with nitrogen during the experiment.

Treatment of Data

Determine the product αN_a using Eqs. (10) and (11); then determine the diffusion coefficient of SO$_2$ in DMSO. α usually has a value of 0.3 to 0.7. Determine the number of electrons in the reduction of SO$_2$.

Plot $i_{pc}/v^{1/2}$ vs. log v, and i_{pa}/i_{pc} vs. log v.

From the shape of the voltammograms you obtained and the results of the graphs plotted above, explain what is happening in this electrochemical process. What is the significance of the second anodic peak you obtain? A plot of i_{pa_1}/i_{pa_2} vs. log v will prove beneficial for this discussion.

Methods for measuring i_{pa}/i_{pc} from a voltammogram are illustrated in Figs. 4-6 and 4-10.

Questions

1 List other aprotic solvents that could be used for this experiment.
2 What problem(s) does an aprotic solvent, such as benzene, that has a very low dielectric constant present for voltammetric experiments?
3 Do you expect traces of water in the DMSO to affect any of the results in this experiment?
4 Write the mechanism for electrochemical reduction of O$_2$ in (*a*) an aprotic solvent and (*b*) water.

EXPERIMENT 4-4

Polarography

Purpose

This experiment involves the study of several of the important aspects of polarography. Among these are the effect of oxygen, the effect of surface-active agents in eliminating maxima, determination of the electrocapillary curve, determination of capillary characteristics, verification of the Ilkovic equation, verification of the equation for the current-voltage relationship (determination of n and of electrode reversibility), as well as a study of methods employed for obtaining quantitative results: the absolute method, the calibration-curve method, and the method of standard additions.

References

1 I. M. Kolthoff and J. J. Lingane, "Polarography," 2nd ed., Interscience, New York, 1952.

2 L. Meites, "Polarographic Techniques," Interscience, New York, 1955.

3 J. J. Lingane, "Electroanalytical Chemistry," 2nd ed., Interscience, New York, 1958, chap. 11.

4 A. J. Bard and L. R. Faulkner, "Electrochemical Methods," Wiley, New York, 1980, chap. 5.

5 A. M. Bond, "Modern Polarographic Methods in Analytical Chemistry," Dekker, New York, 1980.

6 J. Heyrovsky and J. Kuta, "Principles of Polargraphy," Academic Press, New York, 1966.

7 L. Meites in "Treatise on Analytical Chemistry," I. M. Kolthoff and P. J. Elving (eds.), part I, vol. 4, sec. D-2, Interscience, New York, 1963, chap. 46.

Apparatus

Instrument for polarography and recorder
Electrochemical cell
DME

note Mercury is toxic.

SCE reference electrode
Platinum auxiliary electrode
Stopwatch
Nitrogen system for deoxygenating the cell solution (commercially available high purity nitrogen is sufficiently free of oxygen to be used without further purification for all except the most refined measurements)
Volumetric flasks (3), 100 mL
Pipets, one each, 1 mL, 5 mL, 10 mL, 15 mL

Chemicals

6 M nitric acid [HNO_3]
0.001 M potassium chloride [KCl]
Triton X-100 (Rohm & Haas Company, Philadelphia, Pa.), 0.2%
0.02 M cadmium chloride [$CdCl_2$], standard solution
0.2 M hydrochloric acid [HCl] supporting electrolyte (17 mL concentrated HCl per liter)
Unknown: Sample of Cd^{2+} in a 100-mL volumetric flask, about 0.002 M when diluted to the mark
Unknown: Sample of metal ion in a 100-mL volumetric flask, about 0.002 M when diluted to the mark.

Theory

Polarography is the voltammetric technique that utilizes the dropping mercury electrode (DME) as the working electrode in an unstirred solution. In conventional polarography, the potential applied across the DME vs. a references electrode is scanned linearly from an initial to a final potential, and the resulting current is measured to give a polarogram. Such a polarogram for a solution of Cd^{2+} in HCl supporting electrolyte is shown in Fig. 4-11. As the potential is scanned negatively, the onset of Cd^{2+} reduction at the DME is signalled in polarogram B by the increase in cathodic current at the decomposition potential. The current

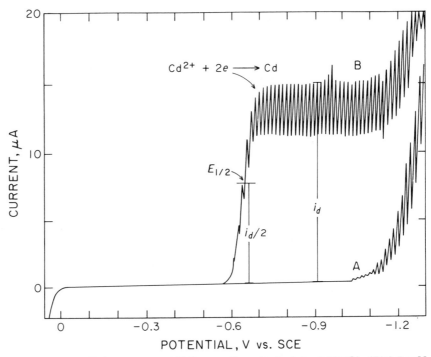

Figure 4-11 / Polarograms. (A) Supporting electrolyte, 1 *M* HCl. (B) 0.5 m*M* Cd²⁺ in 1 *M* HCl. (*From "Experimental Electrochemistry for Chemists," D. T. Sawyer and J. L. Roberts, Jr., copyright © 1974 by Wiley-Interscience, New York, p. 334. Adapted with permission.*)

increases until Cd^{2+} reduction is limited by diffusion of Cd^{2+} to the surface of the expanding drop at which point the current plateaus to give the diffusion current, i_d. The repetitive current fluctuations correspond to growth and loss of each mercury drop on the DME. Polarogram A was recorded with HCl supporting electrolyte alone. This residual current occurring in the absence of Cd^{2+} can be used as a base line for measuring i_d as shown in Fig. 4-11.

The polarographic-wave equation that defines the sigmoid portion of the wave for a reversible system is

$$E = E_{1/2} + \frac{0.0591}{n} \log \frac{i_d - i}{i} \tag{1}$$

where i is the current corresponding to a particular potential E. The half-wave potential $E_{1/2}$ is the potential at which $i = i_d/2$. The $E_{1/2}$ is related to the formal reduction potential $E^{\circ\prime}$ for $Cd^{2+}/Cd(Hg)$

$$E_{1/2} = E^{\circ\prime} + \frac{0.059}{n} \log \frac{D_{Cd(Hg)}^{1/2}}{D_{Cd^{2+}}^{1/2}} - E_{\text{ref}} \tag{2}$$

where D is the diffusion coefficient. As such, $E_{1/2}$ is useful for qualitative identification of the electroactive species. Values of $E_{1/2}$ are tabulated in Appendix Table A-12.

The Ilkovic equation defines the diffusion-limited cathodic current i_d:

$$i_d = knm^{2/3}D^{1/2}t_d^{1/6}C \tag{3}$$

where i_d = diffusion current, μA

           n = number of electron equivalents per molar unit

           D = diffusion coefficient of reducible species, cm^2/s

          m = rate of flow of mercury from the electrode, mg/s

           t = drop time, s

          k = constant (708 for maximum current, 607 for average current)

         C = concentration, mmol/L

As shown by the Ilkovic equation, the magnitude of i_d is directly proportional to the concentration C of electroactive species. As such, i_d is useful for determining the concentration of electroactive species in a solution.

Oxygen dissolved in solution gives a two-wave polarogram such as that shown in Fig. 4-12. Two reduction waves result from O_2 being reduced by $2e$ to H_2O_2 in the first wave and then by $4e$ to H_2O in the second wave. The negative potential limit is defined by reduction of supporting electrolyte; the positive potential limit by oxidation of the mercury electrode. Dissolved O_2 is usually removed by bubbling N_2 through the solution so that the O_2 reduction waves do not interfere with other waves of interest from other species in solution. This gives a smooth base line as shown by polarogram b in Fig. 4-12.

Polarographic maxima can result from hydrodynamic flow of solution around the expanding mercury drop. Maxima are characterized by a large, sometimes erratic, current that diminishes abruptly during the potential scan. One type of maximum occurs at the $E_{1/2}$. Maxima are typically eliminated by addition of a small amount of surfactant, such at Triton X-100, to the solution.

The drop time of the DME varies during the potential scan. Figure 4-13 shows an electrocapillary curve, which is the variation of drop time t_d with potential. The changing drop time reflects the change in surface tension of the mercury

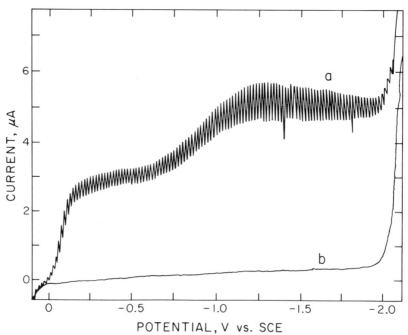

Figure 4-12 / Polarograms of 0.1 *M* KCl. (a) Saturated with air, (b) after deoxygenation by nitrogen bubbling.

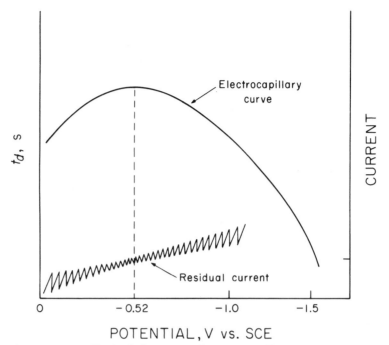

Figure 4-13 / **Typical electrocapillary curve. The magnitude of the residual current is shown on the lower curve.**

drop, depending on the interfacial charge. The potential at which the longest drop time occurs is the electrocapillary maximum. Note how the residual current increases as the potential deviates from the electrocapillary maximum. A more detailed explanation of this phenomenon is found in Ref. 2.

Typical apparatus for polarography is shown in Fig. 4-14. The polarograph (potentiostat) scans the potential applied to the cell, measures the resulting current, and displays it on a recorder. The main feature of the polarographic cell is the dropping mercury electrode. Here mercury is forced by gravity through a very fine capillary to provide a continuous stream of identical droplets, each having a maximum diameter of between 0.5 and 1 mm. Each droplet expands,

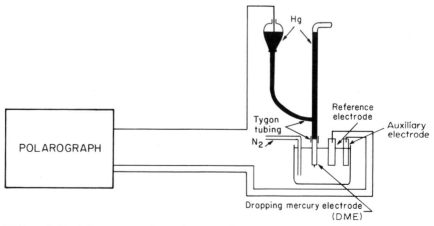

Figure 4-14 / **Apparatus for polarography.**

becomes too heavy to be suspended, and breaks loose from the capillary. The lifetime of a droplet is typically from 4 to 6 s. The actual drop time is determined in part by the rate of flow of mercury through the capillary. This can be varied by altering the height of the mercury head above the capillary. This is accomplished by moving the mercury reservoir up or down. If the reservoir is moved up, mercury flows from the reservoir through the connecting flexible tube to the glass column until the mercury level is equal in the column and the reservoir. In order to increase the lifetime of a droplet, the mercury head should be moved down.

The reference electrode is typically an SCE. A platinum wire usually serves as an auxiliary electrode if the instrument is a three-electrode potentiostat.

A glass tube with a fritted end is also immersed in the sample solution. The tube is connected by a rubber hose to a cylinder of compressed nitrogen. Bubbling nitrogen through the solution removes oxygen. The glass frit acts to disperse the gas into small bubbles that tend to expel the dissolved oxygen more rapidly.

The container of the polarographic cell can be a glass beaker that is fitted with a rubber stopper. Four holes in the stopper accommodate the capillary, the reference electrode, the auxiliary electrode, and the nitrogen tube. The stopper helps keep oxygen out of the cell.

helpful hints 1 *Note: A cardinal rule is that the DME must never be allowed to stand in a solution when the mercury is not flowing.*

Because even traces of dirt or tiny particles will cause erratic behavior of the capillary, or clog it entirely, reasonable care must be exercised to prevent such contamination. Before allowing any solution (or even pure water) to come in contact with the DME, raise the mercury in the column and *check* to see that the mercury has actually begun to flow.

After use, the DME should be withdrawn from the cell, and, with the mercury still issuing from the tip, the capillary should be washed thoroughly with a stream of distilled water from a wash bottle for a minute or two, using 40 to 50 mL of distilled water. The mercury reservoir is then lowered until the mercury flow *just stops* (not further), and the electrode is allowed to stand in the air. An *empty* cell may be raised around the electrode to protect it from laboratory fumes.

2 Contamination of the cell by traces of material left from earlier experiments can cause serious errors. Efficient rinsing of the cell is possible only by removing the rubber stopper and cautiously rinsing the walls with a stream of water from a wash bottle, aspirating out the washings, and repeating several times. An equally cautious technique should be followed in filling the cell with a solution of determinate concentration when the cell is already wet. Errors may be minimized during the investigation of the Cd^{2+} solutions of varying concentration if one works from less concentrated to more concentrated solutions. Naturally the same careful rinsing should be followed when one has finished a day's experimentation.

3 The temperature of the cell should be kept constant at $25 \pm 0.5°C$; even less variation is desirable. A thermostat should be provided to maintain this temperature, but in its absence, best results are obtained by recording calibration curves and unknowns on the same day.

safety caution Mercury vapor is very poisonous. Discard mercury in the waste mercury container provided and clean up any spills immediately. Work over a porcelain tray to catch and recover any spilled mercury conveniently. Notify the instructor in the event of a mercury spill.

Procedures

A. Effect of Oxygen

Maxima

Place 10 mL of 0.1 M KCl in the polarographic cell and record a polarogram of dissolved oxygen by scanning from 0 to -1.9 V vs. SCE. A maximum is usually present in this polarogram.

Suppression of Maxima

Add 2 drops of 0.2% Triton X-100 (maximum suppressor) solution to the 10 mL of 0.1 M KCl, stir well, and record a polarogram. Add a few more drops if the maximum is not entirely suppressed. (*Note:* The addition of too much maximum suppressor will distort the oxygen waves.)

Oxygen Removal (Deoxygenation)

Bubble purified nitrogen gas through the solution remaining in the cell from the "Suppression of Maxima" experiment above. Bubble the gas at a brisk rate for 20 min. Raise the bubbler so that the gas is directed over the surface of the solution. The solution should not be stirred. Record a polarogram of this deoxygenated solution.

Treatment of Data

Write the equations for the electrode reactions of the two oxygen reduction waves.

B. Determination of the Electrocapillary Curve of Mercury

Using the same deoxygenated solution used in Procedure A, set the potential manually at 0.2-V intervals between 0 and -2.0 V and determine the drop time at each setting. Use a stopwatch and measure the time for 10 drops to fall, thus obtaining greater accuracy than if the time for a single drop were measured.

Treatment of Data

Plot drop time t_d vs. E to obtain the electrocapillary curve of mercury in 0.1 M KCl.

C. Measurement of Diffusion Coefficient, $E_{1/2}$, and n for Cd^{2+}

Pipet exactly 5.00 mL of the 0.02 M $CdCl_2$ solution into a 100-mL volumetric flask and dilute exactly to the mark with 0.2 M HCl supporting electrolyte. Deoxygenate this 0.001 M $CdCl_2$ solution and determine its polarogram between -0.1 and -1.2 V vs. SCE. At a fixed potential on the diffusion current of the polarogram, determine, with a stopwatch, the time for 10 drops to form and drop. From this, calculate the time t_d for 1 drop to form and drop. Next, determine the mass of mercury m dropping from the electrode per second. This can be done by collecting the mercury for a known length of time (about 5 min) in a small beaker which contains the supporting electrolyte solution. The capillary is inserted into the solution in the beaker and the mercury is collected for a measured time. The collected mercury is dried with acetone and weighed.

Treatment of Data

Calculate the diffusion coefficient for Cd^{2+} in this supporting electrolyte using the Ilkovic equation. Measure the diffusion current for this calculation, as shown in Fig. 4-15, by extrapolating the residual current portion of the polarogram immediately preceding the rising portion of the polarogram (line A). Take as the diffusion current the difference between this extrapolated line and the current-voltage plateau (B). Measure i_d for the maximum current deflections (see encircled portion of Fig. 4-15) and use 708 as the constant in the Ilkovic equation.

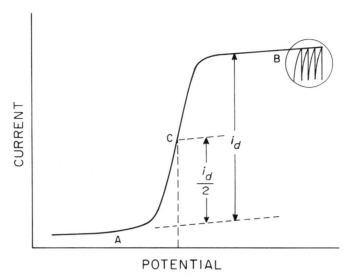

Figure 4-15 / Method for determining diffusion current i_d.

Calculate $E_{1/2}$ for Cd^{2+} by measuring the vertical distance between the residual and diffusion currents, taking half of this value, and extending a line parallel to the residual current through the line of the wave (see Fig. 4-15). From the point of intersection of this parallel line with the wave, point C, drop a perpendicular to the abscissa. (Consult the references for other methods for measuring $E_{1/2}$.)

The equation for a polarographic wave [Eq. (1)] is the equation for a straight line ($y = mx + b$). A plot of E vs. log $[(i_d - i)/i]$ should produce a straight line whose slope is $0.0591/n$ volts. Thus, the value of n can be calculated from the slope. When the value of the logarithmic term becomes zero, the above equation becomes $E = E_{1/2}$, and $E_{1/2}$ is obtained from the point on the plot where the straight line crosses the abscissa. Substances that obey this equation are said to be reversibly oxidized, or reduced.

Plot log $[(i_d - i)/i]$ vs. potential for the polarogram of the 0.001 M $CdCl_2$ above. Select about four or five readings on each side of $E_{1/2}$ from the graphs of current vs. voltage in order to calculate the values of the logarithmic term in the above equation. Indicate on the graph the value of the slope, the calculated value of n, and $E_{1/2}$.

D. Quantitative Determination of Cd^{2+} Concentration in an Unknown

Calibration Curve Method

Prepare solutions of 0.001 M, 0.002 M, 0.003 M, and 0.004 M $CdCl_2$ in 0.2 M HCl by diluting the appropriate amounts of the 0.02 M $CdCl_2$ stock solution with 0.2 M HCl supporting electrolyte. Record polarograms for each standard solution. Obtain an unknown Cd^{2+} solution from the instructor. Transfer exactly 10 mL of the unknown solution to a clean, dry cell; deoxygenate the solution and record a polarogram. Do not vary the height of the mercury column during this series of polarograms.

Method of Standard Additions

While the unknown solution is still in the polarographic cell, pipet a 1.00-mL aliquot of 0.02 M $CdCl_2$ into the cell; deoxygenate and record a polarogram.

Treatment of Data

Calibration Curve Method

Prepare a calibration curve by plotting i_d for the standard solutions at a potential corresponding to a level portion of the limiting current vs. concentration of Cd^{2+}. (See Fig. 4-15 for the measurement of i_d.) Use the calibration curve to calculate the concentration of Cd^{2+} in your unknown from its i_d.

Method of Standard Additions

Calculate the concentration of Cd^{2+} in the unknown from the change in i_d caused by the standard addition. The appropriate equation is

$$C_{unk} = C_{std} \frac{i_d}{i_d' + (i_d' - i_d)(V/v)}$$

where
C_{unk} = concentration of unknown
C_{std} = concentration of standard added
i_d = diffusion current for unknown
i_d' = diffusion current after standard addition
V = volume of unknown
v = volume of standard added

E. Qualitative Identification of an Unknown from $E_{1/2}$

Obtain a sample with an unknown metal ion in 1 M KCl supporting electrolyte from your instructor. Record a polarogram.

Treatment of Data

Measure $E_{1/2}$ and n as illustrated in Procedure C. Determine the identity of the sample by comparison of $E_{1/2}$ with the values listed in Appendix Table A-12. Substantiate your identification by recording a polarogram on a solution of the suspected metal ion in 1 M KCl.

F. Determination of Mixtures (Optional)

To illustrate the possibilities for the analysis of mixtures without the need for separation, record the polarogram of a solution that contains Cu^{2+}, Cd^{2+}, Ni^{2+}, and Zn^{2+} and which is also about 0.1 M in ammonium chloride. The solution should contain 0.002% Triton X-100 to prevent maxima. The waves will occur at about -0.5, -0.8, -1.0, and -1.3 V, respectively, in this medium.

Treatment of Data

Approximate concentrations of each ion can be calculated on the assumption that the diffusion current constant for each ion is the same as that found for cadmium in the previous experiments. This assumption probably is no better than about 90 to 95% valid. (For Cu^{2+} use one-half of this value because n is only 1 for the wave that occurs at about -0.5 V.) In the actual analysis of such a mixture, the constant for each ion would be determined, and one would take into account variations in t_d (which varies markedly with potential, but enters only to the one-sixth power).

Measure i_d for each wave of the polarogram. For waves beyond the first wave, use the extrapolated diffusion current of the preceding wave as a base line. Calculate the approximate concentration of each ion in the mixture using the calibration curve for Cd^{2+} from Procedure D.

Questions

1 The determination of a small amount of Ni^{2+} in the presence of a much larger quantity of Cd^{2+} is hampered by the large Cd^{2+} wave prior to the Ni^{2+} wave.

Propose solution conditions that will improve the analysis by causing the Ni^{2+} wave to precede the Cd^{2+} wave.

2 From an understanding of the Ilkovic equation, suggest experimental measures that could be taken to increase the magnitude of the diffusion current for a given Cd^{2+} sample. Does the increased current also represent a like improvement in detection limit? Justify your answer.

3 Explain the shape obtained for the electrocapillary curve.

4 Explain how the addition of a surfactant to the solution suppresses maxima.

EXPERIMENT 4-5

Differential Pulse Polarography

Purpose

Differential pulse polarography is used to determine Cu^{2+}. The effect of pulse amplitude on peak current, peak potential, and peak width is evaluated.

References

1 A. M. Bond, "Modern Polarographic Methods in Analytical Chemistry," Dekker, New York, 1980, chap. 6.

2 A. J. Bard and L. R. Faulkner, "Electrochemical Methods," Wiley, New York, 1980, chap. 5.

3 D. T. Sawyer and J. L. Roberts, Jr., "Experimental Electrochemistry for Chemists," Wiley-Interscience, New York, 1974.

4 D. E. Burge, *J. Chem. Ed., 47,* A-81 (1970).

5 J. Osteryoung, *J. Chem. Ed., 60,* 296 (1983).

6 J. Osteryoung, *Review of Polarography (Japan), 22,* 1 (1976).

7 S. Borman, *Anal. Chem., 54,* 698A (1982).

8 P. T. Kissinger in "Laboratory Techniques in Electroanalytical Chemistry," P. T. Kissinger and W. R. Heineman (eds.), Dekker, New York, 1984, chap. 5.

Apparatus

Instrument for differential pulse polarography and recorder
DME with drop-knocker (or static mercury drop electrode)

note Mercury is toxic.

Reference electrode (Ag/AgCl or SCE)
Auxiliary electrode (Pt wire)
Nitrogen system for deoxygenating the cell solution
Apparatus for oxygen removal from nitrogen stream (optional) (such as two gas bubbling bottles with medium porosity fritted glass disks, first bottle contains solution of 0.1 M Cr^{2+} in 2.4 M HCl over zinc amalgam, second bottle contains distilled or deionized water)
Pipets, 10 mL, 50 mL
Volumetric flask, 1 L

Chemicals

0.01 M stock solution of Cu^{2+} [dissolve 0.6354 g of copper metal (99.95% pure) in 10 mL of reagent-grade nitric acid (in the hood!). Dilute to 1 L with distilled water]

Standard solutions of Cu^{2+} {dilute the stock solution of 0.01 M Cu^{2+} with 1.0 M sodium nitrate [$NaNO_3$] (supporting electrolyte) to make final concentrations of 1×10^{-4}, 5×10^{-4}, 1×10^{-5}, 5×10^{-5}, 5×10^{-6} M}

Unknown Cu^{2+} [between 1×10^{-4} M and 5×10^{-6} M $Cu(NO_3)_2$]

Optional studies: 1×10^{-7} M and 5×10^{-7} M Cu^{2+} in 1.0 M $NaNO_3$

Theory

Differential pulse polarography is analogous to normal polarography in that the potential of a dropping mercury electrode (DME) is scanned. However, in the pulse technique, a short potential pulse is superimposed on the potential scan as shown in Fig. 4-16. The pulse is sequenced with the dropping mercury electrode so that it occurs once for each drop at exactly the same time after drop dislodgement. The drop lifetime is controlled by a drop-knocker that can be adjusted to give lifetimes that correspond to the pulse repetition period. The pulse is applied immediately before a drop is dislodged when the surface area is at its maximum value. The pulse repetition period and drop time is typically set at 0.5 to 5 s. The pulse magnitude is typically set between 5 and 100 mV.

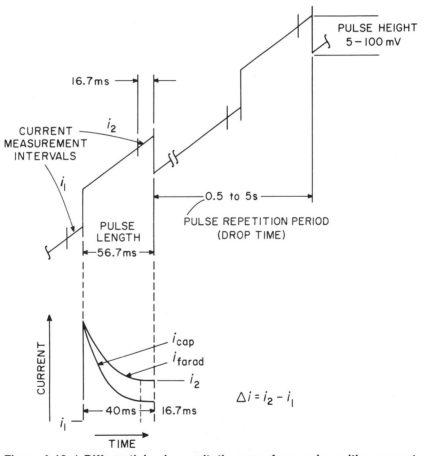

Figure 4-16 / Differential pulse excitation waveform and resulting current-time behavior. [Reprinted with permission from J. B. Flato, Anal. Chem., 44(11), 75A (1972). Copyright © 1972, American Chemical Society.]

A current measurement is made immediately before each pulse i_1 and at the end of each pulse i_2 as indicated in Fig. 4-16. The difference between these two currents ($\Delta i = i_2 - i_1$) is displayed on the recorder for each pulse. The current that accompanies a typical pulse is shown in Fig. 4-16. When the pulse is applied, a surge of current occurs immediately to charge the electrode to the new potential. This capacitance current (i_{cap}) decays rapidly. Another component of current is the faradaic current (i_{farad}) due to electrolysis of the electroactive species being determined in the sample. This current decays more slowly as shown in Fig. 4-16, since new material diffuses to the electrode surface. By measuring the current i_2 at the end of the pulse after i_{cap} has decayed, the interference due to i_{cap} is minimized. This results in a lower detection limit for differential pulse polarography than for normal polarography.

The improved detection limit is illustrated in Fig. 4-17 which shows comparative polarograms for considerably different concentrations of the antibiotic tetracycline. The differential pulse polarogram has a peak shape with the peak potential E_p corresponding to $E_{1/2}$ of the normal dc polarogram. Thus, the greatest value of Δi occurs at the point of steepest slope on a normal polarogram.

The value of E_p is also dependent on the pulse amplitude ΔE as shown in the following equation.

$$E_p = E_{1/2} - \frac{\Delta E}{2} \tag{1}$$

The width of the wave is related to the electron stoichiometry, n. The peak

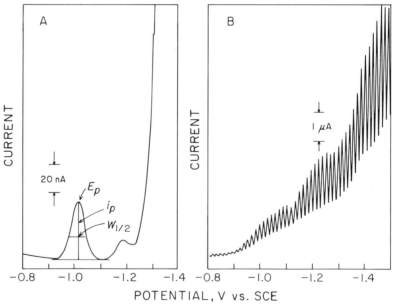

Figure 4-17 / (A) Differential-pulse polarogram. 0.36 ppm tetracycline · HCl in 0.1 *M* acetate buffer, pH 4, dropping mercury electrode, 50-mV pulse amplitude, 1-s drop. (B) DC polarogram. 180 ppm tetracycline · HCl in 0.1 *M* acetate buffer, pH 4, similar conditions. [*Reprinted with permission from J. B. Flato, Anal. Chem., 44(11), 75A (1972). Copyright © 1972, American Chemical Society.*]

half-width ($W_{1/2}$) is 90 mV for $n = 1$ and 45 mV for $n = 2$ for a small pulse amplitude.

The peak height i_p is proportional to concentration. Thus, quantitative determinations are made by measuring i_p.

(See Experiment 4-4 for helpful hints concerning the DME.)

safety caution   Mercury vapor is very poisonous. Discard mercury in the waste mercury container provided and clean up any spills immediately. Work over a porcelain tray to catch and recover any spilled mercury conveniently. Notify the instructor in the event of a mercury spill.

Procedures

A. Calibration Curve and Determination of Unknown Cu^{2+}

Prepare the following standard solutions of Cu^{2+} in 1.0 M $NaNO_3$ by dilution of the 0.01 M Cu^{2+} stock solution: $1 \times 10^{-4}\,M$, $5 \times 10^{-4}\,M$, $1 \times 10^{-5}\,M$, $5 \times 10^{-5}\,M$, $5 \times 10^{-6}\,M$. Obtain an unknown Cu^{2+} solution.

Record differential pulse polarograms of each of the above solutions as well as of the unknown. Use a drop time of 2 s, a pulse amplitude of 50 mV, a scan rate of 5 mV/s, and an initial potential of 0.25 V. After purging the solution with nitrogen (at least 5 min of vigorous bubbling), scan the potential in the negative direction to −0.15 V.

Treatment of Data

Determine E_p for the Cu^{2+} wave from a representative polarogram.

Measure i_p from the polarograms for each concentration of Cu^{2+} and plot i_p vs. concentration of Cu^{2+}. Determine the concentration of the unknown Cu^{2+} solution from the graph.

B. Effect of Pulse Amplitude on i_p, E_p, and $W_{1/2}$

Record differential pulse polarograms on a solution of $5 \times 10^{-6}\,M$ Cu^{2+} in 1.0 M $NaNO_3$ at the following fixed pulse amplitudes: 5, 10, 25, 50, and 100 mV, using a drop time of 2 s and scan rate of 5 mV/s. Use the same potential range as in Procedure A.

Treatment of Data

Determine the effect of pulse amplitude on peak width by plotting $W_{1/2}$ vs. ΔE. Compare your plot with that shown in Ref. 6. Determine from this plot the number of electrons involved in the Cu^{2+} reduction process.

Examine the effect of pulse amplitude on peak potential and peak current by plotting E_p and i_p vs. ΔE. Does the plot of E_p vs. ΔE obey Eq. (1)?

C. Detection of Cu^{2+} Solutions at $10^{-7}\,M$ Concentrations (Optional)

Use a slow scan rate of 1 mV/s or less, a long drop time of 5 s, and a large amplitude pulse of 100 mV, and record differential pulse polarograms of solutions of $1 \times 10^{-7}\,M$ and $5 \times 10^{-7}\,M$ in Cu^{2+}.

Treatment of Data

Examine the peaks obtained on these very dilute solutions. Are they indicative of an accurate determination? Has the limit of detection for Cu^{2+} by this method been reached?

Questions

1   Why does the differential pulse polarogram have a staircase appearance?
2   Show how i_{farad} (see Fig. 4-16) varies as the potential is scanned through a polarographic wave. Use this to explain the shape of the wave in a differential pulse polarogram.
3   Would the same wave shape be observed for a differential pulse voltammogram recorded on a solid electrode such as glassy carbon?

SUPPLEMENTARY EXPERIMENTS

I   Perform the same experiment using pulse polarography.

II Use the method of standard additions to determine the concentration of Cu^{2+} in the unknown (by either differential pulse polarography or pulse polarography).

III Perform any of the above experiments with a different metal ion such as Pb^{2+}, Cd^{2+}, In^{3+}, or Zn^{2+}.

EXPERIMENT 4-6
Polarography of Organic Compounds: Nitrobenzene

Purpose

The effect of pH on the polarographic reduction wave of nitrobenzene is investigated.

References

1   J. E. Page, J. W. Smith, and J. G. Waller, *J. Phys. and Colloid Chem.*, *53*, 545 (1949).
2   I. M. Kolthoff and J. J. Lingane, "Polarography," 2nd ed., Interscience, New York, 1952, chaps. 36–45. Chapter 42 is particularly pertinent.
3   M. J. Allen, "Organic Electrode Processes," Reinhold, New York, 1958.
4   M. Brezina and P. Zuman, "Polarography in Medicine, Biochemistry, and Pharmacy," Interscience, New York, 1958.
5   P. Zuman, "Organic Polarographic Analysis," Pergamon, New York, 1964.
6   P. Zuman, "The Elucidation of Organic Electrode Processes," Academic Press, New York, 1969.
7   P. Zuman and C. L. Perrin, "Organic Polarography," Wiley-Interscience, New York, 1969.
8   C. K. Mann and K. K. Barnes, "Electrochemical Reactions in Nonaqueous Systems," Dekker, New York, 1970.
9   P. Zuman (ed.), "Topics in Organic Polarography," Plenum, London, 1970.
10 A. J. Fry, "Synthetic Organic Electrochemistry," Harper and Row, New York, 1972.
11 M. M. Baizer (ed.), "Organic Electrochemistry," Dekker, New York, 1973.
12 M. R. Rifi and F. H. Covitz, "Introduction to Organic Electrochemistry," Dekker, New York, 1974.
13 N. L. Weinberg (ed.), "Technique of Electroorganic Synthesis," parts I and II, Wiley-Interscience, New York, 1974.
14 M. D. Hawley in "Laboratory Techniques in Electroanalytical Chemistry," P. T. Kissinger and W. R. Heineman (eds.), Dekker, New York, 1984, chap. 17.

Apparatus

Instrument for polarography and recorder
Electrochemical cell
SCE reference electrode

DME

note   Mercury is toxic.

Platinum auxiliary electrode
Nitrogen system for deoxygenating the cell solution

Chemicals   1 m*M* nitrobenzene, pH 3.0 (dissolve 0.123 g nitrobenzene in 105 mL 95%
ethanol. Add the ethanol solution to 900 mL of an aqueous solution
containing 15.7 g citric acid monohydrate and 5.85 g anhydrous
Na_2HPO_4. Add sufficient gelatin to make the solution 0.005% in gelatin)
1 m*M* nitrobenzene, pH 7.0 (dissolve 0.123 g nitrobenzene in 105 mL 95%
ethanol. Add the ethanol solution to 900 mL of an aqueous solution containing 3.71 g citric acid monohydrate and 23.4 g anhydrous Na_2HPO_4.
Add sufficient gelatin to make the solution 0.005% in gelatin)

safety caution   Nitrobenzene is poisonous and is rapidly absorbed through the skin.

note   Commerical aniline contains nitrobenzene as an impurity and may be analyzed
polarographically. Analysis of commercial aniline for nitrobenzene is suggested
as an unknown. The sample should be adjusted to the same solution conditions as
one of the above standard nitrobenzene solutions.

Theory   A large number of organic functional-group compounds can be studied and determined polarographically. These include the following: conjugated unsaturated
compounds; quinones; hydroxylamines; nitro, nitroso, azo, and azoxy compounds; amine oxides; diazonium salts; certain sulfur compounds; certain heterocyclic compounds; peroxides; certain organic halides; and reducing sugars.

In most cases oxidation or reduction of organic compounds proceeds irrevers-
ibly. Nevertheless, in many cases the polarographic wave may be used in the
quantitative determination of these compounds. Reactions of organic compounds
in aqueous solution at the dropping mercury electrode almost always involve hydrogen ion

$$R + mH^+ + me \rightarrow RH_m \tag{1}$$

Consequently, determinations of organic compounds should always be carried out
in solutions with a controlled pH (strongly acidic, strongly basic, or buffered) so
that the concentration of hydrogen ion at the electrode surface is maintained constant. Frequently, different reduction reactions will proceed, depending upon the
acidity of the solution. This is the case for nitrobenzene, which at pH 3 has a two-
step reduction, while at pH 7 it is reduced in only one step.

(See Experiment 4-4 for helpful hints concerning the DME.)

safety caution   Mercury vapor is very poisonous. Discard mercury in the waste mercury container provided and clean up any spills immediately. Work over a porcelain tray to
catch and recover any spilled mercury conveniently. Notify the instructor in the
event of a mercury spill.

Procedure   Record polarograms for each of the two nitrobenzene solutions after they have
been deoxygenated with purified nitrogen (at least 5 min of vigorous bubbling). If
time permits, note the effect of not adding gelatin to the solutions.

A more thorough study is possible if, in addition to the two solutions at pH 3 and pH 7, solutions at other pH's are run. This may be accomplished by adjusting the above two solutions with acid or base to the desired pH with a pH meter.

Record a polarogram of the unknown sample after it has been dissolved in one of the buffer solutions. Prepare a set of standard nitrobenzene solutions by dilution of the 1 mM standard with the appropriate buffer. Record polarograms of these solutions.

Treatment of Data

Interpret the polarographic waves by writing the equations for the reactions taking place at pH 3 and pH 7. Determine the ratio of the diffusion currents for the two waves obtained for the pH 3 solution and compare this ratio with the theoretical value. Compare the total wave height at pH 3 with the wave height at pH 7. Compare the half-wave potentials at the two pH's.

If a series of solutions at various pH's were studied, compare the wave heights for these solutions and note the pH at which two waves are first observed.

If an unknown was analyzed, determine the percentage of nitrobenzene in the sample. Use the polarograms of the standard nitrobenzene solutions to construct a calibration curve for the analysis.

Questions

1 What significance does the change in half-wave potential with pH have?
2 What is the purpose of the ethanol in the preparation of the solutions?
3 What general conclusions can be made concerning the solution equilibria as a result of determining the pH at which there is a transition from one wave to two waves?
4 Is it likely that nitrobenzene can be polarographically reduced at a platinum microelectrode? Would the half-wave potential be different, and if so, more positive or more negative?

EXPERIMENT 4-7

Determination of the Formula and Formation Constant of a Complex Metal Ion by Polarography or Pulse Polarography

Purpose

In recent years voltammetric methods have become more and more prominent tools in the investigation of complex metal ions. The purpose of this experiment is to determine the formula and the formation constant of a complex formed between Pb^{2+} and oxalate.

References

1 K. H. Gayer, A. Demmler, and M. J. Elkind, *J. Chem. Educ., 30,* 557 (1953).
2 J. J. Lingane, *Chem. Revs., 29,* 1–35 (1941).
3 A. E. Martell, and M. Calvin, "Chemistry of the Metal Chelate Compounds," Prentice-Hall, Englewood Cliffs, N.J., 1952, pp. 107–115.
4 I. M. Kolthoff and J. J. Lingane, "Polarography," 2nd ed., Interscience, New York, 1952, chap. 12.

Apparatus

Instrument for polarography (pulse mode required to perform experiment by pulse polarography) and recorder
Electrochemical cell
SCE reference electrode

DME

note Mercury is toxic.

Platinum auxiliary electrode
Nitrogen system for deoxygenating the cell solution
Pipets, 2 mL, 10 mL, 20 mL, 30 mL, 50 mL
Volumetric flasks (5), 100 mL

Chemicals 1.0 M potassium oxalate $[K_2C_2O_4]$ (oxalic acid neutralized with KOH to phenolphthalein end point)

safety caution Potassium oxalate is toxic. If spilled on skin, wash with copious amounts of water.

0.020 M lead nitrate $[Pb(NO_3)_2]$
1.0 M potassium nitrate $[KNO_3]$

Theory The polarographic method for investigating labile metal ion complexes is based on the fact that the characteristic *half-wave potential* ($E_{1/2}$) of a simple metal ion is shifted when the metal ion undergoes complex formation. The extent of this shift in $E_{1/2}$ varies with the concentration of the complexing agent and the formation constant of the complex. By measuring the shift in $E_{1/2}$ as a function of the concentration, it is possible to obtain information concerning both the formula and the stability of the metal complex. Polarograms for Pb^{2+} in the absence of ligand and in the presence of increasing concentrations of ligand X^- are shown in Fig. 4-18. A more negative potential is required to reduce the complex PbX_2 than free (aquated) Pb^{2+}, hence the negative shift in $E_{1/2}$ with added ligand.

A short though not entirely rigorous mathematical derivation is given of the necessary relationships. The derivation assumes the predominance of a single complex species over a fair range of concentration. For chelate complexes this is normally the case, but it is not true for most other metal complexes.

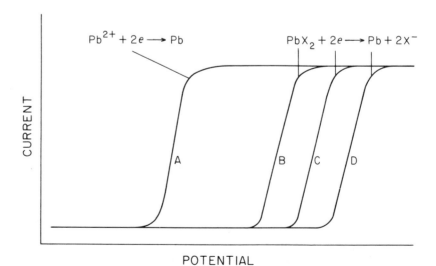

Figure 4-18 / Polarograms for reduction of Pb^{2+} in presence and absence of complexing agent X^-. (A) Pb^{2+} in 0.1 M KNO_3. (B–D) Increasing concentration of X^-.

The equation for the reduction of a metallic ion is:

$$O + ne \rightarrow R \tag{1}$$

and the corresponding Nernst equation is

$$E = E^{\circ\prime} + \frac{0.0591}{n} \log \frac{[O]}{[R]}$$

The complexing agent X forms a complex with the metal ion as follows:

$$O + pX \rightarrow OX_p \tag{3}$$

and therefore

$$K = \frac{[OX_p]}{[O][X]^p} \tag{4}$$

Solving this relationship for [O] and substituting this result into the Nernst equation yields

$$E = E^{\circ\prime} + \frac{0.0591}{n} \log \frac{[OX_p]}{K[X]^p[R]} \tag{5}$$

Because, by definition, the half-wave potential is the potential at which one-half of all the oxidized form which reaches the electrode surface is reduced to the corresponding reduced form, one may say for the reduction of the simple metal ion (assuming equal diffusion coefficients).

$$E_{1/2} = E^{\circ\prime} \qquad \text{when } [O] = [R] \tag{6}$$

and for the reduction of the complex,

$$(E_{1/2})_{\text{complex}} = E^{\circ\prime} + \frac{0.0591}{n} \log \frac{1}{K[X]^p} \tag{7}$$

when

$$[OX_p] = [R] \tag{8}$$

and so the difference in half-wave potentials is given by

$$(E_{1/2})_{\text{complex}} - E_{1/2} = -\frac{0.0591}{n} \log K[X]^p \tag{9}$$

In practice one obtains the $E_{1/2}$ for the simple metal ion and the half-wave potentials for a series of solutions containing a given concentration of metal ion and various concentrations of the complexing agent (e.g., twentyfold and greater excess is employed so that the concentration of X at the electrode reaction will remain essentially constant and of known value) and plots the $E_{1/2}$ against log [X].

A straight line [see Eq. (7)] should be obtained whose slope equals $-0.0591\, p/n$, allowing calculation of p. Once p is determined, one can apply Eq. (9) and evaluate the formation constant K for the complex.

(See Experiment 4-4 for helpful hints concerning the DME.)

safety caution Mercury vapor is very poisonous. Discard mercury in the waste mercury container provided and clean up any spills immediately. Work over a porcelain tray to catch and recover any spilled mercury conveniently. Notify the instructor in the event of a mercury spill.

Procedure Into each of the five volumetric flasks, pipet 2.00 mL of the lead nitrate solution. Into each of these flasks also place

Flask	Substance
1	10 mL of 1.0 M KNO_3
2	10 mL of 1.0 M $K_2C_2O_4$
3	20 mL of 1.0 M $K_2C_2O_4$
4	40 mL of 1.0 M $K_2C_2O_4$
5	80 mL of 1.0 M $K_2C_2O_4$

Dilute each solution to the mark and swirl vigorously. Notice if any precipitate is present. If so, check with instructor before proceeding.

Consult the instructor or the instrument manual instructions on the operation of the particular instrument used in this experiment. Polarograms can be recorded in either the dc or the pulse mode (1-s drop time).

Bubble purified nitrogen through each solution in the polarographic cell and record the polarogram of the solution from -0.2 to -1.0 V vs. SCE at a scan rate of 1 mV/s. Be very careful to start each polarogram *exactly* at -0.2 V, since *accurate* $E_{1/2}$ values are necessary in this experiment.

Treatment of Data Measure the half-wave potential of each curve and test the waves for reversibility by plotting $\log (i_d - i)/i$ vs. E. [See Eq. (1) in Experiment 4-4.] Calculate the value of n.

From a plot of $E_{1/2}$ vs. $\log [C_2O_4^{2-}]$, determine p, the number of oxalates in the lead oxalate complex, from the slope.

Extrapolate the above plot to calculate $E_{1/2}$ for the 1.0 M $C_2O_4^{2-}$ concentration. From this value and from the $E_{1/2}$ obtained in the KNO_3 media, calculate by way of Eq. (9) the formation constant for the complex.

Questions
1 Explain the reason for testing the polarographic waves for reversibility.
2 If (thank goodness it isn't so!) oxalate had acid ionization pK's of 2 and 8, what complications would arise if you recorded your polarograms at a pH of 5? Could you still obtain a value for the formation constant of the complex? *Note:* pK's of oxalic acid are actually 1.2 and 4.3.
3 If you recorded polarograms at pH 11, could you obtain a value for the formation constant of the complex?
4 Would buffered solutions be necessary in question 2? In question 3?
5 Is knowledge of the value for the SCE half-cell potential really necessary? Why?

EXPERIMENT 4-8
Anodic Stripping Voltammetry

Purpose

Anodic stripping voltammetry is used to determine lead and cadmium in unknown samples and lead in pottery.

References

1 W. R. Heineman, H. B. Mark, Jr., J. A. Wise, and D. A. Roston in "Laboratory Techniques in Electroanalytical Chemistry," P. T. Kissinger and W. R. Heineman (eds.), Dekker, New York, 1984, chap. 19.

2 A. J. Bard and L. R. Faulkner, "Electrochemical Methods," Wiley, New York, 1980, chap. 10.

3 E. Barendrecht in "Electroanalytical Chemistry," A. J. Bard (ed.), vol. 2, Dekker, New York, 1967, chap. 2.

4 F. Vydra, K. Stulik, and E. Julakova, "Electrochemical Stripping Analysis," Holsted, New York, 1976.

5 T. R. Copeland and R. K. Skogerboe, *Anal. Chem., 46,* 1257A (1974).

6 W. D. Ellis, *J. Chem. Ed., 50,* A-131 (1973).

7 M. L. Deanhardt, J. W. Dillard, K. W. Hanck, and W. L. Switzer, *J. Chem. Ed., 54,* 55 (1977).

8 *J. Ass. Offic. Anal. Chem., 56*(2), 483 (1973).

9 I. Shain in "Treatise on Analytical Chemistry," I. M. Kolthoff and P. J. Elving (eds.), part I, vol. 4, sec. D-2, Interscience, New York, 1963, chap. 50.

Apparatus

Instrument for linear sweep voltammetry or differential pulse voltammetry (such as Bioanalytical Systems CV-1B, CV-27, or Electrochemical Analyzer; Princeton Applied Research 174 or 173/175; IBM EC/225)

Electrochemical cell

Working electrode [hanging mercury drop electrode (HMDE) or mercury film electrode (MFE)]

note   Mercury is toxic.

Auxiliary electrode (platinum wire)

Nitrogen system for deoxygenating the cell solution

Reference electrode (SCE or Ag/AgCl)

Apparatus for oxygen removal from nitrogen stream (optional) (such as two gas bubbling bottles with medium porosity fritted glass disks, first bottle contains solution of 0.1 M Cr^{2+} in 2.4 M HCl over zinc amalgam, second bottle contains distilled or deionized water)

Micropipets, 10 μL

Pipets (2), 50 mL

Magnetic stirrer and stirring bar

Chemicals (of utmost purity)

Cd^{2+} solutions, standard {cadmium chloride [$CdCl_2$] made up in 0.01 M potassium chloride [KCl] at the following concentrations: $1 \times 10^{-5}\,M$; $3 \times 10^{-5}\,M$; $5 \times 10^{-5}\,M$; $7 \times 10^{-5}\,M$ (Substantially lower concentration ranges can be used in this experiment if the supporting electrolyte is sufficiently pure. See the instructor.)}

Unknown: Cd^{2+} solutions (between 2×10^{-5} and $4 \times 10^{-5}\,M$ made up in 0.01 M KCl)

0.01 M Pb^{2+}, standard {lead nitrate [$Pb(NO_3)_2$] in 4% acetic acid}
Unknown: Pb^{2+} solutions [4×10^{-6} M to 6×10^{-6} M $Pb(NO_3)_2$ in 4% acetic acid]
Samples of pottery (optional—lead glass, soil, water, milk)
Tank of nitrogen (oxygen-free) with regulator valve

Theory

Stripping voltammetry is an electroanalytical technique with an extremely low detection limit. The technique involves an electrochemical preconcentration step in which analyte is deposited into or onto an electrode under controlled conditions. The preconcentrated material is then electrolyzed by a potential sweep during which it is stripped out or off of the electrode. Concentration is determined from the magnitude of the peak current for the voltammetric stripping wave. Detection limits of 10^{-10} M have been achieved for many metal ions.

The remarkably low detection limit of stripping voltammetry is attributable to the preconcentration step. For example, in the determination of Pb^{2+} by anodic stripping voltammetry (ASV), the potential of a mercury electrode is held sufficiently negative to reduce Pb^{2+} by the reaction:

$$Pb^{2+} + 2e \rightarrow Pb$$

This potential, referred to as the deposition potential, is marked (deposit) on the potential axis of Fig. 4-19. Maintaining the potential at this value while stirring the solution results in the accumulation of Pb in the mercury electrode, since Pb is soluble in mercury. If the mercury electrode volume is small relative to the solu-

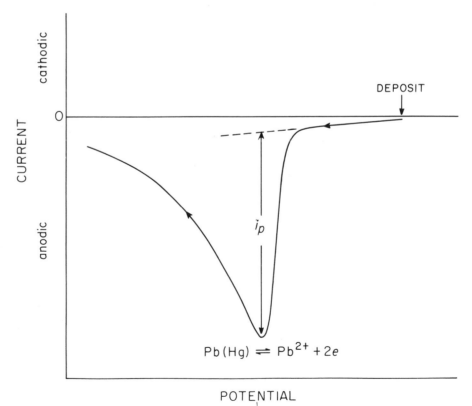

Figure 4-19 / Voltammetric stripping wave for lead.

tion volume, this extraction of Pb^{2+} out of the aqueous solution into mercury as Pb results in a concentration enhancement of lead in the mercury electrode. This concentration step typically takes 1 to 30 min, a longer deposition time being necessary to achieve a lower detection limit. The solution is stirred to maximize the rate of lead deposition. The concentration of lead in the mercury electrode can be enhanced by as much as 10,000, relative to the concentration of Pb^{2+} in the sample solution.

After the deposition step, the solution stirring is stopped and about 30 s is allowed for equilibration. The potential is then scanned positively as indicated by the arrow in Fig. 4-19. When the potential is sufficiently positive, anodic current results from the oxidation of Pb.

$$Pb \rightarrow Pb^{2+} + 2e$$

Thus, Pb is stripped out of the mercury and back into solution as Pb^{2+} with an anodic current, hence the name "anodic stripping voltammetry." A voltammetric wave as shown in Fig. 4-19 results. As the potential is scanned positively, the anodic current increases rapidly until the surface concentration of Pb (in the mercury) approaches zero, causing the current to peak and then decay as Pb is depleted from the vicinity of the electrode surface (inside the mercury). [This is analogous to the behavior of a single sweep in a cyclic voltammogram (see Experiment 4-1).] Qualitative analysis is achieved from the value of the peak potential E_p; quantitative analysis from the magnitude of the peak current i_p. The measurement of these parameters is as indicated in Fig. 4-19.

Two techniques predominate for the stripping step. One is linear sweep voltammetry. In this technique the potential is scanned linearly as a function of time. It is essentially the same as the "forward" scan of a cyclic voltammogram as described in Experiment 4-1. Differential pulse voltammetry is the second commonly used stripping technique. The pulsed waveform and current sampling aspects of this method are described in Experiment 4-5. A lower detection limit can be achieved with differential pulse voltammetry than for linear sweep voltammetry for a given deposition time.

The hanging mercury drop electrode (HMDE) and the mercury film electrode (MFE) are the two most commonly used electrodes for ASV. The larger surface area–to–volume ratio of the MFE compared to the spherical HMDE enables more metal to be concentrated into a given amount of mercury during a specified deposition time. Consequently, a given detection limit can be achieved with a shorter deposition time for the MFE than for the HMDE.

Electrochemical cells of the type described in the text of this chapter are used for ASV.

In stripping voltammetry some fraction of the total metal ion in solution is deposited into the mercury electrode by electrolysis for a given length of time. An exhaustive electrolysis in which all of the metal is deposited into the electrode is time consuming and generally unnecessary, since adequate concentrations can usually be deposited into the electrode to give a satisfactory stripping signal in much shorter times. Since the deposition is not exhaustive, it is important to deposit the same fraction of metal for each stripping voltammogram. The parameters of electrode surface area, deposition time, and stirring must be carefully duplicated for all standards and samples. Deposition times vary from 60 s to 30 min, depending on the analyte concentration, the type of electrode, and the stripping technique. The less concentrated solutions require longer deposition times to give adequate stripping peaks; the MFE requires less time than does the

HMDE; differential pulse voltammetry requires less time than linear sweep voltammetry as the stripping technique.

This experiment can be performed with either the HMDE or the MFE. Although it is written in terms of linear sweep voltammetry as the stripping technique, differential pulse voltammetry can also be used.

safety caution

Mercury vapor is very poisonous. Discard mercury in the waste mercury container provided and clean up any spills immediately. Work over a porcelain tray to catch and recover any spilled mercury conveniently. Notify the instructor in the event of a mercury spill.

Procedures

A. Determination of Cadmium in an Unknown Sample by Direct Calibration Curve

Pipet a 50-mL sample of $1 \times 10^{-5} M$ Cd^{2+} in 0.01 M KCl into the electrolysis cell. Position the electrodes, add a magnetic stirring bar, and deoxygenate for 10 min with oxygen-free nitrogen. At the end of this time, move the nitrogen tube just above the solution to maintain a gentle flow sufficient to prevent oxygen from entering the cell. Turn on the magnetic stirrer, and wait a few seconds for the stirring rate to become constant. Simultaneously apply a potential of -0.8 V and switch on the timer. Electrolyze for 2 min; stop the stirrer and timer at the end of this time and wait 30 s for the solution to become quiescent. (This time allows the amalgam concentration to become more homogeneous and convection in the solution to dampen out.) Obtain a voltammogram in the positive scan direction, using a scan rate of 0.2 V/min until a potential of -0.35 V is reached (or a potential just before the oxidation of mercury). Switch the potentiostat to open circuit (no potential applied to the cell). Repeating the above procedure, record stripping voltammograms for the remaining Cd^{2+} solutions of varying concentration, then record one for the unknown.

For one of the standard solutions, record voltammograms in which (a) the deposition time is varied (e.g., 30 s, 1, 2, 5, 10, and 15 min) and other parameters held constant and (b) the scan rate is varied (e.g., 10, 20, 50, 100, and 200 mV/s) and other parameters held constant.

Treatment of Data

Prepare a calibration curve for the standard Cd^{2+} solutions by graphing peak current vs. concentration. Determine the concentration of the unknown directly from the graph.

Plot i_p vs. deposition time for the series of voltammograms recorded at different deposition times.

Plot i_p vs. scan rate (v and $v^{1/2}$). Which of the two plots is linear? Is this in agreement with the behavior expected for the electrode that you used (HMDE or MFE)?

B. Determination of Lead in an Unknown Sample by the Method of Standard Additions

Pipet a 50-mL sample of the unknown Pb^{2+} solution into the electrolysis cell. Record an anodic stripping voltammogram for this solution. Add 10 μL of standard 0.01 M $Pb(NO_3)_2$ and record a second voltammogram after deoxygenating and stirring the solution. Repeat this procedure with three more 10-μL aliquots of Pb^{2+} standard.

Treatment of Data

Plot i_p vs. concentration of standard Pb^{2+} in the sample after each addition of Pb^{2+}. Draw a line through the data points, extrapolate to zero current, and determine the concentration of the unknown sample from the intercept with the horizontal axis. See Fig. 4-20 for an illustration of the standard addition method.

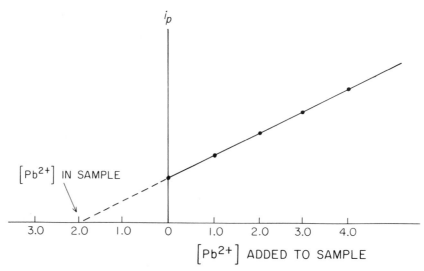

Figure 4-20 / Method of standard additions. Extrapolation to zero current indicates a value of 2.0 for the unknown as obtained from the intercept with the abscissa.

C. Determination of Lead in Pottery (Optional) (FDA Procedure, Ref. 8)

Wash the pottery vessel under test with detergent, rinse thoroughly with distilled water, and fill with 4% acetic acid to near the top (within $\frac{1}{2}$ in), noting the total volume. Cover tightly with plastic wrap such as Saran; let stand for at least 24 h at approximately 25°C. Pipet 50.0 mL of the acetic acid extract in the vessel and proceed as in Procedure B, recording an anodic stripping voltammogram of the extract, and then repeating with additions of 10 μL of standard 0.01 M $Pb(NO_3)_2$ (four additions should suffice). An anodic stripping voltammogram should be run of pure 4% acetic acid also. Occasionally a copper peak may be found at 0.06 V indicating a slight impurity; this small amount will not affect the results at sub-ppm levels for lead.

Treatment of Data

Plot peak current vs. concentration of standard Pb^{2+} added to the sample and determine the unknown concentration as outlined in Procedure B. With the total volume of acetic acid used to extract the lead from the sample, calculate the total amount of lead extracted from the sample.

Questions

1 Discuss the parameters that should be carefully controlled in stripping voltammetry.
2 Explain why the method of standard additions is generally used for the analysis of real samples, rather than a calibration curve obtained from pure metal ion salts dissolved in distilled water with supporting electrolyte.
3 Explain how Cl^- can be determined by *cathodic* stripping voltammetry at a silver working electrode. [*Hint:* See electrode mechanism in Eq. (13) of Chap. 2.]
4 Explain why narrower stripping waves are obtained at an MFE than at an HMDE.

EXPERIMENT 4-9

Spectroelectrochemistry and Thin-Layer Cells†

Purpose

In this experiment $E^{\circ\prime}$, n, and spectra of an organic compound are obtained by thin-layer electrochemical and spectroelectrochemical techniques.

References

1 T. P. DeAngelis and W. R. Heineman, *J. Chem. Ed., 53*, 594 (1976).
2 W. R. Heineman, *Anal. Chem., 50*, 390A (1978).
3 A. J. Bard and L. R. Faulkner, "Electrochemical Methods," Wiley, New York, 1980, chap. 14.
4 W. R. Heineman and P. T. Kissinger in "Laboratory Techniques in Electroanalytical Chemistry," P. T. Kissinger and W. R. Heineman (eds.), Dekker, New York, 1984, chap. 3.
5 T. Kuwana and W. R. Heineman, *Acct. Chem. Res., 9*, 241 (1972).
6 W. R. Heineman, *J. Chem. Ed., 60*, 305 (1983).
7 W. R. Heineman, F. M. Hawkridge, and H. N. Blount in "Electroanalytical Chemistry," A. J. Bard (ed.), vol. 13, Dekker, New York, 1984, chap. 1.
8 I. M. Kolthoff and W. J. Tomsicek, *J. Phys. Chem., 39*, 945 (1935).
9 A. T. Hubbard and F. C. Anson in "Electroanalytical Chemistry," A. J. Bard (ed.), vol. 4, Dekker, New York, 1970, p. 129.

Apparatus

Instrument with scanning capability for cyclic voltammetry and integrator for coulometry (optional)

x-y Recorder

UV-visible absorption spectrophotometer (with large sample compartment)

SCE

Pt wire for auxiliary electrode

Small solution cup (a sawed-off 20-mL beaker will suffice)

Optically transparent thin-layer electrode assembled from following components:

Gold minigrid, 100 wires/in (Buckbee-Mears Co., St. Paul, Minn.) (Approximately 40 cells can be made from one 6 × 6 in piece.)

Teflon tape spacers, 2 mil thick (Fluorofilm DF-1200, Fluorocarbon, Dilectrix Division, Lockport, N.Y.)

Epoxy resin

Microscope slides, 1 × 3 in

Construction of the OTTLE

The OTTLE is constructed from ordinary microscopy slides, 2 mil adhesive Teflon tape spacers, gold minigrid, and epoxy. Strips of Teflon tape approximately 2 mm wide are cut and pressed (adhesive side to the glass) along the periphery of two precleaned microscope slides as shown in Fig. 4-21.

A section of gold minigrid 1 × 3.5 cm is cut and positioned with tweezers within 3 mm of the bottom edge of one microscope slide. Location of the minigrid near the bottom minimizes iR drop (resistance) in the cell. The second microscope slide is then laid on top to form a "sandwich" that is clamped into place. Epoxy applied along the taped edges and allowed to cure at 80°C for 24 h holds the cell together. The two pieces of minigrid extending from the cell edges are used for electrical contact. Since the minigrid is easily torn, these should be

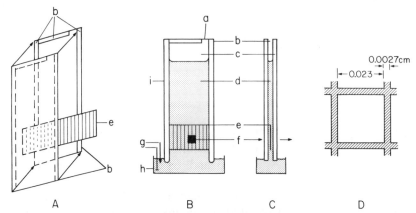

Figure 4-21 / Optically transparent thin-layer electrode. (A) Assembly of the cell. (B) Front view. (C) Side view. (D) Dimension of 100 wires/in gold minigrid. (a) Point of suction application to change solution, (b) Teflon tape spacers, (c) microscope slides (1 × 3 in), (d) solution, (e) transparent gold minigrid electrode, (f) optical path of spectrometer, (g) reference and auxiliary electrodes, (h) solution cup, and (i) epoxy holding cell together. [*Reprinted with permission from T. P. DeAngelis and W. R. Heineman, J. Chem. Ed., 53, 594 (1976). Copyright © 1976, Division of Chemical Education, American Chemical Society.*]

folded over onto the outside of the microscope slide and held in place by a dab of epoxy. After the epoxy dries, a piece of metal foil is folded over the minigrid and the edge of the microscope slides and clamped into place with an alligator clip connected to the potentiostat. (Alternatively, the minigrid extending from the cell can be carefully wrapped around a contact wire and then covered with epoxy.)

Spectroelectrochemical experiments are performed by positioning the OTTLE in a spectrometer so that the light beam passes directly through the minigrid. A spectrometer with a reasonably large sample compartment is desirable. A frame that conveniently suspends the OTTLE above the solution cup and fits into the spectrometer can be fabricated from black Lucite.

Chemicals

4.00×10^{-3} M potassium ferricyanide [$K_3Fe(CN)_6$] in 0.5 M sodium chloride [NaCl]

1.00×10^{-3} M o-tolidine (Eastman Kodak) in 0.5 M acetic acid [HAc], 0.1 M perchloric acid [$HClO_4$]. (To avoid solubility problems the o-tolidine should first be dissolved in concentrated acetic acid and diluted almost to volume with distilled water before adding the perchloric acid.)

Theory

The combination of two quite different techniques, electrochemistry and spectroscopy, has proved to be an effective approach for studying the redox chemistry of inorganic, organic, and biological molecules. Oxidation states are changed electrochemically by addition or removal of electrons at an electrode. Spectral measurements on the solution adjacent to the electrode are made simultaneously with the electrogeneration process. Thus, spectroscopy is used as a probe to observe the consequences of electrochemical phenomena that occur in the solution undergoing electrolysis. Such spectroelectrochemical techniques are a convenient means for obtaining spectra and redox potentials and observing subsequent chemical reactions of electrogenerated species.

A simple spectroelectrochemical cell is the *o*ptically *t*ransparent *t*hin-*l*ayer *e*lectrode (OTTLE). As shown in Fig. 4-21 the OTTLE consists of a transparent gold minigrid electrode sandwiched between two ordinary microscope slides that are separated 0.01 to 0.03 cm by strips of Teflon tape spacers. To use the cell, the bottom edge is dipped into a small cup containing a few milliliters of the solution to be investigated. Reference and auxiliary electrodes are also immersed in this cup. Solution drawn into the OTTLE by application of suction at the top corner maintains its level above the minigrid by capillary action. Electrochemical experiments can then be performed on the thin layer of solution surrounding the minigrid.

An attractive feature of the thin-layer technique is the speed with which coulometric results can be obtained. The solution volume that undergoes electrolysis is the thin layer of solution between the microscope slides that is defined by the area of the minigrid. The volume of this "cell" is only 30 to 50 μL, and complete electrolysis occurs in 30 to 60 s.

Since the minigrid electrode is transparent to light, optical spectra of the solution in the thin-cell volume surrounding the minigrid can be recorded by passing light directly through the minigrid as shown in Fig. 4-21. Spectra of electroactive species in different oxidation states can be obtained by placing the OTTLE in the sample compartment of a spectrometer. Spectra are recorded after the electroactive species has been converted to the desired oxidation state by applying an appropriate potential to the minigrid.

In this experiment, the use of the OTTLE for determining formal redox potentials, n values, and spectra of redox couples is illustrated by measurements on solutions of ferricyanide and o-tolidine using a variety of electrochemical techniques.

Cyclic Voltammetry

Potential scan techniques such as cyclic voltammetry (see Experiment 4-1) are exceedingly useful for locating the redox potentials of electroactive species in a thin-layer cell, just as they are in conventional electrochemical cell configurations. Figure 4-22 shows a cyclic voltammogram for ferricyanide in the OTTLE. The potential scan was initiated at +0.35 V in the negative direction. The onset of cathodic current indicates the reduction of ferricyanide to ferrocyanide.

A rapid drop in current after the peak coincides with the complete electrolysis of ferricyanide in the thin solution layer. Switching the potential scan to the positive direction results in an anodic peak, which signifies the reoxidation of ferrocyanide back to ferricyanide. A formal redox potential for a reversible couple can be determined from the average of the cathodic (E_{pc}) and anodic (E_{pa}) peak potentials

$$E^{\circ\prime} = \frac{E_{pc} + E_{pa}}{2} \tag{1}$$

Coulometry

Coulometry in a thin-layer cell is generally performed by applying a potential that causes complete electrolysis of the electroactive species. Electronic integration of the resulting current gives the total charge consumed by the electrode process. This total charge Q_T contains both (a) the faradaic charge Q_F due to the electrolysis of the reactant and (b) a "blank" charge Q_B, which contains contributions from double layer charging and background reactions resulting from oxidation/reduction of solvent and electrode.

Figure 4-23 shows a charge-time curve in which ferricyanide is reduced to ferrocyanide by a potential step from +0.40 V to 0.00 V. The charge increases rap-

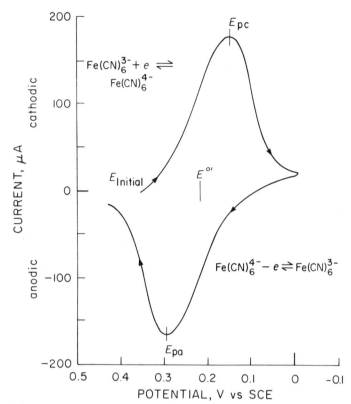

Figure 4-22 / Thin-layer cyclic voltammogram of 4.00 mM Fe(CN)₆³⁻, 0.5 M KCl. Scan rate 2 mV/s. [Reprinted with permission from T. P. DeAngelis and W. R. Heineman, J. Chem. Ed., 53, 594 (1976). Copyright © 1976, Division of Chemical Education, American Chemical Society.]

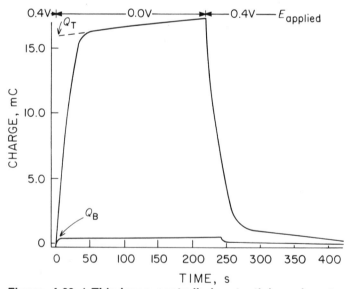

Figure 4-23 / Thin-layer controlled potential coulometry. Charge-time curve for potential step 0.4 V to 0.00 V to 0.4 V vs. SCE. 4.00 mM Fe(CN)₆³⁻, 0.5 M KCl. [Reprinted with permission from T. P. DeAngelis and W. R. Heineman, J. Chem. Ed., 53, 594 (1976). Copyright © 1976, Division of Chemical Education, American Chemical Society.]

idly until all of the ferricyanide in the thin layer is reduced to ferrocyanide at which time the curve levels off. Q_T is usually measured by extrapolating the linear portion of the curve to zero time as shown in the figure. Q_B is measured from the second charge-time curve which was obtained by repeating the experiment on supporting electrolyte only. The desired quantity Q_F, which reflects the amount of reactant electrolyzed, can be calculated by subtracting Q_B from Q_T. The faradaic charge required for complete electrolysis in the thin-layer cell is given by Faraday's Law

$$Q_F = Q_T - Q_B = nFVC \tag{2}$$

where Q = charge, C
         n = number of electrons transferred per molecule, equiv/mol
         F = Faraday's number, 96,485 C/equiv
         V = solution volume of the thin-layer cell, L
         C = concentration of electroactive species, M

The figure also shows the effect of stepping the potential back to the initial value of $+0.40$ V, causing reoxidation of the ferrocyanide.

Calibration of the cell volume V and thickness l of the OTTLE is achieved by potential step coulometry on a solution containing a species, such as ferricyanide, of known C and n.

Determination of $E^{\circ\prime}$, n, and Spectra

Once the cell volume has been calibrated with the standard solution of ferricyanide, the OTTLE can be used to characterize the electrochemistry of an organic compound. A good example is the oxidation of o-tolidine in acidic solution (Ref. 1).

$$H_3N^+ \!-\!\!\!\bigcirc\!\!\!-\!\!\!\bigcirc\!\!\!-\!NH_3^+ \rightleftharpoons H_2N^+ \!=\!\!\!\bigcirc\!\!\!=\!\!\!\bigcirc\!\!\!=^+NH_2 + 2e + 2H^+$$

The $E^{\circ\prime}$ of this reversible couple is easily determined by cyclic voltammetry and n values very close to 2.0 are obtainable by coulometry. The extensive resonance of the oxidized form results in a bright yellow color that is easily seen in the vicinity of the minigrid. The color is sufficiently intense that excellent spectra (see Fig. 4-24, curve a) can be obtained even with the short optical path length of the OTTLE.

A unique spectroelectrochemical approach for examining the Nernst equation is possible with the OTTLE. In an electrochemical cell, the ratio of concentrations of oxidized to reduced forms of the electroactive couple at the electrode surface is determined by the potential which is applied to the electrode as defined by the Nernst equation. (See discussion in Chap. 4.)

$$E_{\text{applied}} = E^{\circ\prime} + \frac{0.0591}{n} \log \frac{[O]}{[R]} \tag{3}$$

In the case of the OTTLE, the potential applied to the minigrid will determine this concentration ratio in the entire solution comprising the thin-layer cell. Upon application of a potential, electrolysis rapidly adjusts the ratio $[O]/[R]$ to the value required to satisfy Eq. (3).

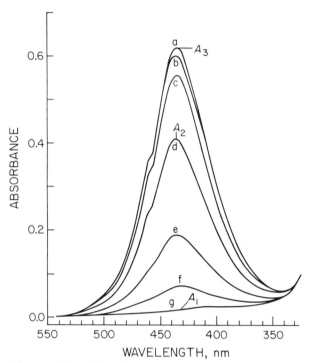

Figure 4-24 / Thin-layer spectra of 0.97 mM o-tolidine, 0.5 M acetic acid, 1.0 M HClO$_4$ for different values of $E_{applied}$. Cell thickness 0.017 cm. (a) 0.800, (b) 0.660, (c) 0.640, (d) 0.620, (e) 0.600, (f) 0.580, (g) 0.400 V vs. SCE. [*Reprinted with permission from T. P. DeAngelis and W. R. Heineman, J. Chem. Ed., 53, 594 (1976). Copyright © 1976, Division of Chemical Education, American Chemical Society.*]

Figure 4-24 shows spectra of o-tolidine in an OTTLE for a series of applied potentials. Curve a was recorded after application of +0.800 V, which caused complete oxidation of o-tolidine ([O]/[R] > 1000). Curve g was recorded after application of +0.400 V, causing complete reduction ([O]/[R] < 0.001). The intermediate spectra correspond to intermediate values of $E_{applied}$. Since the absorbance at 438 nm reflects the amount of o-tolidine in the oxidized form (via Beer's Law), the ratio [O]/[R] which corresponds to each value of $E_{applied}$ can be calculated from the spectra by Eq. (4).

$$\frac{[O]}{[R]} = \frac{A_2 - A_1}{A_3 - A_2} \tag{4}$$

As shown in Fig. 4-24, A_3 is the absorbance when o-tolidine is completely oxidized; A_1, the absorbance when entirely reduced; and A_2, the absorbance for the mixture of oxidized and reduced forms.

Figure 4-25 shows a plot of $E_{applied}$ vs. log ([O]/[R]) for the data in Fig. 4-24. The plot is linear as predicted by the Nernst equation. The slope of the plot is 30.8 mV which corresponds to an n value of 1.92, and the intercept is 0.612 V vs. SCE which corresponds to an $E^{\circ\prime}$ of 0.854 V vs. SHE.

Procedures

A. Cyclic Voltammetry

Locate the redox potential for the Fe(CN)$_6^{3-}$/Fe(CN)$_6^{4-}$ couple by thin-layer cyclic voltammetry on the standard Fe(CN)$_6^{3-}$ solution. Use a scan rate of 4 mV/s. (Scan rates in the OTTLE should be slow to avoid distortion by iR drop.)

Record cyclic voltammograms at scan rates of 0.5, 2, 4, 6, 8, and 10 mV/s.

Treatment of Data

Calculate $E^{\circ\prime}$ for ferricyanide and ferrocyanide from the peak potentials of the cyclic voltammogram by Eq. (1). Convert this value from the SCE to the SHE and compare with reported values obtained under similar solution conditions (Ref. 8).

Plot i_{pa} and i_{pc} against scan rate (v and $v^{1/2}$). Which functionality fits best?

B. Coulometry

Determine Q_T by recording a charge-time curve during a potential step from +0.40 to 0.00 V vs. SCE with the $Fe(CN)_6^{3-}$ solution of known concentration. Determine Q_B by repeating the experiment on a solution containing only supporting electrolyte.

Treatment of Data

Calculate the volume of the thin-layer cell using Eq. (2). From the area defined by the minigrid, calculate the thickness of the thin-layer cell.

C. Determination of $E^{\circ\prime}$, n, and Spectra of o-Tolidine

Assemble the thin-layer apparatus in the cell compartment of a spectrometer. Mask the OTTLE with black tape so the light beam passes only through the center of the minigrid. Locate the o-tolidine redox couple by cyclic voltammetry. Observe the color behavior in the vicinity of the minigrid during reduction and oxidation as a cyclic voltammogram is recorded.

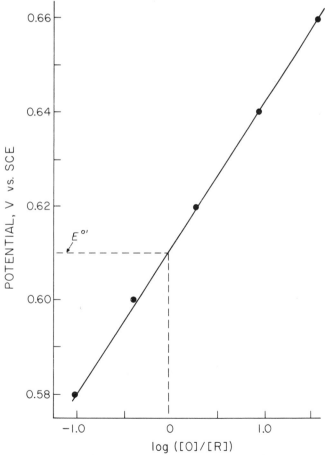

Figure 4-25 / Plot of $E_{applied}$ vs. log ([O]/[R]) from spectra in Fig. 4-24. [*Reprinted with permission from T. P. DeAngelis and W. R. Heineman, J. Chem. Ed., 53, 594 (1976). Copyright © 1976, Division of Chemical Education, American Chemical Society.*]

Record spectra of o-tolidine in its completely reduced and completely oxidized forms. (Procedure D can be performed conveniently at this time.)

Determine Q_T by controlled potential coulometry. A potential step from +0.80 to +0.3 V vs. SCE is suggested. Repeat the potential step coulometry experiment on supporting electrolyte solution to determine Q_B. (The blank experiment can be performed after Procedure D to avoid an additional change of solution.)

Treatment of Data

Calculate $E^{\circ\prime}$ for o-tolidine from the peak potentials of the cyclic voltammogram.

Calculate the molar absorptivity ϵ for λ_{max} of oxidized o-tolidine using Beer's Law and the cell thickness, which was measured coulometrically in Procedure B.

Calculate n for o-tolidine from Q_F by Eq. (2). Is the value of n obtained consistent with the electrode mechanism shown earlier?

D. Nernst Equation: Spectroelectrochemical Study

Assemble the thin-layer apparatus in the cell compartment of a spectrometer. Apply a potential of +0.800 V vs. SCE, and record the spectrum from 325 to 550 nm after electrolysis has ceased. Repeat this procedure for applied potentials of +0.660, +0.640, +0.620, +0.600, +0.580, and +0.400 V.

Treatment of Data

Plot $E_{applied}$ vs. log ([O]/[R]) and determine n and $E^{\circ\prime}$ for the o-tolidine couple.

Questions

1 Explain why the iR drop in the OTTLE is larger than in a conventional cell such as the one shown in Fig. 4-4. What is the effect of the iR drop on E_{pa} and E_{pc} of the cyclic voltammogram?

2 Should the difference in E_{pa} and E_{pc} for a thin-layer cyclic voltammogram of a reversible redox couple be $0.059/n$ volts [see Eq. (4) in Experiment 4-1] or 0 V (see Ref. 9)? Explain.

3 Why does the iR drop not affect $E^{\circ\prime}$ determined by the spectroelectrochemical method?

4 Assuming an OTTLE with an optical path length of 0.20 cm, calculate the concentration needed to perform a spectroelectrochemical experiment if an absorbance change of 0.1 a.u. is required and $\Delta\epsilon$ is 10,000 mol^{-1} cm^{-1} for the compound of interest.

5 Derive Eq. (4) using Beer's Law.

EXPERIMENT 4-10

Amperometric Titrations with Dropping Mercury Electrode

Purpose

The amperometric titration is used to determine Pb^{2+} by a precipitation reaction and Cu^{2+} by a complexation reaction.

References

1 I. M. Kolthoff and Y.-D. Pan, *J. Am. Chem. Soc., 61,* 3402 (1939).

2 R. W. Murray and C. N. Reilley in "Treatise on Analytical Chemistry," I. M. Kolthoff and P. J. Elving (eds.), part I, vol. 4, Interscience, New York, 1963, chap. 43.

3 A. J. Bard and L. R. Faulkner, "Electrochemical Methods," Wiley, New York, 1980, chap. 10.

4 H. H. Bauer, G. D. Christian and J. E. O'Reilly, "Instrumental Analysis," Allyn and Bacon, Boston, 1978, chap. 3

5 H. H. Willard, L. L. Merritt, Jr., J. A. Dean, and F. A. Settle, Jr., "Instrumental Methods of Analysis," 6th ed., Van Nostrand, Princeton, N.J., 1981, chap. 24.

6 J. T. Stock, "Amperometric Titrations," Interscience, New York, 1965.

Apparatus

Instrument for polarography
Electrochemical cell
DME

note   Mercury is toxic.

SCE reference electrode
Platinum auxiliary electrode
Nitrogen system for deoxygenating the cell solution
Rubber stopper drilled for the electrodes, a buret, and a gas delivery tube
Pipet, 10 mL
Buret, 10 mL
Volumetric flasks (2), 100 mL

Chemicals

$0.05\ M$ potassium dichromate $[K_2Cr_2O_7]$
$1\ M$ hydrochloric acid $[HCl]$
Lead nitrate $[Pb(NO_3)_2]$ sample (approximately 5 mmol) in 100-mL volumetric flask (dilute to mark with distilled water)
Acetic acid, sodium acetate buffer, $0.3\ M$ in each
$0.01\ M$ α-benzoin oxime (in 50% ethanol)
Ammonia–ammonium nitrate buffer, $0.3\ M$ in each
Copper(II) nitrate $[Cu(NO_3)_2]$ sample (approximately 0.5 mmol) in 100-mL volumetric flask (dilute to mark with distilled water)

Theory

An amperometric titration is performed by adding successive aliquots of a titrant to a solution of the sample and measuring the diffusion current after addition. This diffusion current may be proportional to the concentration of the substance being titrated, to that of the excess of the reagent, or to that of one of the products of the reaction. Some selectivity is possible by proper choice of the potential applied to the cell. The titration curve is a plot of the diffusion current vs. the volume of reagent added. Amperometric titration curves are similar to the curves of conductometric titrations, and under idealized conditions consist of two straight lines whose intersection corresponds to the end point of the titration. The determination of nonreducible substances such as phosphate and sulfate are made possible by amperometric titrations if a titrant is selected which is polarographically reducible.

Shape of the Titration
Curve

The course of a titration may be followed by employing amperometry as a technique for indicating the concentration of one or more of the species involved in the titration reaction:

$$\underset{\text{Sample}}{X} + \underset{\text{Titrant}}{T} \rightarrow \underset{\text{Product(s)}}{P_1} \tag{1}$$

The shape of the titration curve depends upon which of the species is indicated by the electrode process, which in turn depends upon the potential applied across the cell. Because current either will or will not flow for each of the three species, eight possibilities (2^3) exist and are summarized in the table below. Actually there is an infinite number if exact quantitative values are considered.

Possible Amperometric Titration Conditions

Type	X	T	P_1	Type	X	T	P_1
1	0	0	0	5	0	x	0
2	x	0	0	6	0	x	x
3	x	x	0	7	0	0	x
4	x	x	x	8	x	0	x

x = Active species   0 = Nonactive species

In addition, the active species may give rise to anodic or cathodic currents. Furthermore, the same line of reasoning can be extended to the successive titration of two-component mixtures where Reaction (1) is followed by Reaction (2):

$$\underset{\substack{\text{Sample} \\ \text{(second} \\ \text{component)}}}{Y} + \underset{\text{Titrant}}{T} \rightarrow \underset{\text{Product}}{P_2} \tag{2}$$

The actual shape of the titration curves is further governed by factors such as the equilibrium constants for Reactions (1) and (2), rates of reaction, interferences and impurities, and linearity between the current and the concentration of the active species. This last factor is strongly dependent on the mode of operation employed.

Precipitation Titrations

A reducible ion can be titrated with a nonreducible titrant which is also a precipitating agent, e.g., lead ion with sulfate ion. By using a titrant which is much more concentrated than the sample solution, the volume of the original solution is not appreciably increased during the course of the titration. Under these conditions the diffusion current of the reducible substance will decrease linearly with increasing volume of titrant added. Because the titrant does not yield a diffusion current, the addition of an excess will not affect the measured current, and the titration curve will be of the form shown in Fig. 4-26. The solubility of the precipitate may be appreciable near the end point; thus the experimental points will fall on a curve

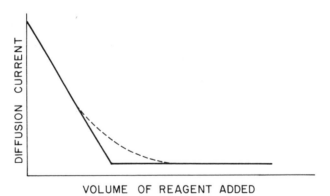

VOLUME OF REAGENT ADDED

Figure 4-26 / Amperometric titration curve for electroactive sample and electroinactive titrant and product.

which lies above the extrapolated lines in this region (dashed line in Fig. 4-26). The presence of a slight excess of either reactant decreases the solubility of the precipitate so that the experimental points approach the straight lines.

In actual practice the dilution of the sample solution by the titrant causes the observed currents to be smaller than those which would be found if the volume remained unchanged during the titration. In order to correct for this dilution, the observed current at each point should be multiplied by the ratio $(V + v)/V$, where V is the volume of the original solution and v is the volume of titrant added.

This correction can be minimized by using a titrant which is much more concentrated than the sample solution. If the ratio of molarities is at least 20:1 the correction can be neglected altogether for most systems.

The advantages and disadvantages of amperometric titrations with the dropping mercury electrode as compared with the rotated platinum electrode are as follows:

Advantages of a DME

1 *Renewed surface.* This is especially desirable for precipitation reactions. As a result of this feature the current values are quite reproducible.
2 *High hydrogen overvoltage of mercury.* Because mercury exhibits a high overvoltage for the evolution of hydrogen but low overvoltage for many other reactions, the scope of amperometric titrations is very wide. This feature makes possible the use of titrants (organic and inorganic) which are not potential-determining but which yield irreversible polarographic waves with well-defined diffusion currents. Platinum electrodes have the advantage of operating over a different voltage range (e.g., $+1.1$ to -0.5 V vs. SCE in 0.5 M HCl) as compared to mercury (e.g., $+0.1$ to -1.1 V vs. SCE in 0.5 M HCl). The exact range depends strongly upon the supporting electrolyte, but the general usefulness of platinum still remains in the positive potential regions.

Disadvantages

1 A practical disadvantage with amperometric titrations using the DME is that such titrations are generally carried out at potentials more negative than -0.2 V vs. SCE and consequently must usually be performed in the absence of oxygen. Considerable time is required to remove dissolved oxygen from the solution prior to titration and after each addition of titrant.
2 *Residual current.* The residual current caused by the charging of the double-layer capacitance of the continually renewed surface of mercury constitutes "noise" superimposed on the diffusion current signal.
3 *Presence of interferences.* The presence of large amounts of materials which yield polarographic diffusion currents at lower potentials than that employed for the titration cause considerable increase in the error in locating the end point. This disadvantage applies to a rotated platinum electrode also. Changing the value of the applied potential, altering the supporting electrolyte, and/or addition of complexing agents may be helpful in some cases.

(See Experiment 4-4 for helpful hints concerning the DME.)

safety caution Mercury vapor is very poisonous. Discard mercury in the waste mercury container provided and clean up any spills immediately. Work over a porcelain tray to catch and recover any spilled mercury conveniently. Notify the instructor in the event of a mercury spill.

Part I. Titration of Lead with Dichromate

Procedure

Pipet a 10-mL aliquot of your sample of lead nitrate into the electrochemical cell and dilute with 15 mL of acetate buffer. Cover the solution with a rubber stopper drilled to accommodate the electrodes, a buret, and a gas delivery tube. *Before* you place the DME capillary in the solution, raise the leveling bulb so that the mercury will be flowing through the capillary. Bubble purified N_2 gas briskly through the solution for 5 to 10 min. Switch the N_2 flow to above the surface of the solution so that the upper part of the cell will be kept in an N_2 atmosphere. Record a polarogram for the lead ions over a potential range from +0.4 to −1.0 V vs. SCE. Record the average current at 0.0 and at −0.6 V vs. SCE.

Add 1.00 mL of the dichromate solution from a microburet, and bubble N_2 through the solution for a minute or two to remove oxygen introduced along with the dichromate solution. Again allow the solution to come to rest by switching the N_2 to flow above the solution. Then measure and record the current at 0.0 and −0.6 V vs. SCE.

Continue in this way until 10 mL of dichromate has been added or until the current is excessive. When excess dichromate is present record a polarogram of this solution over the voltage range of +0.4 to −1.0 V vs. SCE. Discard the solution and rinse the beaker and capillary thoroughly with 1 M hydrochloric acid, then with distilled water. Repeat for duplicate checks if time permits.

Treatment of Data

Plot $i(V + v)/V$ vs. v to secure the two titration curves corresponding to applied potentials 0.0 V and −0.6 V vs. SCE. Mark the end point in each case as well as the potential at which that particular titration curve was obtained. Report the amount (millimoles) of lead in your volumetric flask.

Questions

1 Why is an acid condition desirable for this titration? (See Ref. 1.)
2 Explain how the 1 M hydrochloric acid dissolves the precipitate. Give equations.
3 Which potential, 0.0 or −0.6 V vs. SCE, would you prefer for routine titrations? Why? Consider the effect of oxygen also before you answer.
4 a Could any potential other than 0.0 or −0.6 V vs. SCE be employed?
   b Would any advantages accrue from another choice?
   c Any disadvantages?
   d Would a different titration curve shape be obtained?
5 At the same concentration, which yields the larger diffusion current, Pb^{2+} or $HCrO_4^-$? Why is this to be expected?

Part II. Titration of Copper with α-Benzoin Oxime

Copper in an aqueous ammonia solution yields two polarographic waves. The first step corresponds to the reaction

$$Cu(NH_3)_4{}^{2+} + e \rightarrow Cu(NH_3)_2{}^+ + 2NH_3 \tag{3}$$

and the second wave to the reaction

$$Cu(NH_3)_2{}^+ + nHg + e \rightarrow \underset{\text{Amalgam}}{Cu(Hg)_n} + 2NH_3 \tag{4}$$

The half-wave potentials of the two steps depend upon the concentration of the ammonia—each being more negative at higher ammonia concentrations.

Copper is precipitated by the reagent α-benzoin oxime in alkaline solution to form an olive-green precipitate.

The copper chelate precipitate is very pure and dries easily, allowing excellent gravimetric determinations. However, the precipitate has a somewhat slimy hydrophilic character and consequently filters very slowly. This latter disadvantage is overcome in the amperometric-titration procedure, since filtration is unnecessary. Because α-benzoin oxime can also be reduced at the dropping mercury electrode, the preferred V-shape titration curve may be obtained by proper selection of applied potential.

Procedure

Pipet 10 mL of your copper sample into the electrochemical cell and add 15 mL of ammonia buffer. Cover the solution with a rubber stopper that has been drilled to accommodate the electrodes, a buret, and a gas delivery tube. Record the polarogram of this solution after bubbling N_2 through for 5 to 10 min to remove oxygen. A range of 0.0 to -1.6 V vs. SCE should be satisfactory. Take current readings at closely spaced potential intervals in the range 0.0 to -0.5 V vs. SCE.

Obtain the following data during the course of the titration:

Volume, mL	$i_d(-0.3\ V)$	$i_d(-0.65\ V)$	$i_d(-1.0\ V)$

Titrate using 1-mL increments of reagent until 10 mL of titrant has been added, bubbling the solution with nitrogen between each addition for 1 to 2 min to remove any oxygen introduced by way of the titrant. After 10 mL of titrant has been added, record another polarogram for the range 0.0 to -1.6 V.

Rinse the capillary with 1 M HCl to remove any precipitate and follow with copious quantities of distilled water.

Treatment of Data

Plot the three titration curves (at differing potentials) on different graphs. Label each curve according to its applied potential and indicate the millimoles corresponding to the end point in each case. Report the amount (millimoles) of copper in your volumetric flask.

Questions

1 Give an explanation for each of the two polarograms, including equations for the reduction reactions.
2 Write the reduction reaction taking place at each of the three potentials -0.3, -0.65, and -1.0 V used in the titrations.

EXPERIMENT 4-11

Amperometric Titrations with Rotating Platinum Electrode

Purpose

A rotating platinum electrode is used to study the correlation between amperometric and potentiometric titrations with a polarized electrode.

References

1 R. W. Murray and C. N. Reilley in "Treatise on Analytical Chemistry," I. M. Kolthoff and P. J. Elving (eds.), part I, vol. 4, Interscience, New York, 1963, chap. 43.
2 A. J. Bard and L. R. Faulkner, "Electrochemical Methods," Wiley, New York, 1980, chap. 10.
3 H. H. Bauer, G. D. Christian, and J. E. O'Reilly, "Instrumental Analysis," Allyn and Bacon, Boston, 1978, chap. 3.
4 H. H. Willard, L. L. Merritt, Jr., J. A. Dean, and F. A. Settle, Jr., "Instrumental Methods of Analysis," 6th ed., Van Nostrand, Princeton, N.J., 1981, chap. 24.
5 J. T. Stock, "Amperometric Titrations," Interscience, New York, 1965.

Apparatus

Magnetic stirrer with stirring bar
Instrument for applying potential and measuring current
Beaker, 150 mL [with rubber stopper drilled to accommodate a platinum working electrode, an SCE, and a buret (see Fig. 4-27)]
Graduated cylinder, 10 mL
SCE (with a salt bridge)
Rotating platinum electrode (with synchronous drive 600 rpm)
Pipet, 20 mL
Buret, 50 mL

Chemicals

Concentrated sulfuric acid [H_2SO_4]
0.1 M ferrous ammonium sulfate solution [$Fe(NH_4)_2(SO_4)_2 \cdot H_2O$]
0.1 M ceric ammonium sulfate solution [$(NH_4)_4Ce(SO_4)_4 \cdot 2H_2O$], 1 M in H_2SO_4

Theory

Review the Theory section of Experiment 4-10.

If the conditions for a titration are set up in such a way that either or both the substance being determined and/or the titrant undergoes a reaction at a working electrode, then a current, which is proportional to the concentration of the electroactive species, may occur at a given potential. If the potential across the cell is held constant, then the end point can be determined by the current change during the titration, the titration curve (current vs. volume) usually resolving itself into two straight lines, whose intersection is the end point.

Conversely, one may also perform a titration in which the current is held constant (at zero, or at some finite value). Potential breaks are obtained in each case with the break at zero applied current occurring at the equivalence point, whereas with finite current the break occurs either before or after (depending upon the *direction* of the current) the equivalence point. Helpful ideas concerning the principles underlying this phenomena are given in Experiment 4-10.

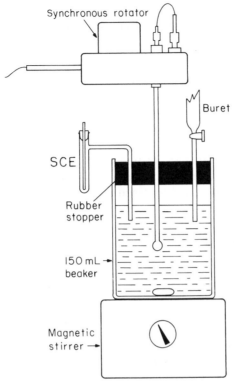

Figure 4-27 / Amperometric titration assembly using a rotated platinum electrode.

Procedure

The following steps are to be taken for this experiment:

1 Clean the platinum electrode by immersing it in hot nitric acid. Rinse both the platinum and calomel electrodes with distilled water. Insert the salt bridge of the SCE into the rubber stopper so that it will make contact with the solution to be placed in the beaker.

2 Into a clean 150-mL beaker pipet 20 mL of the ferrous solution and dilute to a volume such that the electrodes are covered. Add 10 mL of concentrated sulfuric acid to the solution and stir until well mixed. (Check with your instructor to see if an unknown sample is to be issued and if the iron is to be reduced, prior to titration, by the procedure outlined in Experiment 2-3.)

3 Place the solution on the magnetic stirrer, immerse the electrodes, and connect them to the instrument. Insert the buret containing the standard ceric solution into the stopper. Plug in the synchronous motor and turn it on.

4 Adjust the potential applied to the cell until it is at 0.9 V.

5 Take the reading of the current with the potential set at 0.9 V before adding any reagent. When you have this reading, adjust the potential so that the current is 0. Record this value. This is the potential of the cell at zero current. Adjust the potential until the current is at 100 μA (direction + and the same as at 0.9 V above). Record this value. Repeat at 10 μA. Adjust the potential until the current is -100 μA (direction $-$). Record this value. Repeat at -10 μA.

6 Proceed to titrate the sample, taking 2.0-mL increments at the beginning and smaller ones as the end point is approached. Stir the solution after each addition. Take readings until you are past the end point by about 8 mL. After each addition of reagent, repeat step 5. You may conveniently record your readings as follows (place — where readings are experimentally unattainable):

			E			
Volume of Ce^{4+}, mL	$I(0.9\ V)$, μA	0 current	$+100\ \mu A$	$+10\ \mu A$	$-100\ \mu A$	$-10\ \mu A$

Treatment of Data Plot the amperometric titration curve on one graph and the three potentiometric titration curves on a second graph. Determine the end points and the volume of ceric reagent required to reach them in each case and note these values on the graphs. If an unknown was issued, report the grams of iron present.

Questions
1 Describe the effect of an instrument drawing current during a potentiometric titration.
2 Does the direction of the current flow make any difference? Why (explain)?
3 Does the magnitude of the current make any difference? Why (explain)?
4 Write equations for the following electrode processes (major one only):
   a Reaction prior to end point (amperometric procedure)
      (1) At platinum electrode
      (2) In calomel electrode
   b Reaction after end point (amperometric procedure)
      (1) At platinum electrode
      (2) In calomel electrode
   c Reaction at platinum electrode prior to potentiometric end point (1) + current (2) − current
   d Reaction at platinum electrode after potentiometric end point (1) + current (2) − current
5 Why was a value of 0.9 V vs. SCE chosen for the amperometric titration?
6 Describe some reasons for the addition of sulfuric acid.
7 Were any of the electrode reactions "irreversible?" If so, which ones? How can you tell?
8 List the advantages and disadvantages of the DME and the rotating platinum electrode.

EXPERIMENT 4-12
End-Point Detection with Dual Polarized Electrodes

Purpose End-point detection with dual polarized electrodes is used for the titration of Fe^{2+} with Ce^{4+} and arsenite with iodine.

References
1 R. W. Murray and C. N. Reilley in "Treatise on Analytical Chemistry," I. M. Kolthoff and P. J. Elving (eds.), part I, vol. 4, Interscience, New York, 1963, chap. 43.
2 A. J. Bard and L. R. Faulkner, "Electrochemical Methods," Wiley, New York, 1980, chap. 10.

3   H. H. Bauer, G. D. Christian, and J. E. O'Reilly, "Instrumental Analysis," Allyn and Bacon, Boston, 1978, chap. 3.

4   H. H. Willard, L. L. Merritt, Jr., J. A. Dean, and F. A. Settle, Jr., "Instrumental Methods of Analysis," 6th ed., Van Nostrand, Princeton, N.J., 1981, chap. 24.

5   J. T. Stock, "Amperometric Titrations," Interscience, New York, 1965.

Apparatus

Polarograph (two-electrode)
Platinum-wire electrodes (2), identical
Beaker, 250 mL
Magnetic stirrer with stirring bar
Pipet, 10 mL

Chemicals

0.1 M ceric sulfate $[Ce(SO_4)_2]$
0.1 M ferrous ammonium sulfate $[Fe(NH_4)_2(SO_4)_2 \cdot 6H_2O]$
Concentrated hydrochloric acid $[HCl]$
0.025 M arsenite solution (arsenic trioxide $[As_2O_3]$ in 1 M sodium bicarbonate $[NaHCO_3]$ buffer)
0.05 M iodine solution $[I_2]$

Theory

The electrochemical methods that are usually employed for detecting the end point in a titration are generally classified as *amperometric, potentiometric,* or *conductometric,* each referring to the electrical quantity measured. The variations observed during the titration depend quite obviously on the electrode systems and the electric circuits employed. The following classification can be made:

A   Potentiometric titrations
   1   One indicator electrode (e.g., Pt) and one reference electrode (e.g., calomel)
     a   *At zero current.* This is the classical potentiometric technique.
     b   *At constant current.* In this procedure, a small constant current is passed through the indicator circuit (solution is stirred at a constant rate). The potential break at the end point is often found to be *larger* or *smaller* than in the case where no current is passed. The *direction* of current flow determines this order.
   2   Two indicator electrodes (e.g., two Pt electrodes)
     a   *At zero current* (seldom employed). Thermodynamics would predict no difference of potential in this solution. Actually, a small difference is usually observed depending on the previous history of the electrodes (surface effects).
     b   *At constant current.* In this case, potential differences are theoretically expected, and the method is quite useful. Current passing *into* one electrode causes an equal current to flow *out* of the other electrode. A potential difference then occurs between the two electrodes which is attributed to the potentials required at each electrode for supporting this current. Two examples are given; one illustrates the potentials resulting from two reversible systems, the other is concerned with one irreversible system titrated by a reversible system.
       (1)   *Two reversible systems.* For example, imagine that a Fe^{2+} solution is to be titrated with Ce^{4+}. *At the initial point* in the titration, the probable electrode reactions are

$$\text{Cathode:} \quad 2H_2O + 2e \rightarrow H_2 + 2OH^-$$
$$\text{Anode:} \qquad\qquad Fe^{2+} \rightarrow Fe^{3+} + e$$

Because these reactions are forced (by the constant-current condition) to take place, and they require quite different potentials, a rather large potential difference is expected at this point in the course of the titration.

Half way to the end point the solution contains equal quantities of Fe^{2+}, Fe^{3+}, and Ce^{3+}. The following electrode reactions are then expected:

$$\text{Cathode:} \quad Fe^{3+} + e \rightarrow Fe^{2+}$$
$$\text{Anode:} \qquad\quad Fe^{2+} \rightarrow Fe^{3+} + e \quad \text{(Oxidation of } Ce^{3+}$$
$$\text{would be more difficult.)}$$

Because these two reactions take place at nearly the same potential, a potential difference of close to zero is expected at this halfway point.

At the end point, equal quantities of Fe^{3+} and Ce^{3+} are present, and the expected electrode reactions become

$$\text{Cathode:} \quad Fe^{3+} + e \rightarrow Fe^{2+}$$
$$\text{Anode:} \qquad\quad Ce^{3+} \rightarrow Ce^{4+} + e$$

These reactions occur at widely different potentials, and consequently the potential difference at the equivalence point in the titration would be large. This maximum value of ΔE is taken to be the end point as shown in Fig. 4-28.

Past the end point, excess Ce^{4+} has been added, and the electrode reactions become

$$\text{Cathode:} \quad Ce^{4+} + e \rightarrow Ce^{3+} \quad \text{(Reduction of } Fe^{3+} \text{ is}$$
$$\text{more difficult.)}$$
$$\text{Anode:} \qquad\quad Ce^{3+} \rightarrow Ce^{4+} + e$$

The potential difference would therefore be small. Figure 4-28 illustrates the titration of Fe^{2+} with Ce^{4+} using a constant-current potentiometric end point.

(2) *Titration of irreversible couple with a reversible couple.* The titration of arsenite with iodine is an example of this type. Because arsenite is oxidized to arsenate at an electrode surface with difficulty (large overvoltage) and because the reduction of arsenate to arsenite is also irreversible, a large potential difference is expected up to the end point with the following predominant reactions

$$\text{Cathode:} \quad H_2O + e \rightarrow \tfrac{1}{2}H_2 + OH^-$$
$$\text{Anode:} \qquad\quad H_2O \rightarrow \tfrac{1}{2}O_2 + 2H^+ + 2e \quad \text{(Initial condition)}$$
$$I^- \rightarrow \tfrac{1}{2}I_2 + e \quad \text{(Remainder of titration)}$$

Past the end point, excess I_2 is added, and the following reactions (reversible) take place

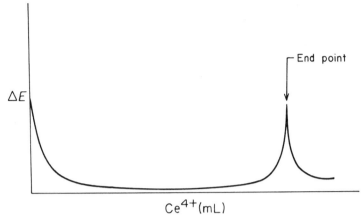

Figure 4-28 / Titration of Fe^{2+} with Ce^{4+} with a constant-current potentiometric end point with two indicator electrodes.

$$\text{Cathode:} \quad \tfrac{1}{2}I_2 + e \rightarrow I^-$$
$$\text{Anode:} \quad \quad I^- \rightarrow \tfrac{1}{2}I_2 + e$$

The potential difference then becomes small.

This titration is illustrated in Fig. 4-29.

B Amperometric titrations

    1 One indicator electrode (e.g., rotating platinum indicator electrode or DME vs. SCE). This is the classical amperometric procedure. Because this subject has been discussed in Experiments 4-10 and 4-11, a discussion of its principle of operation is not repeated here.

    2 Two indicator electrodes (e.g., two identical platinum wire electrodes). In this technique, a constant *potential* difference is applied between the two electrodes, and the resulting current is then measured during the course of the titration.

        a *Titration of a reversible system with a reversible system.* Again let us consider Fe^{2+} being titrated with Ce^{4+}. The potential applied across the two electrodes is generally quite small, and this usually restricts the

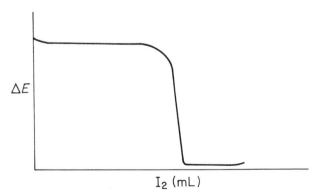

Figure 4-29 / Titration of arsenite with I_2 with a constant-current potentiometric end point with two indicator electrodes.

electrode reactions to the reverse of one another. That is, if one electrode reaction is

Anode: $Fe^{2+} \rightarrow Fe^{3+} + e$

the other reaction must become

Cathode: $Fe^{3+} + e \rightarrow Fe^{2+}$

The current is a measure of the overall rate of electrode reaction, which in turn generally is limited (because of low concentrations of reactants or irreversibility in the electron transfer process) either by the anode *or* cathode reactions (both cathodic and anodic currents are equal).

(1) *Initial point.* At the beginning of the titration only Fe^{2+} is present, and therefore only the anodic reaction is possible. Therefore the current at this point is zero (or slightly higher due to residual currents).

(2) *During titration.* As Fe^{3+} increases in concentration during the titration, the current also increases. Up to the half-way mark, the concentration of Fe^{3+} is the factor that limits the current. Past the half-way mark, the concentration of Fe^{2+} becomes less than that of Fe^{3+}, and the current begins to decrease, since the current now is limited by the Fe^{2+} concentration. At the end point the current approaches zero, since only Fe^{3+} and Ce^{3+} are present. Past the end point, excess Ce^{4+} is present. Because the concentration of Ce^{3+} is larger than that of Ce^{4+}, the current just after the end point is proportional to the concentration of Ce^{4+} and thus increases with added reagent. This titration is illustrated in Fig. 4-30.

The same principles can be used to explain the shape of the arsenite-iodine and iodine-arsenite titrations as shown in Figs. 4-31 and 4-32.

Procedures

note   In these experiments, try to keep the rate of stirring at the same level throughout the titration.

A. Titration of Fe^{2+} with Ce^{4+}

Place 10 mL Fe^{2+} stock solution in the beaker, add 10 mL concentrated HCl, and then sufficient water to cover the electrodes by about $\frac{1}{2}$ in. Titrate with Ce^{4+}, collecting the following data during the course of the titration:

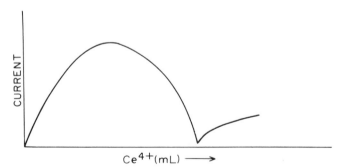

Figure 4-30 / Titration of Fe^{2+} with Ce^{4+} with an amperometric end point with two indicator electrodes.

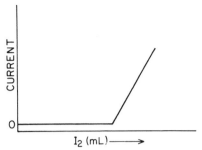

Figure 4-31 / Titration of arsenite with I_2 with an amperometric end point with two indicator electrodes.

	E			i		
Volume, mL	$i = 2\ \mu A$	$i = 20\ \mu A$		$E = 0.05\ V$	$E = 0.2\ V$	$E = 0.4\ V$

note Take all five readings after each addition of titrant, changing the applied potential or current as may be necessary.

B. Titration of Arsenite with Iodine

Place 10 mL arsenite stock solution in the beaker and add sufficient water to cover the electrodes by $\frac{1}{2}$ in. Titrate with iodine, collecting the data as directed above.

Treatment of Data

Plot the following on four sheets of graph paper:

1 Data on potentiometric titration of Fe^{2+} with Ce^{4+} at applied currents of 2 and 20 μA
2 Data on potentiometric titration of arsenite with iodine at applied currents of 2 and 20 μA
3 Data on amperometric titration of Fe^{2+} with Ce^{4+} at applied potentials of 0.05, 0.2, and 0.4 V
4 Data on amperometric titration of arsenite with iodine at applied potentials of 0.05, 0.2, and 0.4 V

Questions

1 In your own words, explain the reasons for the four titration curve shapes obtained in the four graphs.
2 Which method and conditions do you prefer for the titration of Fe^{2+} with Ce^{4+}? Why?

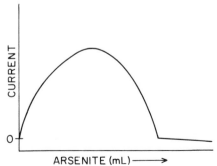

Figure 4-32 / Titration of I_2 with arsenite with an amperometric end point with two indicator electrodes.

3 Which method and conditions do you prefer for the titration of arsenite with iodine? Why?

4 Explain in detail why the different current values caused the effect illustrated in graph 1. [*Hint:* Sketch the hydrodynamic voltammograms expected for the solution at (*a*) the beginning of the titration, (*b*) the middle of the titration, (*c*) just before the end point, (*d*) the end point, (*e*) just past the end point, and (*f*) far past the end point.]

5 Does there seem to be any difference in the degree of reversibility between the Fe^{3+}, Fe^{2+} couple and the Ce^{4+}, Ce^{3+} couple? What is meant by the "degree of reversibility" of a couple?

6 What titration curve would be expected if I_2 were titrated with arsenite?

EXPERIMENT 4-13
Glucose Determination with an Immobilized Enzyme Electrode†

Purpose

Glucose is determined by the amperometric detection of H_2O_2 with an immobilized enzyme electrode.

References

1 G. Sittampalam and G. S. Wilson, *J. Chem. Ed., 59,* 70 (1982).

2 D. N. Gray, M. H. Keyes, and B. Watson, *Anal. Chem., 49* (12), 1067A (1977).

3 N. Lakshminarayanaiah, "Membrane Electrodes," Academic Press, New York, 1976.

4 G. G. Guilbault in "Methods in Enzymology," K. Mosbach (ed.), vol. XLIV, Academic Press, New York, 1976, p. 579

5 R. K. Kobos in "Ion-Selective Electrodes in Analytical Chemistry," H. Freiser (ed.), vol. II, Plenum, New York, 1980, chap. 1.

6 H. V. Malmstadt and H. L. Pardue, *Anal. Chem., 33,* 1040 (1961).

Apparatus

Enzyme electrode†
YSI Model 25 oxidase meter‡
Strip chart recorder (with adjustable full scale)
7- or 4-dram capacity vials (5)
Magnetic stirring bars (6), $\frac{1}{2} \times \frac{1}{4}$ in
Pipets (2), 2- and 10-mL capacity calibrated in $\frac{1}{10}$ mL
Beaker, 50 mL
Magnetic stirrer
Wash bottle, 100 mL

Chemicals

Phosphate buffer (pH 7.3), 250 mL (in wash bottle)
100 mg/dL glucose stock solution (made in pH 7.3 phosphate buffer), 50 mL
Unknown solution of glucose
Deionized distilled water

---

†From an experiment developed at the University of Arizona and Bowling Green State University, courtesy of G. S. Wilson, G. Sittampalam, and V. Srinivasaan. Reprinted in part with permission from Ref. 1, copyright 1982, Division of Chemical Education, American Chemical Society.

‡Both the amperometric probe with the membrane kit and the oxidase meter can be purchased from Yellow Springs Instrument Co., Box 279, Yellow Springs, OH 45387.

Theory

Techniques utilizing immobilized biological molecules, such as enzymes and antibodies, have become increasingly popular in routine chemical analysis. Among these, immobilized enzymes have been used with considerable success, especially in clinical and environmental analyses.

This experiment is designed to use the enzyme glucose oxidase. β-D-glucose, which is a specific substrate for this enzyme, undergoes the following enzymatic reaction:

$$\beta\text{-D-glucose} + O_2 \xrightarrow[\text{pH 7.3, phosphate buffer}]{\text{glucose oxidase}} \text{gluconic acid} + H_2O_2 \qquad (1)$$

To follow the above reaction, one can detect the formation of one of the two products or the depletion of one of the reactants. In monitoring the products, it is convenient to measure the concentration of H_2O_2 generated, which will be directly proportional to the glucose concentration in the sample solution. H_2O_2 concentration can be monitored either by a potentiometric measurement of iodine produced by the reaction between H_2O_2 and excess I^- in the sample (Ref. 6) or by a direct anodic oxidation of H_2O_2 at a fixed applied potential. The latter approach is used in this experiment.

The enzyme is "immobilized" or trapped between an inner cellulose acetate and an outer collagen or polycarbonate membrane, which is cast at the tip of the amperometric probe. The probe consists of a platinum anode (working electrode) and a silver cathode between which a constant potential of +0.700 V is applied. This potential is applied when the probe is plugged into the oxidase meter, which also measures the current due to the oxidation of H_2O_2 at the platinum anode.

When the probe is immersed in a glucose solution, glucose diffuses through the outer membrane into the enzyme layer and is converted to gluconic acid with the concomitant production of H_2O_2 as shown in Eq. (1). The H_2O_2 diffuses *selectively* through the cellulose acetate membrane toward the platinum anode, where it undergoes an oxidation, while the O_2 dissolved in solution reduces at the silver cathode as shown below:

$$\text{Pt anode:} \qquad\qquad H_2O_2 \rightarrow O_2 + 2H^+ + 2e \qquad (2)$$

$$\text{Ag cathode:} \quad \tfrac{1}{2}O_2 + 2H^+ + 2e \rightarrow H_2O \qquad (3)$$

The steady-state current from the anode reaction will be directly proportional to the H_2O_2 generated by the enzymatic reaction, which in turn will be proportional to the glucose concentration in the sample. Thus, the steady-state current is a direct measure of glucose concentration in the sample solution.

The experimental setup is shown in Fig. 4-33.

Glucose oxidase is immobilized on the electrode by trapping between cellulose acetate and collagen membranes as shown in Fig. 4-34.

Procedure

The experimental setup should always be left as shown in Fig. 4-33 with the oxidase meter ON and the enzyme probe immersed in fresh buffer solution.

Turn on the stirrer motor and set the knob at a slow, but constant stirring speed. Make sure the electrode is immersed in fresh buffer solution.

Zero the oxidase meter and recorder using appropriate control knobs. Set the chart speed at 3 cm/min. Set full scale to 250 to 500 mV on the recorder. Set the selector switch on the oxidase meter to "nanoamp" mode. Now the full scale on the meter will read 100 nA.

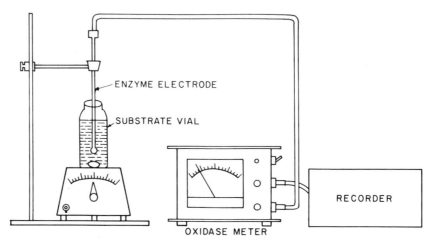

Figure 4-33 / Experimental apparatus for glucose determination.

Prepare 2 mg/dL, 4 mg/dL, 6 mg/dL, 8 mg/dL, and 10 mg/dL glucose solutions in labeled vials with stirring bars. (10 mL of these standard solutions are conveniently made using the vials provided.) Use the 100-mg/dL stock solution; the appropriate dilutions are made using pH 7.3 phosphate buffer (dL ≡ deciliter ≡ 100 mL).

Immerse the probe in each of the above standard solutions and record the "steady-state" current for 1 min. A typical current response is shown in Fig. 4-35. The probe should be washed with fresh buffer solution in between the introduction of each substrate solution.

Pipet 0.50 mL of unknown solution into a separate vial with a stirring bar. Dilute to 10.00 mL using pH 7.3 phosphate buffer and record the steady-state current.

Remove the probe from the unknown solution and leave it immersed in fresh buffer solution. Turn off the stirrer motor.

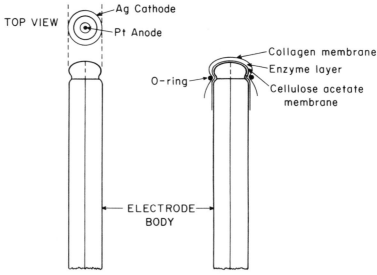

Figure 4-34 / Enzyme electrode.

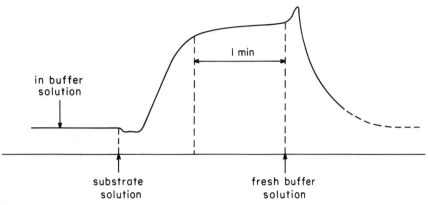

Figure 4-35 / Response of glucose electrode.

Treatment of Data

Plot a standard curve of the steady-state current vs. glucose concentration using the data for the glucose standards. From the standard curve calculate the concentration of glucose in the unknown.

Questions

1 Briefly discuss the probable roles of the cellulose acetate and collagen membranes.

2 Explain the slight slope observed in your steady-state current response. (*Hint:* Consider the possible directions of diffusion of H_2O_2 produced in the enzyme layer!)

3 Explain why the current measured by the oxidase meter in this experiment is proportional to $[H_2O_2]$ at the cathode, rather than $[O_2]$ at the anode.

5
Electrolytic Methods†
and Controlled-Current Methods

Whereas voltammetric methods in most cases are nondestructive and do not alter the sample solution, electrolytic methods are based on the principle of complete conversion of the sample to another form and oxidation state. The quantity of the sample species can be determined either by weighing the electrolysis product (usually the free metal) or by measuring the number of coulombs of electricity to effect the complete electrolysis of the sample. The latter requires that the electrolysis proceed with 100% current efficiency.

The most familiar electrolytic method is the electrogravimetric determination of metals by plating onto a weighed platinum electrode and reweighing this electrode after complete electrolysis has been accomplished. An example of such a method is the determination of copper with two platinum electrodes in a solution containing $0.1\ M\ Cu^{2+}$ and $0.1\ M\ H_2SO_4$. Because the electrodes are identical the cell potential is zero. However, if an external potential is applied to this cell so that electrons are driven into the left electrode, then cupric ions and hydrogen ions migrate toward the left (negative) electrode, and sulfate ions migrate to the right (positive) electrode (see Fig. 5-1).

The possible electrode reactions are

$$\text{Cathode:} \quad Cu^{2+} + 2e \rightarrow Cu \qquad E° = 0.34\ \text{V} \tag{1}$$

$$2H^+ + 2e \rightarrow H_2 \qquad E° = 0.00\ \text{V} \tag{2}$$

†From a discussion by Dr. R. L. Pecsok, University of California, Los Angeles.

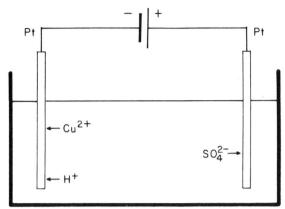

Figure 5-1 / Electrochemical cell for electrolytic methods.

Anode: $2HSO_4^- \rightarrow S_2O_8^{2-} + 2H^+ + 2e$ $E° = -2.05$ V   (3)

$$2H_2O \rightarrow O_2 + 4H^+ + 4e \qquad E° = -1.23 \text{ V} \qquad (4)$$

For the reactions as written, the standard potentials indicate that Reactions (1) and (4) have the greater tendency to take place. The actual potentials calculated from the Nernst equation should be used rather than the standard potentials. As the Cu^{2+} concentration is decreased, the potential of the cathode becomes more negative until it is sufficiently negative to cause the reduction of H^+.

As soon as the electrolysis commences, the cell is modified and becomes

$$Cu \quad \bigg| \quad Cu^{2+} \quad H^+ \quad SO_4^{2-} \quad O_2 \quad \bigg| \quad Pt$$

with a copper electrode and an oxygen electrode. Assuming a partial pressure for oxygen of 1 atm and that the Cu^{2+} and H^+ concentrations are both $0.1\ M$, then the potential of this cell can be calculated to be 0.86 V, with the copper electrode being negative.

$$E_{cell} = E°_{O_2,H_2O} E°_{Cu^{2+},Cu} + \frac{0.059}{4} \log \frac{P_{O_2}[H^+]^4}{[Cu^{2+}]^2} = 0.86V \qquad (5)$$

This represents the tendency for the cell reaction and indicates that the copper electrode would spontaneously go to Cu^{2+}. Thus, to prevent dissolution of copper an applied potential of 0.86 V must be applied to the cell with the negative terminal of the applied potential attached to the copper electrode. This potential is referred to as the equilibrium decomposition potential and is that potential that will just counterbalance the tendency for the cell reaction to take place in the opposite direction. As electrolysis takes place the concentration of the active species decreases in the immediate vicinity of the electrode. This causes the decomposition potential to increase and is referred to as concentration polarization.

A potential just greater than 0.86 V would be expected to cause the electrolysis of copper; however, in many cases it will take a considerably larger potential than the equilibrium decomposition potential because of overpotential effects (irreversible electrode reactions). The magnitude of the overpotential depends upon a number of variables:

1 Kind and condition of the surface of the metal used for the electrode.
2 Nature of the electrode reaction. (It is high for reactions involving a gas discharge and usually low for the deposition of metals.)
3 Temperature. (Increased temperature reduces the overpotential.)
4 Current density. (Increased current density increases the overpotential.)

The overpotential, regardless of whether it occurs at the anode or at the cathode, acts to increase the magnitude of the applied potential necessary to maintain a given rate of electrolysis.

In addition to the equilibrium decomposition potential and the overpotential, the applied potential must also overcome the resistance of the entire circuit if any current is to flow. Thus, we have the overall equation

$$E_{applied} = E_{cell} + E_{oc} + E_{oa} + iR \tag{6}$$

where E_{oc} is the overpotential of the cathode, E_{oa} is the overpotential of the anode, and iR is the ohmic voltage. The necessary applied potential is equal to the equilibrium decomposition potential (using the concentration of the active species in the immediate vicinity of the electrode) plus the overpotential plus the ohmic voltage. Equation (6) is presented graphically in Fig. 5-2. As current flows, the

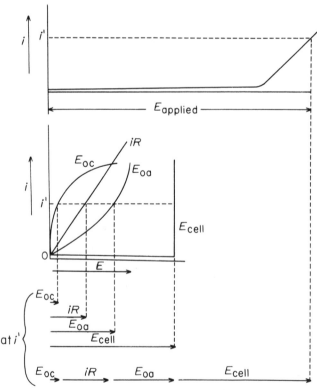

Figure 5-2 / Graphical presentation of Eq. (6).

electrode reactions take place and the chemical composition of the electrolyte is continually changing. This has a number of effects:

1 The electrode potentials are changed and the equilibrium decomposition potential increases.
2 The initial electrode reactions may be replaced by others as the first substances are depleted.
3 The resistance changes because of the replacement of one kind of ions by another. Frequently a metal is deposited at the cathode, and an equivalent amount of H^+ is formed at the anode; this decreases the resistance.
4 The temperature is increased due to the passage of current through the electrolyte. This decreases the overpotentials and decreases the resistance of the electrolyte.

It is apparent that if the applied potential is kept constant during an electrolysis the cathode potential may change considerably due to changes in concentrations. As the current and resistance change, the overpotentials also are affected, and all of these indirectly cause the cathode potential to change through Eq. (6). It is normal for the cathode potential to become more negative. The more negative the cathode potential, the more likely it is that other reactions can occur. It would obviously be more desirable to control the cathode potential rather than the total applied potential. (The same considerations apply to the anode in case one is interested in that reaction.)

In order to measure the cathode potential, a stable reference electrode is required. The cell anode will not suffice, since its potential is also changing as reaction takes place. A third electrode is inserted for this purpose, usually an SCE. The potential between the cathode and the reference electrode is held constant by adjusting the total applied potential as required. This technique is valuable in dealing with mixtures where several different electrode reactions might occur at the cathode if its potential were not held constant.

For the analysis of mixtures some degree of selectivity may be achieved through adjustment of the electrolyte in the sample solution. For example, if a solution is made acidic those metal ions above hydrogen in the electromotive series will not be electrolyzed. Thus, in an acid solution Cu^{2+} may be electrolyzed selectively in the presence of Ni^{2+} because H^+ is more easily reduced than Ni^{2+}. In a similar fashion, the presence of nitrate or other electroactive ions will permit selectivity for specific mixtures.

A number of useful analytical techniques are derived from these principles:

1 *Electrogravimetric determinations.* Metals are deposited and weighed on the cathode, or compounds of the metals, such as lead dioxide, are deposited and weighed on the anode. This technique is illustrated by Experiment 5-1.
2 *Electroseparations.* One or a group of metals may be deposited, leaving others in solution. The deposited metals may then be stripped from the cathode by reversing the current or dissolving in acids. Mercury cathodes can be used for this purpose and the metals recovered upon evaporation of the mercury.
3 *Coulometric analysis.* If the electrode reaction can be restricted to a single reaction operating with 100% current efficiency (usually by controlling the cathode potential), the amount of charge passed is a measure of the substance to be determined. Alternatively, with a constant current the time required to complete the reaction is a measure of the substance. The latter method

requires some independent method of determining the completion of the desired reaction. The technique of constant-current coulometry is illustrated by Experiment 5-2.

EXPERIMENT 5-1
Electrogravimetric Analysis of a Copper-Nickel Mixture†

Purpose

This experiment illustrates the utility of the electrolytic method and the control of the electrode reaction that is made possible by control of the pH and the solution media.

References

1 D. A. Skoog and D. M. West, "Principles of Instrumental Analysis," 2nd ed., Saunders, Philadelphia, 1980, chap. 20.
2 H. A. Laitinan and W. E. Harris, "Chemical Analysis," 2nd ed., McGraw-Hill, New York, 1975, chap. 14.
3 J. J. Lingane, "Electroanalytical Chemistry," 2nd ed., Interscience, New York, 1958, chap. 10.
4 L. Meites and H. C. Thomas, "Advanced Analytical Chemistry," McGraw-Hill, New York, 1958, chap. 7.
5 H. H. Willard, L. L. Merritt, Jr., J. A. Dean, and F. A. Settle, Jr., "Instrumental Methods of Analysis," 6th ed., Van Nostrand, Princeton, N.J., 1981, chap. 25.
6 H. A. Strobel, "Chemical Instrumentation," Addison-Wesley, 1973, chap. 27.
7 N. Tanaka in "Treatise on Analytical Chemistry," I. M. Kolthoff and P. J. Elving (eds.), part I, vol. 4, sec. D-2, Interscience, New York, 1963, chap. 48.

Apparatus

Beakers (2), 200 mL, tall form, lipless
Split watch glasses (2)
Platinum-gauze electrodes (2 pairs) (cathode approximately 100 cm²)
DC power source with outlets for two electrolysis assemblies
Magnetic stirrers with stirring bars (2)
Pipet, 25 mL
Graduated cylinder, 100 mL
Volumetric flask, 100 mL (containing unknown copper-nickel sample‡; approximately 0.5 to 2 g of each metal as the nitrate or sulfate; 1 mL of HNO_3)
Conical flasks (2), 500 mL

Chemicals

3 M sulfuric acid [H_2SO_4]
16 M nitric acid [HNO_3]
6 M HNO_3
18 M H_2SO_4
15 M ammonium hydroxide [NH_4OH]
Acetone or methanol
Dimethylglyoxime in alcohol (1% solution)

---

†From a similar experiment at the University of California, Los Angeles, courtesy of Dr. R. L. Pecsok.
‡An alloy of Cu and Ni can be used; however, if a precipitate forms due to Fe or Al it must be filtered before the electrolysis.

Theory

The general principles of electrogravimetric analysis have been reviewed in the text of the chapter.

Procedure

Clean the platinum electrodes in dilute nitric acid, rinse in distilled water, then rinse in acetone or methanol and dry for a few minutes in the oven. Cool and weigh to the nearest 0.1 mg, or the limit of accuracy.

Dilute the unknown sample in the 100-mL volumetric flask precisely to the mark, and pipet a 25-mL sample into each of the two 200-mL tall-form beakers. Neutralize each of the samples with ammonium hydroxide, then add 6 mL of 3 M H_2SO_4 and 2 mL of 16 M HNO_3, and dilute to a final volume of about 100 mL. Position the platinum-gauze electrodes, making sure that the large outer electrode is the cathode and that the electrodes do not touch! Place a magnetic stirrer below each pair of electrodes, put a stirring bar in each beaker, and position each beaker on a magnetic stirrer so that the electrodes are centered and extend about 2 mm above the solution. Start the stirrer and adjust the applied voltage to 2 to 4 V in order to obtain a current of 3 to 4 A. Continue the electrolysis for 45 to 60 min or until the blue color of the Cu^{2+} has entirely disappeared. With electrolysis continuing, rinse the watch glass and the sides of the beaker until the cathode screen is completely immersed. Continue the electrolysis for another 15 min. If no copper has deposited on the newly immersed part of the electrode, the electrolysis is complete; otherwise continue the electrolysis after immersing an additional portion of the cathode. With the voltage still applied, remove the stirrer and the watch glass, and slowly lower the beaker, while playing a stream of wash water on the electrodes. Any acid allowed to remain on the deposit will dissolve it. Be sure to retain the remaining solution for the nickel determination. Rinse the cathode in water and acetone or alcohol, dry it for a *few minutes* in an oven, cool, and weigh. (Prolonged drying of the copper deposit will cause oxidation.) After weighing remove the copper deposit with nitric acid.

The remaining solution in the beaker should be evaporated *almost* to dryness in the hood to remove all nitrates; a completely dry solid will be difficult to dissolve. The evaporation can be done on a hot plate, or preferably the solution is transferred to a 500-mL conical flask and fumed vigorously over a flame. *Swirl continuously to avoid bumping.* Add 5 mL of 18 M sulfuric acid and fume a second time.

Cool the residue, carefully add 25 mL water, and neutralize with ammonium hydroxide. If the evaporation was done in a conical flask, transfer the solution back to the tall-form beaker. Add 15 mL of 15 M ammonia, dilute to 100 mL, and electrolyze as for copper. Test the solution for completeness of deposition by withdrawing 1 mL and adding it to 2 mL of 1% dimethylglyoxime solution; formation of a red precipitate indicates the presence of Ni^{2+}. Rinse, dry, and weigh the cathode.

Treatment of Data

Report the weight of Cu and of Ni in the sample.

Questions

1 Why is it necessary to remove the NO_3^- prior to the electrolysis of nickel?
2 What will happen if insufficient ammonium hydroxide is added prior to the nickel determination?
3 Outline a scheme for the determination of Cu, Cd, and Pb in the same alloy sample.
4 Suggest a procedure for the electrolytic determination of Cu and Ni in a mix-

ture which would not require weighing the deposit. Indicate the necessary equipment and solution conditions.

5 Calculate the theoretical time required for plating the nickel in one of your samples assuming an average current of 3.5 A. Compare this with the actual time and give a reason for discrepancy.

SUPPLEMENTARY EXPERIMENT

Electrogravimetric Determination of Copper in Brass and Copper Ore

Purpose

This experiment illustrates the use of the electrogravimetric method for the analysis of real samples. Copper is determined in brass and/or copper ore.

Apparatus

Electroanalyzer and platinum electrodes
Conical flask, 250 mL
Funnel and filter paper
Beaker, 300 mL, tall form
Graduated cylinder, 25 mL
Magnetic stirrer and stirring bar
Beakers, 400 mL, 250 mL
Volumetric flask, 250 mL
Pipet, 50 mL
Fluted watch glass

Chemicals

Unknown brass and copper ore samples
1:1 nitric acid:water
Concentrated sulfuric acid
Nitric acid
Ammonium sulfate, solid
Acetone or ethyl alcohol

Theory

The general principles for electrogravimetric analysis are discussed in the text of the chapter.

Procedures

A. Determination of Copper in Brass

Weigh out a 1-g sample accurately to the nearest tenth of a milligram, and transfer it to a 400-mL beaker. Add 20 mL of 1:1 nitric acid; when solution is effected, add 10 mL of concentrated sulfuric acid and boil vigorously until the brown fumes cease.

safety caution

The dissolution process should be done in a hood!

Dilute the sample to approximately 75 mL, filter into a 250-mL volumetric flask, wash the residue with 4 to 5 portions of distilled water, and dilute to the 250-mL mark. Transfer a 50-mL aliquot to a 300-mL tall-form beaker, add 2 g of ammonium sulfate, and dilute to about 200 mL total volume.

With all switches in the OFF position, connect the anode to the positive pole of the electroanalyzer and a previously cleaned and weighed cathode to the negative pole. Drop the magnetic stirring bar into the sample beaker and mount the beaker in position, immersing the electrodes to within approximately $\frac{1}{4}$ in of the tops and taking care that they clear the magnet. Turn on the stirrer and power and electrolyze with a current of 2 A for 35 min. Raise the liquid level, electrolyze 15 min more. If no more copper has deposited, stop the electrolysis. With the voltage still applied to the cell, slowly lower the beaker away from the electrodes, directing a stream of water from a wash bottle against the deposited copper as it is exposed. Turn off the current *after* the electrodes are exposed. Dip the cathode in a beaker of distilled water, then in acetone or alcohol, and dry at 110°C for 2 to 3 min. (Prolonged drying of the copper deposit will cause oxidation.) Cool and weigh the cathode. Clean the electrode in warm concentrated nitric acid, rinse with water and acetone, and dry.

Treatment of Data   Calculate the percentage of copper in the sample.

B. Electrolytic Determination of Copper in Copper Ore   The procedure is similar to that for the copper in brass except for the sample preparation, which is herein described.

Weigh a 1-g sample into a tall-form 300-mL beaker, cover, and add rapidly 10 mL of nitric acid. Cover with a fluted watch glass and heat the mixture below boiling to a volume of 5 mL. Cool, add 10 mL of sulfuric acid, and evaporate cautiously to strong white SO_3 fumes. Cool, dilute to 100 mL, and warm for 20 min. Filter, wash well with hot distilled water, evaporate to a volume of 100 mL, add 5 mL nitric acid, and electrolyze at 0.5 A and 2 V for 1 h.

Treatment of Data   Calculate the percentage of copper in the sample.

Questions
1 What is the purpose of adding ammonium sulfate to the sample?
2 Why is the voltage applied to the cell left on when the electrodes are removed from the sample after electrolysis is complete?
3 What is the limit of detection for copper by this technique? What is the limiting factor?

EXPERIMENT 5-2
Constant-Current Coulometry: Titration of Arsenic

Purpose   The objective of this experiment is the determination of the amount of arsenic in a sample by titration of As(III) with I_3^-, which is generated electrochemically by a constant current.

References
1 A. J. Bard and L. R. Faulkner, "Electrochemical Methods," Wiley, New York, 1980, chap. 10.
2 H. H. Bauer, G. D. Christian, and J. E. O'Reilly, "Instrumental Analysis," Allyn and Bacon, Boston, 1978, chap. 4.
3 D. A. Skoog and D. M. West, "Principles of Instrumental Analysis," 2nd ed., Saunders, Philadelphia, 1980, chap. 20.
4 D. J. Curran in "Laboratory Techniques in Electroanalytical Chemistry," P. T. Kissinger and W. R. Heineman (eds.), Dekker, New York, 1984, chap. 20.
5 H. H. Willard, L. L. Merritt, Jr., J. A. Dean, and F. A. Settle, Jr., "Instrumental Methods of Analysis," 6th ed., Van Nostrand, New York, 1981, chap. 25.

6 D. D. DeFord and J. W. Miller in "Treatise on Analytical Chemistry," I. M. Kolthoff and P. J. Elving (eds.), part I, vol. 4, sec. D-2, Interscience, New York, 1963, chap. 49.

Apparatus

Pipets, 3 mL, 25 mL
Beaker, 250 mL
Volumetric flasks (2), 100 mL
Graduated cylinder, 100 mL
Magnetic stirrer
Large foil platinum electrode (foil 1×2 cm)
Platinum-wire electrode
Constant-current coulometer
Electrochemical cell

Chemicals

Starch solution (3%) [3 g starch plus 30 mL cold (room temperature) formamide, stir. Pour into 65 mL hot (110°C) formamide; do this in the hood.]
0.006 M arsenic trioxide [As_2O_3] stock solution (1.2 g As_2O_3 per liter, slightly acid for stability, dissolve in base first)
Composite potassium iodide [KI] and sodium bicarbonate [$NaHCO_3$] buffer solution (240 g KI, 41 g $NaHCO_3$, and 1 mL arsenite stock solution per 4 liters)
Unknown: Solution of As_2O_3

Theory

A coulometric titration involves the electrochemical generation of a titrant by current in an electrochemical cell. This titrant then reacts with the species to be determined as in the case of a conventional titration. The equivalence point of the titration is signaled by an indicator such as a chemical indicator that undergoes a color change as the end point or an instrumental method of end-point detection. The amount of titrant generated during the titration is calculated from the charge passed through the electrochemical cell.

A typical electrochemical cell for a coulometric titration is shown in Fig. 5-3. The cell consists of a platinum generator electrode with a large surface area and a second electrode to complete the electrochemical cell. The second electrode is usually a coiled platinum wire that is isolated in a separate solution from the sample solution by a sintered glass disk, which prevents the products that are formed at this electrode from interfering with the titration reaction in the main cell compartment. The cell is positioned on a stirring motor to enable stirring during the titration.

The cell is connected to a coulometer. The coulometer applies a constant current to the cell for a precisely measured time interval. The magnitude of the current is adjustable—the larger the current passing through the cell, the more rapidly the generation of titrant. The current ON/OFF switch is analogous to the stopcock of a buret for a conventional titration.

The amount of titrant generated to reach the end point of a titration is calculated from the charge passed through the cell. The coulometric titration is based on Faraday's Law which states that 96,485 coulombs (C) of electricity will oxidize or reduce 1 g equiv of an electroactive substance. The quantity of electricity is measured in terms of the coulomb and the faraday. For a constant current of i amperes flowing for t seconds, the number of coulombs Q is given by the equation $Q = it$. The faraday is equal to 1 mol (6.02×10^{23}) of electrons, which is

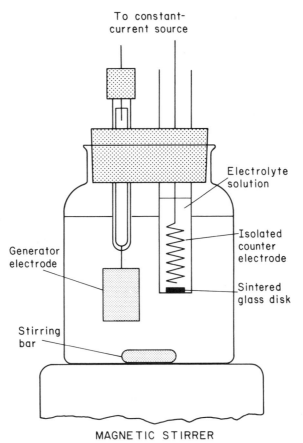

Figure 5-3 / Electrochemical cell for coulometric titration.

equal to 96,485 C. The weight of substance oxidized or reduced at an electrode can be calculated from the number of coulombs by the expression

$$W = \frac{QM}{nF}$$

where W = weight of substance oxidized or reduced, g
Q = number of coulombs passed through the cell, C
M = formula weight of the substance
F = faraday, 96,485 C
n = number of equivalents per mole (the number of electrons added or removed per ion or molecule)

In this experiment a platinum generating electrode is placed in a solution of iodide, and iodine is generated by the following reaction:

Anode reaction: $3I^- \rightarrow I_3^- + 2e$

The iodine so formed can be used to carry out titrations. Thus, if an unknown quantity of As(III) is present in the same solution, it will be oxidized by the electrochemically generated I_3^- to As(V) in a manner analogous to a conventional redox titration. Under the pH conditions in this experiment, the titration reaction is

Solution reaction: $\quad I_3^- + H_3AsO_3 + H_2O \rightleftharpoons 3I^- + HAsO_4^{2-} + 4H^+$

When all of the As(III) has reacted, an excess of I_3^- appears which then reacts with a starch indicator, turning it purple. It is necessary that the actual amount of iodine generated be known. This can be calculated from Faraday's Law if the number of coulombs consumed is known. The most *convenient* method of measuring this value is to employ a constant generating current and simply to measure the time necessary to reach the starch end point. Knowing the current and the time, the number of ampere-seconds (A · s) or coulombs is known, and the number of equivalents can be calculated.

The advantages of the coulometric titration are several:

1 Standard solutions are generated electrolytically rather than by adding them from a buret. Actually, electric current is a most versatile reagent, and a constant-current supply can be used to generate acids, bases, oxidizing agents, reducing agents, precipitating agents, and complexing agents in almost any strength desired. In other words, the coulomb becomes a primary standard replacing a host of chemical standards.

2 Because reagents may be generated and used on the spot, unstable reagents that are not useful for volumetric work may be employed because time for them to decompose or evaporate is considerably decreased. An excellent example of this is the use of chlorine (generated very simply from a saline solution).

3 Ease of microaddition of reagent. The amperage of the current (analogous to the strength of a volumetric reagent) can be set accurately to almost any level, and extremely small quantities can be easily added near the equivalence point. The electric switch takes the place of the stopcock of a buret.

4 The fact that addition of electrons as a reagent causes no dilution is important in certain cases.

5 Because of the electrical mode of generation, coulometric procedures are ideal for remote operation and for automatic procedures.

Procedure

It is advisable to clean the Pt electrodes before assembling the cell. This can be done by placing the electrodes in warm concentrated HNO_3 for a few minutes followed by a thorough rinsing with distilled water. Plug the timer, stirrer, and coulometer into the electric outlet and turn the power switch of the coulometer to the ON position. Check to see that the large platinum-foil (generating) electrode is connected to the positive terminal and that the small platinum-wire electrode is connected to the negative terminal.

Into the 250-mL beaker place 75 mL of $NaHCO_3$–KI solution and 4 to 8 drops of 3% starch solution. Place the stirring bar in the beaker and set the beaker on the stirrer. Start the stirrer and cautiously lower the electrodes into the beaker, being careful to avoid bending the large platinum foil by letting it hit the bottom of the beaker or the stirrer.

See the instructor for any special instructions for adjusting or calibrating the instrument. Turn the control switch to the ON position. The counter should begin to run. Turn the control switch to OFF. You will probably notice that the solution is dark blue because the generation of I_2 has formed the blue iodo-starch compound. Add a small amount of the arsenite stock solution drop-wise until the blue color disappears. Then flip the control switch to ON and back to OFF *until a light blue color develops in the entire solution*. This color will be taken as the end point for the titration. Set the counter back to zero.

Pipet 25 mL of a diluted stock arsenite solution (diluted 25 mL to 100 mL in a volumetric flask) into the beaker and turn the control switch ON. Let the titration continue with rapid stirring until the light blue color develops (about 500 s at 30 mA), then stop the titration and record the time in seconds.

Return the timer to zero, pipet in 25 mL more of the diluted arsenite solution and repeat the coulometric titration. Several runs may be made in the same beaker. Repeat this until *t* values are obtained which agree to within ±1 s.

Dilute your unknown arsenite in the 100-mL volumetric flask to the mark and titrate 25-mL portions as you did the diluted stock solution.

The control switch may be thought of as an electronic stopcock, for it permits you to add, in this case, small amounts of I_2. As you near the end point you will want to stop the titration to allow the I_2 formed (seen as blue-colored streaks streaming from the large generating electrode) to react before addition of the next portion. This is a good procedure and is recommended.

Treatment of Data

Report (1) molarity of the original stock arsenite solution and (2) milligrams of As_2O_3 in the 100-mL volumetric flask (your unknown sample).

Questions

1 What is the purpose of the KI? (*Hint:* Where do the electrons go at first? Ultimately?)
2 What is the purpose of the bicarbonate buffer solution?
3 Why is the anode a *large* platinum foil and the cathode a *small* platinum wire?
4 Why use platinum electrodes instead of copper, zinc, or mercury electrodes?
5 Why doesn't the cathodic process counteract the anodic process?
6 What is the fundamental requirement for a coulometric titration?

EXPERIMENT 5-3
Chronopotentiometry †

Purpose

The fundamental parameters of chronopotentiometry are investigated and the rate constant of the chemical reaction of an electrogenerated species is determined.

References

1 A. J. Bard and L. R. Faulkner, "Electrochemical Methods," Wiley, New York, 1980, chap. 7.
2 W. R. Heineman and P. T. Kissinger in "Laboratory Techniques in Electroanalytical Chemistry," P. T. Kissinger and W. R. Heineman (eds.), Dekker, New York, 1984, chap. 4.
3 D. T. Sawyer and J. L. Roberts, Jr., "Experimental Electrochemistry for Chemists," Wiley, New York, 1974, chap. 8.
4 W. H. Reinmuth, *Anal. Chem., 32,* 1514 (1960).

†Adapted with permission from an experiment developed at Case Western Reserve University by T. Kuwana and H. N. Blount, based on Ref. 5.

156 / Electrochemical Methods

5 A. C. Testa and W. H. Reinmuth, *Anal. Chem., 32,* 1512 (1960).
6 I. M. Kolthoff and W. J. Tomsicek, *J. Phys. Chem., 39,* 945 (1935).

Apparatus

Instrumentation for chronopotentiometry
Recorder with time base
Pt working electrode
Pt auxiliary electrode
SCE
Volumetric flasks (2), 100 mL

Chemicals

0.1 *M* sodium chloride [NaCl]
2 m*M* potassium ferrocyanide [$K_4Fe(CN)_6$] in 0.1 *M* NaCl
0.1 *M* sulfuric acid [H_2SO_4]
0.2 *M* H_2SO_4
p-aminophenol

Theory

Chronopotentiometry is the electrochemical technique in which a constant current is impressed across an electrochemical cell containing an unstirred solution. This step-current excitation signal is shown in Fig. 5-4A. The resulting potential is measured as a function of time, giving a chronopotentiogram as shown in Fig. 5-4B. The change in measured potential reflects the change in surface concentrations of electroactive species during the electrolysis. The transition time τ corresponds to the time required for the surface concentration of the electroactive species to reach zero. At this point the potential shifts rapidly so that electrolysis of another solution component can occur to support the constant current.

The equation relating potential and time for a reversible system is

$$E = E_{\tau/4} + \frac{RT}{nF} \ln \frac{\tau^{1/2} - t^{1/2}}{t^{1/2}} \tag{1}$$

where RT/nF have their usual significance, t is time and τ is the transition time. The value $E_{\tau/4}$, the quarter-wave potential, is analogous to the polarographic half-wave potential and is defined by

$$E_{\tau/4} = E^{\circ\prime} + \frac{RT}{nF} \ln \frac{D_R^{1/2}}{D_O^{1/2}} \tag{2}$$

where D is the diffusion coefficient and $E^{\circ\prime}$ is the formal electrode potential.

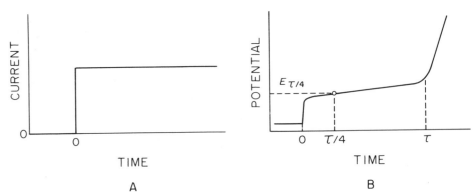

Figure 5-4 / Chronopotentiometry.   (A) Excitation signal.   (B) Chronopotentiogram.

The relationship between the applied current and τ is shown by the Sand equation (for a planar electrode),

$$i\tau^{1/2} = \frac{nFAD^{1/2}\pi^{1/2}C}{2} \tag{3}$$

where
i = applied current, A
τ = transition time, s
A = electrode area, cm^2
D = diffusion coefficient of electroactive species, cm^2/s
C = concentration, mol/cm^3
n = number of electrons transferred per ion or molecule
F = Faraday's number, 96,485 C/equiv

It is apparent that the quantity $i\tau^{1/2}$ is a constant for a given concentration of electroactive species.

If the current is reversed at or before the transition time τ is reached, a reverse-transition time τ_r will be obtained as shown in the chronopotentiogram in Fig. 5-5.

In the absence of any coupled chemical reactions, the reverse-transition time τ_r will be one-third of the forward electrolysis time t_f.

$$\tau_r = \tfrac{1}{3}\, t_f \tag{4}$$

In many cases a chemical reaction in solution is coupled to the heterogeneous electron transfer reaction at the electrode. For example, a typical electrode mechanism is the following chemical reaction (or EC mechanism):

$$R \underset{}{\overset{-e}{\rightleftharpoons}} O \overset{k}{\rightarrow} Y \tag{5}$$

in which the electrogenerated species O undergoes a solution reaction to form Y at a rate characterized by rate constant k. This particular mechanism is amenable to investigation by current-reversal chronopotentiometry. In this technique the amount of species O generated during an initial anodic current can be probed by reversing to cathodic current. The following chemical reaction of O to form Y

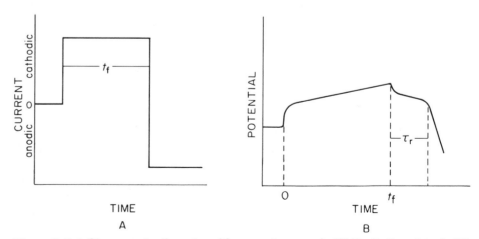

Figure 5-5 / Chronopotentiometry with current reversal. (A) Excitation signal. (B) Chronopotentiogram.

causes τ_r to be less than $\frac{1}{3}t_f$ by an amount which depends on the magnitude of the rate constant k.

In this part of the experiment, the electrode mechanism for the oxidation of *p*-aminophenol (PAP) is investigated by current-reversal chronopotentiometry. The electrode mechanism for this oxidation involves the electrochemical generation of quinoneimine (QI), which reacts with water to form quinone as shown below.

Electrode reaction:

$$\text{(6)}$$

Solution reaction:

$$\text{(7)}$$

Procedures

A. Fundamental Parameters for Chronopotentiometry

In this part of the experiment, the Sand equation is verified by chronopotentiometric oxidation of ferricyanide

$$\text{Fe(CN)}_6^{4-} \rightleftharpoons \text{Fe(CN)}_6^{3-} + e \tag{8}$$

at a platinum electrode. The relationship of τ_r to t_f in current-reversal chronopotentiometry is also examined.

1 Record the anodic current chronopotentiogram of a solution of 2 mM Fe(CN)_6^{4-} in 0.1 M NaCl using a platinum working electrode. The current setting for this experiment should be adjusted to give a transition time of approximately 5 s. Repeat this step in triplicate. (Between each run stir the solution for 30 s and then allow it to stand for 2 min to reestablish initial conditions at the electrode surface.)

2 Vary the current and record triplicate chronopotentiograms with transition times of approximately 3, 7, 10, 15, 20, and 30 s.

3 Using the same solution in the same cell, repeat steps 1 and 2 but reverse the current just before the first transition time is reached.

4 Empty, clean, and dry the cell. Wash your electrodes thoroughly with distilled water.

Treatment of Data

For steps 1 and 2, plot $i\tau^{1/2}/C$ vs. . Also calculate $i\tau^{1/2}/C$ for each chronopotentiogram. Does $i\tau^{1/2}/C$ show a trend with time? Use the overall average value of $i\tau^{1/2}/C$ to evaluate $AD^{1/2}$ for your system from the Sand equation. If the electrode area is known, calculate $D_{\text{Fe(CN)}_6^{4-}}$.

For step 3, calculate τ_r/t_f for each chronopotentiogram. Does τ_r/t_f show a trend with time? How well does it agree with the expected value for a diffusion-controlled process under conditions of equal anodic and cathodic currents?

Calculate $E^{\circ\prime}$ for the ferrocyanide-ferricyanide couple (assume $D_O = D_R$). Does your value agree with the literature values of $E^{\circ\prime}$ for this system? (See Ref. 6.) Why or why not? Explain.

B. Determination of
Rate Constant of
Chemical Reaction of
Electrogenerated
Quinonimine

In this part of the experiment the rate constant for the reaction of electrogenerated quinonimine [Eq. (7)] is determined by current-reversal chronopotentiometry, as follows:

1 Accurately weigh (to the nearest 0.1 mg) two 10-mg samples of *p*-aminophenol (PAP). Transfer each sample to a clean, *dry,* 100-mL volumetric flask.

note Do not dilute at this point. See below.

2 Dilute only one PAP sample to 100 mL with 0.10 *M* H_2SO_4. Make sure that all the PAP is dissolved and that the solution is thoroughly mixed.

3 Fill the clean electrolysis cell with the PAP solution. Record duplicate current-reversal chronopotentiograms (anodic current step first) at forward electrolysis times of 5, 10, 15, 20, 25, and 30 s. Make sure that for each experiment the anodic and cathodic currents are equal. (Note the two transitions on the reverse step.)

4 Dilute the second PAP sample to 100 mL with 0.20 *M* H_2SO_4.

5 Repeat step 3 using this second sample.

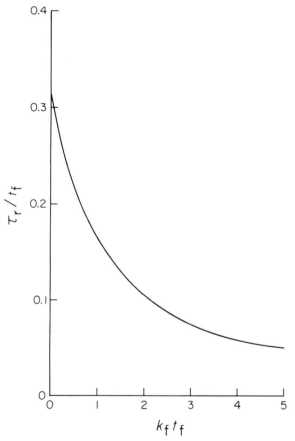

Figure 5-6 / Working curve for the calculation of the rate constant k_f for a following chemical reaction from current-reversal chronopotentiometry. [*Reprinted with permission from A. C. Testa and W. H. Reinmuth, Anal. Chem., 32, 1512 (1960). Copyright © 1960, American Chemical Society.*]

Treatment of Data

Using the working curve in Fig. 5-6 calculate the rate constant for the succeeding chemical reaction at each acid concentration. How well does your rate data agree with that of Testa and Reinmuth (Ref. 5)?

Questions

1 Write the electrode reaction for the second wave on the reverse step of the chronopotentiogram for PAP. (*Hint:* Is the product of the reaction electroactive?)

2 Does oxygen dissolved in solution interfere with either the ferrocyanide or the PAP experiment? Explain.

3 What might cause $i\tau^{1/2}/C$ to increase at long electrolysis times?

4 Explain how the reaction of electrogenerated quinoneimine could be studied by cyclic voltammetry. (See Experiment 4-1.)

II

Methods Based on Electromagnetic Radiation

6
Analytical Ultraviolet-Visible Absorption Spectroscopy

The absorption of electromagnetic radiation by ions and molecules serves as the basis for numerous analytical methods of analysis, both *qualitative* and *quantitative*. In addition, studies of absorption spectra provide knowledge concerning the formula, structure, and stability of many chemical species as well as establish the most favorable conditions and wavelength for an analysis.

The experiments in this chapter deal with photon absorption. Because the energy of a molecule is the sum of many individual types of energy,

$$E_{\text{molecule}} = E_{\text{electronic}} + E_{\text{vibrational}} + E_{\text{rotational}} + E_{\text{zero-point vibration}} + E_{\text{translational}} + \text{other types}$$

absorption of a photon can increase molecular energy in a variety of ways. Figure 6-1 illustrates the multiplicity of energy transitions and their quantized nature.

When a photon is absorbed, the energy of the absorbing species is increased by an amount proportional to the frequency of the photon

$$E'' - E' = h\nu$$

where $E'' - E' =$ the increase in energy of the molecule, ergs
                $h = 6.62 \times 10^{-27}$ erg-s
                $\nu =$ the frequency of the photon, s^{-1}

The frequency ν is related to the wavelength of the photon λ by the expression

$$\nu = \frac{c}{\lambda}$$

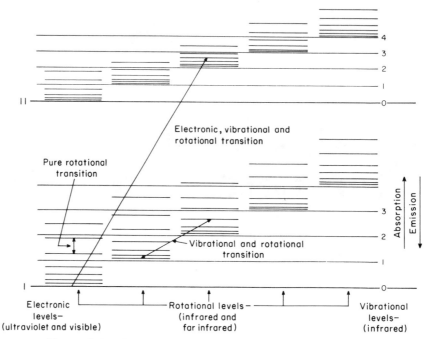

Figure 6-1 / Energy levels and spectroscopic transitions.

where c is the velocity of light, 3×10^{10} cm/s. Frequency often is reported in units called wavenumbers (Kaysers) $\bar{\nu}$ which are defined as

$$\bar{\nu} = \frac{1}{\lambda}$$

where λ is in centimeters. Figure 6-2 summarizes the forms of photon absorption and their relation to the types of energy transitions.

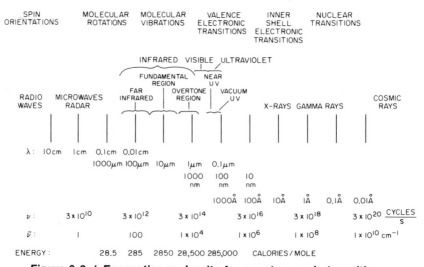

Figure 6-2 / Energetics and units for spectroscopic transitions.

To determine the concentration of a solution by measuring the amount of light it absorbs requires a quantitative relationship. This is provided by the Beer-Lambert Law, sometimes known more simply as "Beer's Law," which is a modification of an earlier law, derived by Lambert. The latter is concerned with the transmission of monochromatic light by homogeneous solids.

Lambert concluded that each unit length of material through which light passes absorbs the same fraction of entering light. If P (or I) represents the power (or intensity) of transmitted light and P_0 (or I_0) represents the power of incident light, then the change in P is proportional to the power of incident light multiplied by the change in thickness b of the material through which the light passes. Mathematically,

$$dP = -kP \, db$$

The proportionality constant is k, and the negative sign arises from the fact that P becomes smaller when b becomes larger, as illustrated in Fig. 6-3. Rearranging and integrating,

$$\frac{dP}{P} = -k \, db \qquad \text{and} \qquad \int_{P_0}^{P} \frac{dP}{P} = -k \int_{0}^{b} db$$

which yields $\ln (P/P_0) = -kb$ or $\log (P/P_0) = -(k/2.303)b$.

Beer modified the law to apply to solutions. He found that doubling the concentration of light-absorbing molecules in a solution produced the same effect as doubling the thickness. The modified law (the Beer-Lambert Law mentioned above) may be expressed mathematically as

$$\log \frac{P}{P_0} = -\epsilon b C$$

In this expression, the concentration C of the solution is conventionally expressed in moles per liter; ϵ is then called the *molar absorptivity* (molar extinction coeffi-

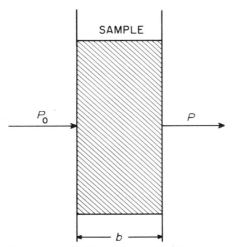

Figure 6-3 / Absorption of radiant energy.

cient) and is a constant (corresponding to $k/2.303$ in the former expression) characteristic of the absorbing substance and of the particular wavelength of light used. The cell width b is expressed in centimeters and is essentially constant in most experimental work. Thus, $\log (P/P_0)$ is directly proportional to concentration.

If $\log (P/P_0)$ is plotted against concentration for a solution which obeys the Beer-Lambert Law, a straight line results whose slope is $-\epsilon b$. P/P_0 is called the *transmittance* of the solution.

More often, in experimental work dealing with optical methods of analysis, the terms *percent transmittance* ($\%T = P/P_0 \times 100$) and *absorbance* [$A = \log (P_0/P) = 2 - \log \%T$] are used. (Absorbance is also called *optical density*.) As with $\log (P/P_0)$, both absorbance and $\log \%T$ are directly proportional to concentration when a solution obeys the Beer-Lambert Law. This may be shown by slightly modifying the Beer-Lambert equation given above and then observing the direct proportionality with concentration. For absorbance, the desired equation results simply by multiplying through by the factor (-1); that is,

$$A = \epsilon b C$$

Applications of this law will be demonstrated in experiments concerning visible, ultraviolet, and infrared spectrophotometry. Additional experiments are concerned with absorption spectra and their use for determining the formulas of complex ions, the structure of organic molecules, and the evaluation of equilibrium constants. A general discussion of the pertinent theory will be given with the specific experiment.

The general principles of spectrophotometry are discussed and demonstrated in detail as a part of the experiments in this chapter. Although other spectrophotometers can be used in most of these experiments, specific directions are limited to the Bausch and Lomb Spectronic 20 Spectrophotometer. The latter is not a precision spectrophotometer, but it is eminently satisfactory for demonstrating basic principles and is simple to use in a classroom situation.

The spectrophotometer is an important analytical instrument that makes possible a quantitative measurement of the light passing through a clear solution. The first step in such an analysis is to determine the optimum wavelength (i.e., color of light) to use in the analysis.

The wavelength that is chosen must be appreciably absorbed by the substance under analysis, for otherwise a measurement of transmitted light would not be a significant measure of the concentration of the desired substance. On the other hand, the substance must not absorb too much of the wavelength chosen, for the transmitted light might be of too weak an intensity to measure accurately. It can be shown mathematically that the best compromise between too much and too little absorption comes in the region between 15 and 65 $\%T$ and precisely at 36.8 $\%T$. [That is the region where the error in reading the transmittance is least, in comparison with the actual transmittance of the substance for readout-limited instruments (e.g., the Spectronic 20).]

This, however, is not the only consideration. The wavelength of light that is chosen must fall in a region where the $\%T$ is not changing rapidly with change in wavelength. This is because the spectrophotometer cannot isolate a single wavelength but, rather, isolates a band of wavelengths. However, if all the wavelengths in this narrow band are absorbed to nearly the same extent by the substance, then the situation is nearly the same as when one is able to isolate a single wavelength.

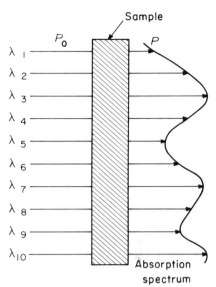

Figure 6-4 / Absorption of light as a function of wavelength.

Thus, flat portions of the $\%T$ vs. wavelength plot (absorption spectrum) are selected for analysis, e.g., λ_3, λ_5, λ_7, λ_9, or λ_{10} in Fig. 6-4.

To obtain the information required by the above considerations, the $\%T$ of a solution is measured (and plotted) as a function of wavelength of the incident light across all of the spectrum. Such a plot is called an absorption spectrum.

Some of the nomenclature in the references is different from that used in these experiments. Symbol I is often used instead of P to indicate light intensity. Symbol a (absorptivity when C is in grams per liter) corresponds to ϵ multiplied by the molecular weight of the substance being measured, and symbol l is often used to refer to cell length, herein indicated by b. *Transmittance* is T. These symbols and their synonyms are summarized in Table 6-1.

Table 6-1 / Summary of Symbols and Their Synonyms

Accepted Symbol	Meaning	Accepted Name	SYNONYMS† Symbol	SYNONYMS† Name
T	P/P_0	Transmittance		Transmission
A	$\log(P_0/P)$	Absorbance	O.D., D, E	Optical density, extinction
a	A/bc, $c =$ grams per liter	Absorptivity	k	Extinction coefficient, absorbancy index
ϵ	A/bM, $M =$ moles per liter	Molar absorptivity	a_M	Molar extinction coefficient, molar absorbancy index
b	$A/\epsilon c$	Sample path length, cm	l, d	
nm	10^{-9} m	Nanometer	$m\mu$	Millimicron
μm	10^{-6} m	Micrometer	μ	Micron
λ_{max}	Wavelength of maximum absorption			

†Not officially recommended, but are used in the older literature and in some nonchemical disciplines.

EXPERIMENT 6-1

Spectrophotometry in the Visible Region: Absorption Spectra, Beer's Law, and the Simultaneous Analysis of a Two-Component Mixture

Purpose

This experiment is designed to introduce the principles of spectrophotometry as well as the operating characteristics of a spectrophotometer.

References

1 H. H. Willard, L. L. Merritt, Jr., J. A. Dean, and F. A. Settle, Jr., "Instrumental Methods of Analysis," 6th ed., Van Nostrand, Princeton, N.J., 1981, chaps. 2 and 3.

2 G. W. Ewing, "Instrumental Methods of Chemical Analysis," 4th ed., McGraw-Hill, New York, 1975, chaps. 2 and 3.

3 H. H. Bauer, G. D. Christian, and J. E. O'Reilly (eds.), "Instrumental Analysis," Allyn and Bacon, Boston, 1978, chaps. 6 and 7.

4 E. D. Olsen, "Modern Optical Methods of Analysis," McGraw-Hill, New York, 1975, chap. 2.

5 D. A. Skoog and D. M. West, "Principles of Instrumental Analysis," 2nd ed., Saunders, Philadelphia, 1980, chaps. 6 and 7.

6 C. K. Mann, T. J. Vickers, and W. M. Gulick, "Instrumental Analysis," Harper and Row, New York, 1974, chaps. 11 and 16.

7 E. J. Meehan in "Treatise on Analytical Chemistry," P. J. Elving, E. J. Meehan, and I. M. Kolthoff (eds.), 2nd ed., part I, vol. 7, Interscience, New York, 1981, chaps. 1, 2, and 3.

8 G. D. Christian, "Analytical Chemistry," 3rd ed., Wiley, New York, 1980, pp. 411–413.

9 A. E. Harvey, Jr., J. A. Smart, and E. S. Amis, *Anal. Chem., 27,* 26 (1955).

Apparatus

Spectrophotometer operating between 375 and 625 nm (Bausch and Lomb Spectronic 20 Spectrophotometer)

Cuvettes (3), matched, one special cuvette for observing light-path color (prepare by placing a piece of chalk in a regular cuvette as shown in Fig. 6-5)

Volumetric flasks (5), 25 mL

Pipets, 5 mL, 10 mL, and 20 mL

Beakers, two 100 mL, five 50 mL

Chemicals

0.0500 M chromium(III) nitrate solution [$Cr(NO_3)_3$]

0.1880 M cobalt(II) nitrate solution [$Co(NO_3)_2$]

Theory

Spectrophotometers and Absorption Spectra

In a typical spectrophotometer white light that emanates from a tungsten lamp passes through an entrance slit and is dispersed by a diffraction grating or prism. Of the dispersed beam a narrow band of similar-wavelength light (ideally, monochromatic) passes through a second slit into the sample solution being measured. Any of this light which is not absorbed by the sample solution, but which passes through the solution, falls upon the phototube of the instrument, where the intensity of the transmitted light is measured electronically.

The most common diffraction grating is a flat sheet of plastic-coated glass. The surface of the plastic is a replica of a surface that has been ruled with many fine, parallel grooves: 600 or more accurately spaced grooves to the millimeter. The plastic surface is aluminized to make it reflect. The white light falling upon the

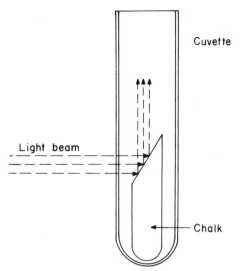

Figure 6-5 / Cell system for viewing the color of visible light.

grating is dispersed into a horizontal fan of beams with the short wavelengths (violet and ultraviolet) at one end and the long wavelengths (red and infrared) at the other.

A schematic diagram of the Bausch and Lomb Spectronic 20 Spectrophotometer is shown in Fig. 6-6. The following discussion describes the operation of this particular instrument. However, the operation in general is similar to other instruments in the same region of the spectrum.

The spectrum of light falls on a dark screen with a slit cut in it. Only that portion of the spectrum which happens to fall on the slit goes through into the sample, and any desired part of the spectrum can be projected onto the slit by turning the grating with the wavelength control knob. Attached to this knob is a dial calibrated in wavelengths of nanometers or millimicrons (1 nm = 1 mμ = 10 $\mathring{A}$ = 10^{-6} mm). The slit of the instrument passes a band of wavelengths that is 20 nm in width. Because of the linear dispersion of the grating, this bandwidth of 20 nm is constant over the entire wavelength region. (Although the Spectronic 20 has a fixed bandwidth of 20 nm, other instruments may have a different fixed value

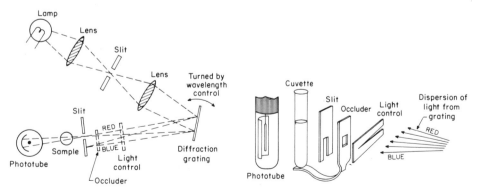

Figure 6-6 / Schematic layout for the Bausch and Lomb Spectronic 20 Spectrophotometer.

or a variable slit control.) The spectrophotometer is turned on by rotating the left-hand knob clockwise. This should be done at least 20 min before measurements are made. After the instrument has warmed up, the knob is used to adjust the phototube amplification so that the meter needle will read "0" on the percentage transmission scale when no light is striking the phototube (which is the case when a cuvette is *not* present in the sample holder; without an inserted cuvette a shutter blocks the optical path).

The right-hand knob regulates the amount of light passing through the second slit to the phototube. The need for this light-control knob arises because the light source does not emit light of equal intensities at different wavelengths and the phototube is not equally responsive to light of varying wavelengths. In addition, the "blank" solution (the medium in which the substance being measured is located) may itself absorb light of certain wavelengths. In order to measure the absorbance due to only a particular species in solution, side effects which affect the $\%T$ reading must be compensated for. Therefore, after the spectrophotometer has been zeroed (by means of the amplifier control knob), a blank solution is placed in the light path, and the light-control knob is rotated until the dial reads $100\ \%T$ to achieve this desired compensation. If a sample solution is now placed in the light path, any change in the $\%T$ reading is due to the particular light-absorbing species in the sample, and the $\%T$ reading is a measure of the quantity of that species present.

The handling of the cuvettes is extremely important. Often two or more cuvettes are used simultaneously, one for the blank solution and the others for the samples to be measured. Yet any variation in the cuvette (such as a change in cuvette width or the curvature of the glass, stains, smudges, or scratches) will cause varying results. To avoid significant experimental errors several rules must be followed:

1   Do not handle the lower portion of a cuvette (through which the light beam will pass).
2   Always rinse the cuvette with several portions of the solution before taking a measurement.
3   Wipe off any liquid drops or smudges on the lower half of the cuvette with a clean lint-free wiper (Kim-wipes, Scott wipers, or other lens paper) before placing the cuvette in the instrument. *Never* wipe cuvettes with towels or handkerchiefs. Inspect to insure that no lint remains on the outside and that small air bubbles are not present on the inside walls.
4   When inserting a cuvette into a sample holder
    a   To avoid any possible scratching of the cuvette in the optical path, insert the cuvette with the index line facing toward the *front* of the instrument.
    b   After the cuvette is seated, line up the index lines *exactly*. The cuvette should be removed in the reverse manner (pointing the cuvette index toward the front of the instrument *before* withdrawing it).
5   When using two cuvettes simultaneously, use one of the cuvettes *always* for the blank solution and the other cuvette for the various samples to be measured. Mark the tubes accordingly and do not interchange the cuvettes during the remainder of the course. If absorption spectra are being run on several different samples at the same time, matched cuvettes must be used. To match three or more tubes, half-fill each *clean* tube with $CoCl_2$ solution [containing 23 g of $CoCl_2$ per liter in 1% (by volume) of HCl]. Set the wavelength scale to 510 nm, set the zero with the amplifier control, and place one tube in the spectrophotometer. Adjust the light control so the meter reads 50 $\%T$.

Check the other tubes and record the $\%T$ of each. Use the three tubes which match the closest.

Part I. Operation and Response of the Spectrophotometer

Theory

The phototube in the Spectronic 20 Spectrophotometer is a cesium-antimony surface photoemissive cell (type S-4). The relative response of the phototube to a beam of monochromatic light of constant intensity is

Wavelength, λ	Response (relative), %
350	90
375	98
400	100
425	98
450	91
475	81
500	68
512	61
525	53
550	37
575	21
600	10
612	7
625	5

The phototube is much more sensitive to light of wavelength 400 nm than to light of wavelength 600 nm. This means that the phototube will require a greater flux of 600-nm than of 400-nm monochromatic light for the same $\%T$ reading to be registered upon the spectrophotometer dial.

Procedures

A. Spectrophotometric Response

Plug in and turn on the spectrophotometer. Adjust the amplifier control knob (left front) until the meter needle reads 0 $\%T$, and allow the instrument to warm up for 20 min. After the instrument has warmed up, rezero the instrument if necessary. Be sure the sample holder is in place by pressing straight down on the holder. Shut the top of the holder whenever you adjust the instrument or take readings to eliminate as much stray light as possible.

Put approximately 3 mL of distilled water into a rinsed cuvette, and wipe dry with a clean lint-free wiper (such as a Kim-wipe or Scott wiper). (*Note the five rules* concerning the use of cuvettes.) Turn the light-control knob (right front) *counter*clockwise as far as it will turn to diminish the amount of light passing to the phototube. Insert the water-filled cuvette into the sample holder and align the index lines exactly.

Turn the wavelength control knob to 510 nm.

Rotate the light-control knob clockwise until the meter needle registers about 90 on the $\%T$ scale.

By rotating the wavelength knob, scan the visible spectrum and note how the response (measured by the position of the meter needle) varies with wavelength.

Determine the wavelength of light to which the instrument is most responsive (it will be near 510 nm). Adjust the light-control knob until the $\%T$ reads 100 at this wavelength. Then, *without readjusting* either the amplifier control knob or the light-control knob, run a spectral curve for the relative response of the spectrophotometer, reading from the $\%T$ scale. Take readings at the following wavelengths: 350, 375, 400, 425, 450, 475, 500, 512, 525, 550, 575, 600, 612, and 625 nm.

B. The Visible Spectrum

Now place the special cuvette into the sample holder to observe the color of the light beam (Fig. 6-5). Rotate the cuvette until the light path strikes the sloping surface of the chalk. Observe and record the color of the beam every 50 nm from 650 to 350 nm. It may be necessary to rotate the light-control knob to increase or decrease the intensity of the light. Do not allow the meter needle to read off scale.

Adjust the wavelength to 600 nm and note the variation in color across the band of light. (See Question 4 at the end of the experiment.) What range of wavelengths of light comprises the band you see?

Adjust the wavelength to 550 nm. Rotate the light-control knob and observe the change in light intensity. The amount of light passing into the sample is regulated by a movable metal strip into which a narrow V-shaped aperture has been cut. As the strip is moved back and forth in the light path (by turning the light-control knob), more or less light is allowed to pass between the sides of the "V." Note the variation in light intensity across the band and that the far edge of the light band has the greatest intensity. (See Question 5 at the end of the experiment.)

Treatment of Data

On a piece of fine-ruled graph paper (10×10 to the centimeter) plot "relative overall response" of your spectrophotometer vs. wavelength, using the data you obtained in the laboratory. Plot the wavelength horizontally.

On the *same* graph, plot a curve representing the "relative response of the phototube" as a function of wavelength by use of the data in the discussion at the beginning of Part I. Along the top of the graph indicate the colors observed for the various wavelengths.

While the relative response of the phototube is great toward light of wavelength 400 nm, the relative response of the entire instrument at this wavelength is low. The spectrophotometer has a much greater relative response at 525 nm than would be expected from a consideration of the phototube response alone. The difference is due primarily to the light source.

From the two curves of your graph, calculate the relative intensity of the spectrophotometer's lamp emission (plus a small factor due to the optics) over the visible range of the spectrum. At each wavelength studied, divide the "relative overall response" by the "relative phototube response." This gives a series of numbers which indicate "relative lamp intensity" at various wavelengths, the largest of numbers being in the neighborhood of 3.0.

To convert these relative numbers to a scale where the maximum is approximately 100, multiply each of the numbers by the factor $(\frac{100}{3})$. Thus

$$\text{Relative lamp intensity} = \frac{\text{instrument response}}{\text{phototube response}} \times \frac{100}{3}$$

(The exact value obtained for a particular wavelength is unimportant; the importance lies in how the value *changes* with wavelength.) Finally, plot a curve on your graph that represents the "relative lamp intensity" as a function of wavelength.

Questions

1 To what color (and wavelength) of light is the instrument most responsive? What color of light is emitted most strongly by the tungsten lamp?

2 In what way do the various knobs on the spectrophotometer affect the light beam which is passed through the sample holder?

3 The effective bandwidth is 20 nm on the Spectronic 20 Spectrophotometer and is constant over the entire wavelength region. What is meant by the term *bandwidth?*

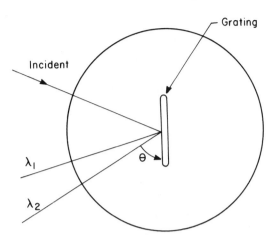

Figure 6-7 / Dispersion of light by a grating.

4 When light is diffracted from a grating, the sine of the angle of diffraction is directly proportional to wavelength. That is $\sin \theta = \alpha \lambda$, where α is a constant (see Fig. 6-7). Which color light is diffracted at a greater angle—red light or yellow light?

5 If both red light and yellow light pass through the second slit of the spectrophotometer at the same time, the intensity of which color light will be affected most by the V shape of the light-control aperture? Why? (The diffraction grating is so arranged in the Spectronic 20 to give horizontal dispersion of light, i.e., the color of the dispersed light varies horizontally but not vertically.)

6 When the solution is red, does the solution absorb red light strongly or transmit red light strongly?

7 In calculating the relative lamp intensity of the spectrophotometer at various wavelengths, the water and the glass of the blank cuvette were assumed to absorb none of the light (in the region of wavelengths used). Do you think this assumption is justified? Why?

8 (This one is just to think about.) "The effective bandwidth . . . is constant over the entire wavelength region." For this statement to be true, the spectrum produced by the diffraction grating must be *linear* (equal angles between consecutive wavelengths), and the light intensity must be regulated by varying the vertical height of the light beam rather than by changing the width of the beam. How are those two conditions consistent with the original statement?

Part II. Absorption Spectra

Theory

A graphical plot of absorbance vs. wavelength is referred to as an *absorption spectrum*. A discussion of the theory of absorption spectra is given in the introductory section of Chap. 7.

Procedure

Prepare the following solutions:

1 0.0200 *M* Cr(III), by pipetting 10 mL of the 0.0500 *M* stock solution of $Cr(NO_3)_3$ into a 25-mL volumetric flask and diluting to the mark. Mix well by inverting the flask about 15 times.

2 0.0752 *M* Co(II), by pipetting 10 mL of the 0.1880 *M* stock solution of $Co(NO_3)_2$ into a 25-mL volumetric flask and diluting to the mark. Mix.

Obtain three matched cuvettes (see rule 5 under the discussion of spectrophotometer operation for the procedure to match cuvettes). Set the wavelength dial at 375 nm, and adjust the instrument to read 0 %T with no cuvette and 100 %T when the water-filled cuvette is in the sample holder. (If the instrument will not adjust at this wavelength, set it at the shortest wavelength at which it will adjust.) For this, as well as the Beer's Law and Two-Component System experiments, be sure to use the same cuvette for the distilled water blank, a second cuvette for the chromium solutions, and a third one for the cobalt solutions.

Place approximately 3 mL of the 0.0200 M Cr(III) solution in your second cuvette. Place 3 mL of the 0.0752 M Co(II) solution in your third cuvette. Wipe and insert the Cr(III) solution into the sample holder. Read the %T of the solution from the dial. Replace this cuvette with the Co(II) solution and read the %T.

Although the meter is electronically protected from burnout, it is best to turn the light control counterclockwise before changing to another wavelength.

Turn the wavelength dial to 400 nm. Again, set the "0" and the "100" using the blank solution. Place the Cr(III) solution in the holder again and read the %T at this wavelength. Replace the Cr(III) with the Co(II) and read the %T of this solution. Continue this procedure at 5-nm intervals from 405 through 625 nm.

Empty and rinse the cuvettes thoroughly with water (never with cleaning solution). Be sure to keep the cuvettes distinguished so that the same cuvette will be used for the blank, a second one for all the Cr(III) solutions, and a third for the Co(II) solutions.

Treatment of Data

Convert all readings of %T to absorbance ($A = 2 - \log \%T$). Using a sheet of fine-ruled graph paper (10 × 10 to the centimeter) plot the absorption spectra of Cr(III) and Co(II). Save this graph for use in the next two parts of the experiment.

Question

1 By referring to your Cr(III) absorption spectrum suggest a desirable wavelength for analyzing Cr(NO$_3$)$_3$ solutions that have a concentration between 0.02 and 0.04 M. Explain briefly why this same wavelength might be undesirable for the determination of a Cr(NO$_3$)$_3$ solution with a concentration appreciably greater or less than 0.02 to 0.04 M.

Part III. Beer's Law

Theory

See the introduction of this chapter.

Beer's Law Plots

Graphs of absorbance vs. concentration, or $\log \%T$ vs. concentration, are known as Beer's Law plots. These are prepared by measuring the light absorbed by a series of solutions with different known concentrations. The cell width and the wavelength of the light are maintained constant. If a linear plot is obtained [the Beer-Lambert relationship holds for the solution (and instrument) at that wavelength], the plot may be used with confidence to determine the concentrations of unknown solutions.

Unfortunately, the condition of truly monochromatic light upon which the Beer-Lambert Law is based is only approximated in the laboratory. Because more than one wavelength of light passes through a solution at the same moment, apparent deviations from the law are observed to give nonlinear Beer's Law plots. The object, therefore, of much preliminary work in optical analysis is to find a suitable wavelength band where the apparent deviation from the law will be only slight or negligible.

The spectral curves obtained in Part II serve as guides in the search for a desirable wavelength for Cr(III) and Co(II) determinations. The Beer's Law plots from the present experiment will illustrate the relative merit of different wavelengths for quantitative measurements. The choice of a wavelength for a given determination is to some extent dependent upon the approximate concentration of the solution.

Procedure

The spectrophotometer should be turned on at least 20 min before any measurements are made.

Obtain approximately 75 mL of the 0.0500 M Cr(III) stock solution in a small beaker. Make up solutions that are 0.0100, 0.0200, 0.0300, and 0.0400 M in chromium by pipetting 5, 10, 15, and 20 mL, respectively, into four 25-mL volumetric flasks and diluting each to the mark. These solutions plus a portion of the 0.0500 M stock solution may then be transferred to five 50-mL beakers and covered with Parafilm® until measured with the spectrophotometer. Clean the four 25-mL volumetric flasks.

Obtain approximately 75 mL of the 0.1880 M Co(II) stock solution in a small beaker. Make up solutions that are 0.0376, 0.0752, 0.1128, and 0.1504 M in cobalt by pipetting 5, 10, 15, and 20 mL, respectively, into four 25-mL volumetric flasks and diluting to the mark. The stock solution provides a fifth concentration.

By referring to your plots of absorbance vs. wavelength (Part II) for the 0.0200 M Cr(III) solution select six wavelengths at which to study absorbance as a function of concentration. Use the same wavelengths for your study of the Co(II) solution. Include in your selection:

1 Wavelengths at the two maxima of the curve for Cr(III), which are near minima (410, 575 nm) for Co(II)
2 A wavelength where A is at a maximum for Co(II) (510 nm)
3 A wavelength corresponding to a steeply rising portion of the curve for Co(II) and near the intersection of the Co and Cr curves (450 nm)
4 A wavelength corresponding to a steeply descending portion of the curve for Co(II) near the intersection of the Co and Cr curves (540 nm)
5 The 625-nm wavelength

Obtain Beer's Law plots for the Cr(NO$_3$)$_3$ and for the Co(NO$_3$)$_2$ solutions for each of the selected wavelengths.

Turn the wavelength dial to the shortest of the chosen wavelengths.

Adjust the spectrophotometer for 0 and 100 %T, using the same cuvette for the distilled water blank as in Part II. Remove the cuvette from the sample holder and see if the meter returns to 0 (after 10 s). If not, reset the 0 and then repeat the adjustment for 100 %T. Continue this until 0 and 100 are obtained with the cell out and in, respectively.

Measure and record the %T and absorbance (optical density) of the 0.01 M Cr(III) solution. Then measure and record the 0.0376 M Co(II) solution using a matched cuvette. Adjust the spectrophotometer for the next longest wavelength. Read the 0.01 M Cr(III) and the 0.0376 M Co(II) at this wavelength. Continue until all six wavelengths have been set and measured. Discard the 0.01 M Cr(III) and rinse the cuvette several times with small portions of the 0.02 M solution. Read these two solutions at the selected wavelengths. Continue with each successively more concentrated solution for Cr(III) and for Co(II) at each of the six wavelengths. Be sure to set the instrument to 0 and 100 %T with the blank after *each* change of wavelength.

Obtain an unknown Cr(III) solution in a 25-mL volumetric flask and dilute to the 25-mL mark if not previously diluted by the instructor. Determine the $\%T$ and A of this solution at each of the six wavelengths by use of the same cuvette which was used for the previous Cr(III) solutions. After all determinations are made, rinse the cuvettes with distilled water. Before leaving the laboratory, consider your answer to Question 3 at the end of the experiment.

Treatment of Data

Plot log $\%T$ vs. concentration and absorbance vs. concentration for each set of data obtained at each of the six wavelengths. Although absorbance readings can be made directly from the spectrophotometer dial, these readings are often erroneous because of poor meter markings; it is better to convert the $\%T$ readings to absorbance by calculator (or a log table) and then to use these calculated values for plotting. Use *one* sheet of graph paper for all of the Cr(III) readings, and another sheet for all of the Co(II) readings of absorbance. Use one sheet of semi-log paper for all of the Cr(III) readings and another sheet of semi-log paper for all of the Co(II) readings; plot $\%T$ on the vertical axis.

Draw a horizontal line across your Beer's Law plots at 65 $\%T$ and 15 $\%T$. Measurements of $\%T$ should be restricted within these two limits to avoid excessive errors in the estimated concentration.

From an appropriate Beer's Law curve, determine the concentration (in moles per liter) of your unknown Cr(III) solution.

These Beer's Law plots will be used for reference in Part IV.

Questions

1 Briefly account for the appearance of the Beer's Law plots that you obtained in this experiment (i.e., where do the plots adhere to or deviate from Beer's Law). Why? (Summarize for each of the six wavelengths.)

2 At which of the six wavelengths is the molar absorptivity the smallest for the Cr(III) solutions? For the Co(II) solutions?

3 Estimate how accurately you can read $\%T$ from the spectrophotometer dial (i.e., $\pm$ how great a $\%T$). Estimate how accurately you can determine the wavelength of light passing through the sample (i.e., $\pm$ how many nanometers).

4 Assuming that you could reproducibly read the $\%T$ scale to ± 1 $\%T$ at each of the six selected wavelengths, how accurately could you report the concentration of a 0.0150 M Cr(III) solution at each wavelength (e.g., 0.0150 $M \pm 0.0050$ at 968 nm), of a 0.0550 M solution? (*Hint:* Use the Beer's Law plots of $\%T$ vs. concentration that are plotted on semi-log paper.)

5 What wavelengths should be used for the determination of Cr(III) at an approximate concentration of 0.015 M and of 0.055 M? Why? Consider both adherence to Beer's Law and inherent error in the final concentration that is reported.

6 (A question just to think about.) Starting with the Beer-Lambert relationship

$$A = \log \frac{P_0}{P} = \epsilon bc$$

a Show that $A = 2 - \log \%T = -\log T$.

b Show that, in order for the relationship to be dimensionally correct, the molar absorptivity ϵ must have the dimensions of liters per mole per centimeter.

c Indicate how to determine the slope that a Beer's Law plot would have at some given wavelength if you have at your disposal only an absorption spectrum of a solution of known concentration and the knowledge that at the given wavelength Beer's Law is obeyed. (Assume a constant cuvette length.)

Part IV. Simultaneous Analysis of a Two-Component Mixture

Theory

When a solution of two colored (light-absorbing) substances is prepared, the presence of the second component often causes a change to occur in the light-absorbing properties of the first component. With such solutions the absorbance by the components is not additive, which precludes a simple quantitative determination of their concentrations.

However, there are many instances in which the two components do not react or interact with one another, and, thus, neither affects the light-absorbing properties of the other. The absorption of light by such components is additive. That is to say, the total absorbance of the two-component solution is just the *sum* of the absorbances which the two substances would have individually, if the substances were in separate solutions under similar conditions and had the same concentrations as in the mixture. When this is true, their individual concentrations can be determined from spectrophotometric measurements.

Because interactions often do arise, sometimes when they are least expected, the nature of the absorption spectra of the components must be investigated when they are separate *and* when they are in solution together. Such an investigation is to be carried out for the Co(II)–Cr(III) system.

If there were some wavelength of light at which cobalt does not absorb, and chromium absorbs strongly, and some other wavelength of light where the converse were true, then these two wavelengths could be used separately to determine each of the two components in the same way that one does one-component analyses, i.e., just as if the other component were not there. Unfortunately, this is not usually the case.

However, if at one wavelength light is absorbed weakly by the cobalt while it is absorbed strongly by the chromium, and at any other wavelength the converse is true, their concentrations can be determined because the *absorbances are additive*. Such a pair of wavelengths does exist for the Co(II)–Cr(III) system: one near 510 nm and the other near 575 nm. That these wavelengths also occur at flat portions of the spectra is added good fortune.

In the introduction to the chapter, the proportionality relationship between the absorbance of a substance and its concentration was developed

$$A = \epsilon b C$$

By using the same cuvette for each sample to be analyzed, the factor b is kept constant, so that the two constants ϵ and b can be combined into a single absorptivity constant k,

$$A = kC$$

which represents a proportionality factor that relates A and C for some particular substance at some particular wavelength. For a Beer's Law plot of A (vertically) vs. C (horizontally), k is the *slope* of the line.

For a solution that contains n light-absorbing components whose absorbances are additive, the total absorbance of the solution at some wavelength i is the sum of the individual absorbances

$$A_i = \sum_{j=1}^{n} k_{ij} C_j \qquad i = 1$$

The subscript j refers to components just as i refers to wavelengths of light used.

This general equation may be written more explicitly as

$$A_1 = k_{11}C_1 + k_{12}C_2 + \cdots + k_{1n}C_n \qquad (1)$$

For a second wavelength ($i = 2$) the explicit equation is

$$A_2 = k_{21}C_1 + k_{22}C_2 + \cdots + k_{2n}C_n \qquad (2)$$

which expresses the fact that the total absorbance at the second wavelength (A_2) is equal to the absorbance by substance 1 with concentration C_1 plus the absorbance by substance 2 with concentration C_2, etc.

When a two-component solution is analyzed, only the first two "kC" terms are relevant. The use of two different wavelengths of light provides two equations in two unknowns. By solving the equations simultaneously, the concentrations of the two components can be determined. The various k's are determined from Beer's Law plots for the separate components at the two wavelengths chosen for the analysis.

Procedure

Prepare a Cr(III)–Co(II) mixture by pipetting 10 mL of 0.0500 M Cr(III) stock solution and 10 mL of 0.1880 M Co(II) stock solution into a 25-mL volumetric flask and diluting to the mark. The resultant solution contains 0.0200 M Cr(III) and 0.0752 M Co(II).

Obtain an unknown solution containing Cr(III) and Co(II) in a 25-mL volumetric flask and dilute to the mark if it has not been diluted already by the instructor.

Determine the absorption spectra of the two-component Cr(III)–Co(II) solution and the unknown mixture using matched cuvettes and taking readings at 5-nm intervals from 375 nm to 625 nm. Although the values of absorbance A are required, read the values of %T and convert these values to A later. Readings of %T must be used to obtain the higher values of A. (The spacings between numbers on the A scale of the instrument are too close for accurate readings.)

Treatment of Data

Convert all %T readings to A values.

Plot the absorption spectra for the two solutions on the same graph prepared in Part II, which contains the spectra for individual solutions of Cr(III) and Co(II).

For each wavelength, add the values of A for the single-component curves and plot these points on the same graph as above. These points, which represent the sum of the first two curves, should have a similar shape to the curve for the mixture. This confirms that the noninteraction condition is fulfilled.

Determine the k's from the Beer's Law Plots of Part III

By referring to the chromium and cobalt absorption spectra, find two desirable wavelengths at which to carry out an analysis of a Cr–Co mixture. These should lie near 510 and 575 nm. From your Beer's Law plots of Part III, determine the slopes (k's) for Cr(III) at 510 and 575 nm and the slopes for Co(II) at 510 and 575 nm. The calculated k values correspond to k_{11}, k_{12}, k_{21}, and k_{22} in Eqs. (1) and (2). (These k values represent the relative molar absorptivities for each of the components at the two wavelengths, because the cell length is maintained constant.) If the cell thickness is equal to 1 cm, then $k = \epsilon$. An alternative to the evaluation of slopes is to calculate each k or ϵ for all of the concentrations at a particular wavelength and take their average.

Calculate the concentration (in moles per liter) of each of the components in the unknown by setting up simultaneous equations and solving for the two un-

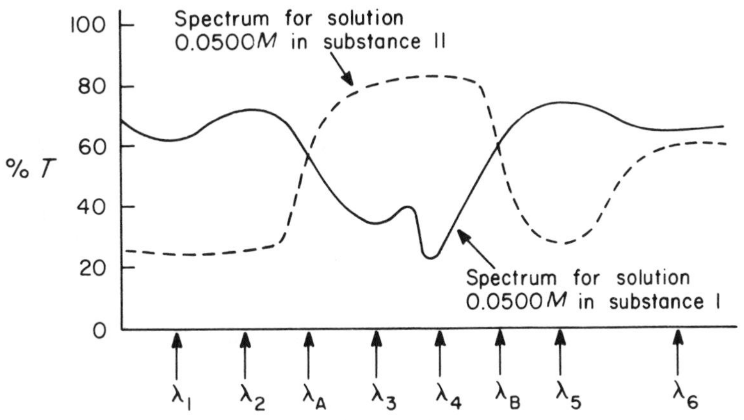

Figure 6-8 / Absorption spectra.

knowns. One method of calculation is indicated in Question 3. Another is as follows:

Let C_1 = Co(II) concentration

C_2 = Cr(III) concentration

k_{11} = absorptivity constant for Co(II) at 510 nm

k_{12} = absorptivity constant for Cr(III) at 510 nm

k_{21} = absorptivity constant for Co(II) at 575 nm

k_{22} = absorptivity constant for Cr(III) at 575 nm

A_1 = absorbance of Co(II) plus Cr(III) at 510 nm = $k_{11}C_1 + k_{12}C_2$

A_2 = absorbance of Co(II) plus Cr(III) at 575 nm = $k_{21}C_1 + k_{22}C_2$

Solving the equations for A_1 and A_2 simultaneously:

$$C_1 = \frac{k_{22}A_1 - k_{12}A_2}{k_{11}k_{22} - k_{12}k_{21}} \quad \text{and} \quad C_2 = \frac{k_{11}A_2 - k_{21}A_1}{k_{11}k_{22} - k_{12}k_{21}}$$

Questions

1   If a Cr(III) solution contained a high concentration of a pure-blue material, e.g., saturated $Cu(NO_3)_2$, suggest how one might accurately analyze the solution for Cr(III) in the concentration range from 0.0300 to 0.0600 M.

2   Figure 6-8 illustrates the absorption spectra from 400 to 600 nm for two substances (I and II) which do not react or interact in any way.

   a   If you were to analyze an unknown mixture of I and II of approximately equal concentrations to those above, why would it be unwise to choose for your analysis λ_1 and λ_2? λ_1 and λ_4? λ_2 and λ_5?

   b   If you were to analyze a solution of substance I in which a small quantity of substance II was present as an impurity, what wavelength(s) would you consider first in planning the analysis? Why?

   c   Why is a two-component analysis possible by use of λ_2 and λ_A but not possible with λ_A and λ_B?

3   (Optional) The simultaneous equations for total absorbance A_1 given in the discussion may be rewritten in a form to give the concentrations directly. For two components ($j = 1$ and $j = 2$) and two wavelengths ($i = 1$ and $i = 2$) the general form of the rewritten equation is

$$C_j = \sum_{i=1}^{2} P_{ji}A_i \quad j = 1, 2$$

where the P_{ji} factors may be calculated from the k_{ij} factors in the previous expressions.

a Write out the two equations explicitly.

b Find algebraic expressions for the P_{ji} in terms of the k_{ij}. (One way to do this is to solve the earlier equations simultaneously for the C_j and then to equate equivalent coefficients.)

c Compute numerical values for the P_{ji} terms by use of the results of step **b** and the experimental values for the k_{ij} terms. Substitute these numerical values into the equations in step **a** and solve directly for the concentrations of cobalt and chromium in your unknown.

d In qualitative terms, summarize what has been done in steps **a** and **c**.

SUPPLEMENTARY EXPERIMENTS

I. Determination of a Mixture of Cobalt and Nickel

Chemicals

0.150 M cobalt(II) nitrate solution [$Co(NO_3)_2$]
0.150 M nickel(II) nitrate solution [$Ni(NO_3)_2$]

Procedure

For Beer's Law plots, make dilutions to concentrations of 0.1125, 0.075, and 0.0375 M. Use appropriate dilutions and determine the absorption spectra. Next, undertake the study of a component solution in a manner similar to that used for the chromium-cobalt mixture in Part IV of the experiment.

Treatment of Data

Treat the data in a manner similar to that described in Part IV. (Cobalt has an absorption maximum at 510 nm; nickel at 660 nm and 395 nm.) Report the concentration (in moles per liter) of each of the components in the unknown.

II. Determination of Chromium (as $Cr_2O_7^{2-}$) and Manganese (as MnO_4^-) Mixtures

Chemicals

0.020 M potassium permanganate [$KMnO_4$] in 0.5 M sulfuric acid [H_2SO_4] (plus 2 g KIO_4 per liter)
0.020 M potassium dichromate [$K_2Cr_2O_7$] in 0.5 M H_2SO_4 (plus 2 g KIO_4 per liter)

Procedure

For Beer's Law plots, make dilutions to concentrations of 0.0008, 0.0016, 0.0024, 0.0032, and 0.0040 M. Use the same procedure as described in Part IV for the two-component analysis. Use 440 nm as the absorption maximum for $Cr_2O_7^{2-}$ and 545 nm for MnO_4^-. To obtain stable solutions of $KMnO_4$, add 0.5 g of solid KIO_4 and boil the solution. For details, see Ref. 8.

Treatment of Data

Treat the data in a manner similar to that described in Part IV.

III. Simultaneous Determination of Fe(II) and Fe(III)

Chemicals

Make up the following in separate 25-mL flasks:

5.0×10^{-5} M Fe(II) and 2.0×10^{-4} M 1,10-phenanthroline
5.0×10^{-5} M Fe(III) and 2.0×10^{-4} M 1,10-phenanthroline
2.0×10^{-4} M 1,10-phenanthroline

Procedure

Scan each solution between 700 and 350 nm with a recording spectrophotometer. From an examination of the resulting curves, select two wavelengths for study. Convert the %T's at these two wavelengths and calculate the molar absorptivities of the Fe(II) and Fe(III) complexes. Record an absorption spectrum of an unknown solution that contains Fe(II) and Fe(III) and measure the absorbances at the two selected wavelengths.

Treatment of Data

Calculate the concentration of the two ions in the unknown solution. See Ref. 9.

EXPERIMENT 6-2

"Precision" Spectrophotometry

Purpose

The principles of precision spectrophotometry are illustrated in this experiment by the determination of chromium(III).

References

1   R. Bastian, *Anal. Chem., 21,* 972 (1949).
2   H. H. Willard, L. L. Merritt, Jr., J. A. Dean, and F. A. Settle, Jr., "Instrumental Methods of Analysis," 6th ed., Van Nostrand, Princeton, N.J., 1981, chap. 3.
3   G. W. Ewing, "Instrumental Methods of Chemical Analysis," 4th ed., McGraw-Hill, New York, 1975, chap. 3.
4   D. A. Skoog and D. M. West, "Principles of Instrumental Analysis," 2nd ed., Saunders, Philadelphia, 1980, chap. 7.
5   E. D. Olsen, "Modern Optical Methods of Analysis," McGraw-Hill, New York, 1975, chap. 1.
6   E. J. Meehan in "Treatise on Analytical Chemistry," P. J. Elving, E. J. Meehan, and I. M. Kolthoff (eds.), 2nd ed., part I, vol. 7, Interscience, New York, 1981, chap. 2.

Apparatus

Spectrophotometer (Bausch and Lomb Spectronic 20)
Cuvettes (2)
Pipets, 1 mL, 2 mL, 3 mL, 4 mL, 5 mL, and 10 mL
Beaker, 150 mL
Graduated cylinder, 100 mL
Volumetric flasks (9), 25 mL (including two for unknowns A and B)

Chemicals

0.2500 M chromium(III) nitrate [$Cr(NO_3)_3$]
Unknowns: Usually 0.10 to 0.15 M in chromium (when diluted to the mark)

Theory

The ordinary spectrophotometric method, although more widely used than any other method of analysis, has two drawbacks: (1) its operation is limited to samples that transmit between 15 and 65% of the incident light, and (2) it has lower precision than volumetric and gravimetric techniques. These two limitations can be circumvented by a differential approach to the spectrophotometric measurement, the accuracy of which approximates that of volumetric procedures. This experiment illustrates such an approach.

The four types of spectrophotometric measurements may be classified according to the "standards" used to establish the "0" and "100" points on the percent transmittance scale. These standards actually correspond to known concentrations (C_1 or C_2) of the sample species: C_1 is used to set the "0" and C_2 to set the "100"; C_x corresponds to the unknown concentration of the sample species whose value is to be determined by the procedure.

Method	"0" %T Standard	"100" %T Standard
I. Ordinary	Shutter ($C_1 = \infty$)	Solvent ($C_2 = 0$)
II. High absorbance	Shutter ($C_1 = \infty$)	Solution of definite concentration ($C_2 < C_x$)
III. Low absorbance	Solution of definite concentration ($C_1 > C_x$)	Solvent ($C_2 = 0$)
IV. Ultimate precision	Solution of definite concentration ($C_1 > C_x$)	Solution of definite concentration ($C_2 < C_x$)

The *high-absorbance method* is a modification of the ordinary spectrophotometric method whereby solutions of high concentration (or more properly, high absorbance) can be measured with considerably greater accuracy. In the ordinary spectrophotometric method, two cuvettes (cells) are used, one containing the sample solution under analysis and the other containing pure solvent (the 100 %T blank). Arrangements are made instrumentally so that the absorbance of the solvent and cuvette is subtracted from the absorbance of the sample, the difference being the absorbance of the desired compound. While this method is applicable to solutions of moderate absorbance, solutions of high absorbance cannot be measured accurately. To allow accurate measurements of high-absorbance solutions, a known solution of the substance to be measured is employed as the 100 %T blank, rather than the pure solvent. The known solution to be used as the blank is so chosen as to be slightly more dilute than the unknown. By setting the 100 %T with such a known solution, one effectively increases the size of the %T scale (what had once been 10 %T, for example, may now read 100 %T, although no change has occurred with the 0 %T). Accordingly, greater accuracy is possible in the %T readings and the concentration determinations (see Fig. 6-9a). In this high-absorbance method, the absorbance "read" becomes the difference between the absorbance of the sample and the absorbance of the standard, or (where k represents ϵb):

$$A = kC_x - kC_2 = k(C_x - C_2)$$

The *low-absorbance method* applies to the other extreme in the concentration range. In this precision method the 100 %T is set, as usual, with distilled water. The 0 %T, however, is set by use of a known solution of the substance being measured, with a concentration that is somewhat greater than that of the unknown solutions. The effective size of the %T scale is thereby increased, except that this time weakly absorbing solutions are favored (see Fig. 6-9b). With the low-absorbance method, however, a calibration curve (log %T vs. concentration or A vs. concentration) must be prepared because C_x and log %T are not directly proportional.

The method of *ultimate precision* involves a combination of the two preceding methods. The 0 %T is set with a standard solution somewhat more concentrated than the unknown, and the 100 %T is set using a standard solution somewhat less concentrated than the unknown (see Fig. 6-9c). Because of nonlinearity, a calibration curve also is necessary with this method.

Cell Corrections

Because of the nonidentity of the sample and reference cells, a correction (β-factor) is evaluated and applied. A brief description of the β-factor is given as a guide

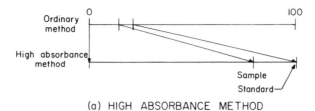

(a) HIGH ABSORBANCE METHOD

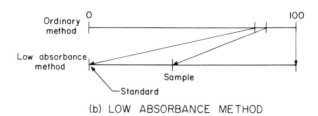

(b) LOW ABSORBANCE METHOD

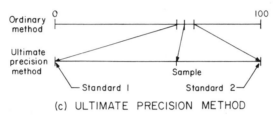

(c) ULTIMATE PRECISION METHOD

Figure 6-9 / Examples of "precision" spectrophotometry.

for use in evaluating the experimental data and should not be considered a full explanation.

Consider two cuvettes which are identical only in the fact that they contain the *same* solutions [i.e., $0.0300\,M$ Cr(III)]. Further denote one of the cuvettes as the sample cuvette by the subscript s and the other as the reference or blank cuvette by the subscript r. If Beer's Law holds, $A_s = \epsilon_s b_s C_s$ and $A_r = \epsilon_r b_r C_r$ at a given wavelength for the sample cuvette and the reference cuvette, respectively.

But suppose that $A_s \neq A_r$, as will be the case in most optical work due to the difference in the optical properties of the two cuvettes. Because the two solutions in the cuvettes are identical, $C_s = C_r$ and $\epsilon_s = \epsilon_r$. Hence,

$$\frac{A_r}{A_s} = \frac{\epsilon_r b_r C_r}{\epsilon_s b_s C_s} = \frac{b_r}{b_s} = \beta$$

To correct a value of A_s' (transformed from the $\%T$ read) for some other concentration—assuming ϵ is constant and, thus, that Beer's Law is valid—multiply by β

$$A_s'\left(\frac{A_r}{A_s}\right) = A_s'\,\beta = A_s^*$$

where A_s^* is the corrected absorbance.

**Procedure
(High-Absorbance
Method)**

Observe the operating procedure for the spectrophotometer.

note In order to achieve the expected high accuracy inherent in the high-absorbance method, *careful* technique in every operation is mandatory.

In the volumetric flasks carefully prepare 15 solutions by diluting appropriate volumes of the 0.2500 *M* Cr(III) stock solution with distilled water to give concentrations of 0.0100, 0.0200, 0.0300, . . . , 0.1400, and 0.1500 *M*. Take about 150 mL of the stock solution to use for preparing all the solutions in this experiment and for rinsing the pipets. These solutions may be prepared in sets of five, as they are needed. Do not discard the 0.0500 *M* or the 0.1000 *M* solution until you have completed the experiment.

It is *important* in this experiment that the blank cuvette be kept as the blank. There are times when both cuvettes will contain Cr(III) solutions [the blank consisting of a Cr(III) solution rather than of distilled water], and some intermixing of cuvettes may result unless care is used. The cuvettes must be placed in the sample holder in exactly the same position. The analytical wavelength should be about 550 nm. (Do not use the wavelength of approximately 575 nm that was determined in Experiment 6-1.) This new wavelength is selected by considering all of the data of Experiment 6-1—a combination of the λ of strong absorption and λ of high instrumental response; the latter is a major consideration.

Preliminary Evaluation of Cuvettes

Set 100 %*T* on the instrument, using distilled water in the blank cuvette (λ = 550 nm). Empty the cuvette and fill it with 0.0300 *M* Cr(III) solution and record the %*T*. Now fill the "sample" cuvette with the 0.0300 *M* solution and immediately read the %*T* obtained with this cuvette. The %*T* readings should be near 37 %*T*. Repeat both of the readings several times to obtain a statistical average for the %*T* for each cuvette.

During the remainder of the experiment, use the cuvette that has the *greater* %*T* reading as the blank cuvette (mark the cuvette accordingly).

Calibration Curves

Using distilled water as the blank (with due attention to the proper cell and its position), measure and record %*T* of each solution from 0.100 through 0.0500 *M*.

Now rinse and fill the blank cuvette with the 0.0500 *M* solution. (*Note:* Do not discard the 0.0500 *M* solution.) Using this solution as the blank or reference, measure the %*T* of Cr(III) solutions in the concentration range 0.0500 through 0.1000 *M*.

Make similar measurements for the 0.1000 to 0.1500 *M* solutions with the 0.1000 *M* solution as the blank or reference. If the instrument cannot be set to 100 %*T* with the 0.1000 *M* solution as reference, use the solutions of next lower concentration until one is found which can be set at 100 %*T*. With high-concentration solutions, errors due to fluctuations in the %*T* readings may be diminished by making rapid consecutive readings on the reference and the sample solutions. (*Note:* Do not discard the solution that was used as the reference.)

This process theoretically could be continued indefinitely, but in practice solutions with absorbances that are greater than that for 0.13 *M* Cr(III) preclude setting the blank to 100 %*T* with the Spectronic 20.

Unknowns

Dilute the unknowns (A and B) to the mark with distilled water and, by the use of proper blank solution(s), determine the %T of each.

Treatment of Data

Preliminary Evaluation of Cuvettes

From the %T readings obtained in this part of the experiment, calculate the β-factor to be used with the cuvettes.

Calibration Curves

Convert all the %T readings recorded in this part of the experiment to absorbance and correct these A values by use of your β-factor. Plot the correct A values vs. concentration on a graph. The resulting plot should consist of several sloping lines which are approximately straight and parallel to one another.

Unknowns

From the %T values and by use of the β-factor, determine a corrected absorbance value for each unknown.

note The β-factor must be determined *anew* each day that %T readings are made; thus, any %T readings obtained on the second day should be corrected with the β-factor that was determined on that day.

Determine the molarity of the two unknowns. Report your results to *four* decimal places, e.g., 0.1254 M and 0.0975 M for A and B, respectively, etc.

Analysis of Error (Ordinary and High-Absorbance Methods)

Under the assumption that the system obeys Beer's Law and that the error may be attributed solely to inaccuracy in reading the %T scale, a treatment of the expected error in the analysis can be made. This consists of calculating the expected error at each experimental point on the calibration curve and then noting how the expected error varies from point to point on the curve.

The first step in the treatment is to evaluate the error in concentration dC that results from an error in reading the percent transmittance scale by a given amount dT. The value of dC is the absolute error and depends strongly upon the portion of the transmittance scale where the reading is made; in the high-transmittance region the absolute error is smaller than in the lower-transmittance region. The quantitative relationship is easily derived from Beer's Law, as follows:

$$A = -\log T = \epsilon b(C_x - C_2)$$
$$-A = \log T = -k(C_x - C_2)$$
$$0.43 \ln T = -k(C_x - C_2)$$

Therefore,

$$dC_x = -\frac{0.43}{kT}\,dT$$

When T reads 1 (absorbance = 0), then $dC_x/dT = -0.43/k$. Because T cannot exceed unity, dC_x/dT can never be smaller than $-0.43/k$. Note also that the k is the slope of the line for a plot of absorbance vs. concentration.

Determine the three k values corresponding to the three straight lines on your absorbance vs. concentration graph. (The three k values should be similar.)

By use of these k values and the preceding relationship, calculate the absolute error for each of your 17 experimental points, assuming an error of 1% in reading the percentage transmission scale ($\Delta T = 0.01$).

Of more significance to the analyst, however, is the *relative error*, which is the absolute error in concentration divided by the actual concentration [and often then multiplied by 100 for relative percent error, i.e., $(dC/C) \times 100$]. Calculate the relative percent error (for 1% error in reading the percentage transmission scale) for each experimental point, and record each of these values at the proper points on your A vs. concentration graph.

Questions

1 a Calculate the relative percent error caused by misreading the $\%T$ scale by 1% at the point of minimum error for the ordinary colorimetric method (i.e., at 36.8 $\%T$).

   b Find on your calibration curve (high-absorbance technique) the point where the relative percent error is lowest. How much greater precision (i.e., twofold, 8.2-fold, . . .) results from operating near this point with the high-absorbance technique, rather than diluting the sample so as to operate in the 36.8 $\%T$ (optimum) region with the ordinary spectrophotometric technique?

   c Which of your 15 standard solutions should be used as the 100 $\%T$ reference solution, in order to obtain an optimal analysis of an unknown Cr(III) solution, which happens to be 0.096 M? Of *your* two unknown Cr(III) solutions?

2 If you were measuring the concentrations of aqueous Cr(III) solutions by use of 550-nm light, which of the four methods of spectrophotometry would you use for solutions in the following concentration ranges? Why?

   a Range of 10^{-3} to 10^{-2} M
   b Range of 10^{-2} to 10^{-1} M
   c Range of 10^{-1} to 1 M
   d Range of 0.04 to 0.05 M

SUPPLEMENTARY EXPERIMENTS

I. Low-Absorbance Method

Chemicals

Stock solution: 0.01 M chromium(III)
Standard solutions prepared by dilution of stock solution: 0.002, 0.004, 0.006, 0.008, 0.010 M chromium(III)
Unknown: 0.002 to 0.01 M in chromium(III)

Procedure

Measure the $\%T$ of the chromium solutions in the usual manner at 575 nm by setting the instrument scale reading to zero transmittance with no cuvette, and to 100 $\%T$ by use of distilled water.

Next, adjust the zero setting with the most concentrated solution. Readjust the 100 $\%T$ with distilled water as blank. Recheck the zero setting and the 100 $\%T$ setting, and read the $\%T$ for each of the other solutions, as well as that of the unknown.

Treatment of Data

Convert the %T values that were obtained for all the solutions to absorbance. Make two plots of absorbance vs. concentration, one for the data recorded in the usual manner and one for the data recorded with the zero setting adjusted with the most concentrated solution. Determine the concentration of the unknown from the latter plot.

II. Ultimate Precision Method

Chemicals

Standard solutions of chromium(III): 0.0200, 0.0300, 0.0350, 0.0400, 0.0500 M (or if the high-absorbance method was performed, use solutions prepared for that experiment)

Unknown: 0.02 to 0.04 M chromium(III)

note

Other solutions that can be used for these studies and the appropriate wavelengths include: cobalt nitrate at 510 nm; nickel sulfate at 395 nm; potassium dichromate at 440 nm; and potassium permanganate at 525 nm.

Procedure

Measure the %T of each solution (including the unknown) in the usual manner (i.e., by setting the zero with no cuvette and the 100 %T with a distilled water blank at 575 nm).

Using the least concentrated solution (0.0200 M) set the 100 %T. Using the most concentrated solution (0.0500 M), set the zero adjustment. Recheck the 100 %T setting and zero setting. Measure the transmittance of the remaining solutions and the unknown.

Treatment of Data

Plot transmittance vs. concentration and absorbance vs. concentration for each set of data. Determine the unknown concentration from each calibration curve.

EXPERIMENT 6-3
Photometric End-Point Detection and Spectrophotometry in the Near-Infrared Region

Purpose

The photometric detection of an end point is illustrated with a complexometric titration that is monitored in the near-infrared region of the spectrum.

References

1 H. H. Willard, L. L. Merritt, Jr., J. A. Dean, and F. A. Settle, Jr., "Instrumental Methods of Analysis," 6th ed., Van Nostrand, Princeton, N.J., 1981, chaps. 2 and 3.

2 G. W. Ewing, "Instrumental Methods of Chemical Analysis," 4th ed., McGraw-Hill, New York, 1975, chaps. 2 and 3.

3 E. D. Olsen, "Modern Optical Methods of Analysis," McGraw-Hill, New York, 1975, chap. 2.

4 D. A. Skoog and D. M. West, "Principles of Instrumental Analysis," 2nd ed., Saunders, Philadelphia, 1980, chaps. 6 and 7.

5 C. K. Mann, T. J. Vickers, and W. M. Gulick, "Instrumental Analysis," Harper and Row, New York, 1974, chaps. 11 and 16.

6 E. J. Meehan in "Treatise on Analytical Chemistry," P. J. Elving, E. J.

Meehan, and I. M. Kolthoff (eds.), 2nd ed., part I, vol. 7, Interscience, New York, 1981, chaps. 1 and 2.

7 P. B. Sweetser and C. E. Bricker, *Anal. Chem., 24,* 1107 (1952).

8 R. F. Goddu and D. N. Hume, *Anal. Chem., 26,* 1679, 1740 (1954).

Apparatus

Spectrophotometer with blue-sensitive phototube (type S-4)

Cuvettes (3), including special cuvette for observing light-path color (for method of preparation, see Experiment 6-1)

Phototube, red-sensitive (1P 40)

Red filter

Stirring bar

Magnetic stirrer

Ring stand

Buret clamp

Buret, 10 mL

Beaker, 250 mL

Graduated cylinder, 100 mL

Pipets (2), 5 mL

Chemicals

0.250 *M* EDTA (ethylenediaminetetraacetic acid, disodium salt) solution

Cupric nitrate [$Cu(NO_3)_2$] and calcium nitrate [$Ca(NO_3)_2$], each 0.05 to 0.1 *M*, as an unknown solution in a 100-mL volumetric flask (dilute to mark)

Ammonia [NH_3] buffer (350 mL concd NH_3 and 54 g ammonium chloride [NH_4Cl] diluted to 1 L) *or* ethanolamine buffer (200 mL ethanolamine and 60 mL 6 *M* hydrochloric acid [HCl] per liter)

Theory

In the photometric method of end-point detection for titrations, use is made of the difference in the molar absorptivities (at the analytical wavelength selected) of the various species present. The appearance or disappearance of an absorbing species will give a linear, or concentration-dependent, change in absorbance which will yield two straight lines that intersect at the end point. In this respect photometric titration curves resemble in shape the curves that are obtained in conductimetry and amperometry (in which the conductance and the current, respectively, also vary linearly with concentration).

The selection of the analytical wavelength requires much care, for there are at least *three* components present which may absorb light: the original substance, the titrant, and the resulting product or products. The usual procedure is to select some wavelength at which only one component absorbs. However, the mere fact that only one component absorbs at a particular wavelength does not necessarily mean that that particular wavelength should be selected for the analysis. If the absorbance is intense, the %*T* readings may be limited to the undesirable 0 to 20 %*T* region, where comparatively large errors in measuring absorbance would overshadow the inherent accuracy of the photometric titration.

For a successful photometric titration it is necessary that the measured species adhere roughly to Beer's Law, and the necessary chemical and instrumental precautions must be observed to maintain the relation $A = \epsilon bC$.

Some typical photometric titration curves are shown in Fig. 6-10 for the reaction

$$X + T \rightarrow P$$

where T represents the titrant, X the component to be determined, and P the prod-

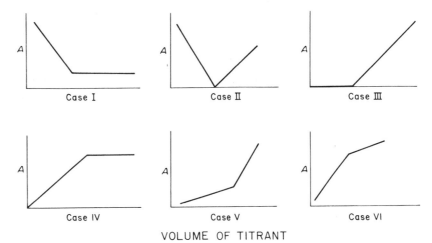

Figure 6-10 / Photometric titration curves. Case I: X and P both absorb. Case II: X and T both absorb. Case III: Only T absorbs. Case IV: One (or more) of the products P absorbs. Case V: P and T absorb and $\epsilon_T > \epsilon_P$. Case VI: P and T absorb and $\epsilon_T < \epsilon_P$.

uct(s) of the reaction. Case II in Fig. 6-10 yields the greatest accuracy due to the acute angle of intersection, which facilitates locating the end point.

Procedures

A. The Red and Near Infrared Region of the Spectrum

The Second-Order Visible Spectrum; Effect of Red Filter

In the sample holder of the spectrophotometer place the special cuvette for observing the light path (see Experiment 6-1). Making observations at 50-nm intervals, record the color of light observed from 750 to 1000 nm. The intensity of the beam should be set as high as possible to observe these colors.

Compare the colors at 400 and 800 nm, 475 and 950 nm, and 500 and 1,000 nm.

note The instrument is much more sensitive to the shorter-wavelength light; adjust the right-front knob accordingly.

The colors seen at the above pairs of wavelengths may appear somewhat different, because of the much lower intensity of the longer-wavelength light; therefore try to compare only the average color observed.

Now hold the red filter over the "special cuvette" and, looking at the chalk surface through the red filter, observe the visible spectrum from 350 to 650 nm.

note Do not touch the filter except by its edge; otherwise difficult-to-remove fingerprints will remain.

Record the range of wavelengths obscured by the filter and the range not obscured. Carefully replace the filter in the lens paper and temporarily set it aside.

Instrument Response with Red-Sensitive Phototube

Remove the special cuvette and unplug the instrument. Carefully tilt the instrument back and unscrew the backplate. Remove the original (blue-sensitive) phototube—*not* the light bulb—by carefully pulling on the tube while wiggling it

back and forth. Note that the inner surface of the curved "target" in the phototube is directed toward the sample holder. Insert the red-sensitive phototube
(1P 40) in place of the blue-sensitive tube. Any fingerprint smudges should be
wiped off the phototubes with a clean tissue. Store the blue-sensitive tube in
the box which originally contained the 1P 40 phototube. After making sure that the
phototube is firmly seated in the spectrophotometer, shut the backplate, right the
instrument, and plug it in again.

Record the instrument response from 350 to 1000 nm at 50-nm intervals, except between 600 and 700 nm, where a 25-nm interval should be used. The instrument response should be determined as in Experiment 6-1. The intensity of
the light beam is reduced (by turning the right-front knob), and the spectrum is
scanned to determine the wavelength of maximum sensitivity. At this wavelength
the instrument is set to 100 %T with distilled water in the cuvette and the remaining readings are taken without changing the intensity control.

Instrument Response with Red-Sensitive Phototube and the Red Filter

Again unplug the instrument and tilt it back. Unscrew the backplate and place the
red filter carefully in position. Replace the backplate, taking care that the filter
does not drop out of its slot.

Again record the instrument response, as without the red filter, from 350 to
1000 nm.

The red-sensitive phototube and the red-filter combination are to be used in
Procedure B of this experiment.

Treatment of Data

Plot a graph (%T vs. λ) of the response of the spectrophotometer:

1 With only the red-sensitive phototube
2 With both the red-sensitive phototube and the red filter

Include the spectral region obscured by the red filter.

B. Photometric
Titration of Ca–Cu
Mixture

Because the copper–EDTA complex and copper ion have a maximum difference
in absorption at 755 nm, this wavelength should be used as the analytical wavelength. (See absorption spectrum data given in Question 3 at the end of this experiment.) The other solute molecules involved during the titration are all either
nonabsorbing or only absorb slightly with respect to 755-nm light.

Rinse and fill your buret with the 0.250 M EDTA stock solution.

Pipet 10 mL of Cu^{2+}–Ca^{2+} unknown solution and add 25 mL of ammonia
buffer (or 50 mL ethanolamine buffer) to a clean 250-mL beaker. Add 65 mL of
water with ammonia buffer or 40 mL of water with ethanolamine buffer, *keep a
record of the exact amount which you add*. Mix well, using the mechanical stirrer,
before taking a %T measurement on the solution.

note   The solution *must be mixed* thoroughly before every new measurement; at least
15 s of stirring time should be allowed.

This experiment should be performed reasonably rapidly to avoid loss of ammonia, if ammonia buffer is used. Otherwise sharp end points will not be observed.

Using a 5-mL pipet (pouring directly out of the beaker is not recommended),
rinse your sample cuvette thoroughly with the unknown solution in the 250-mL
beaker, returning the rinsings to the beaker. Then refill your sample cuvette and

read the %T of this solution at 755 nm against a solution of untitrated copper sample (1 mL of sample diluted with 2 mL of ammonia buffer and 8 mL of H_2O *or* 5 mL ethanolamine buffer and 5 mL of H_2O).

Repeat the 100 %T adjustment and the %T reading on the sample until suitable agreement is found. This is important.

Return the sample to the beaker. Add a 0.5-mL increment of 0.250 M EDTA, and *allow thorough mixing.* Measure %T as above, repeating the measurement until suitable agreement is obtained.

Repeat the addition of EDTA increments and %T measurements until a total of 9.5 mL of EDTA has been added. Return all rinsings and samples to the original beaker of solution being titrated.

If time permits, the above titration may be repeated for a duplicate check.

Finally, unplug the spectrophotometer, carefully remove the red filter and the red-sensitive phototube, and replace with the original blue-sensitive phototube.

Treatment of Data

Plot the data obtained in the photometric titration (absorbance vs. milliliters of 0.250 M EDTA). Three straight lines should result which give two intersections corresponding to a copper end point and a calcium end point.

Calculate the number of moles of Ca(II) and Cu(II) in the unknown Cu–Ca solution.

note While no direct statement has been made regarding which inflection point corresponds to Cu(II) and which corresponds to Ca(II), this should be evident from the preceding discussion and the shape of the titration curve. Be sure you associate the correct volume with the proper metal ion.

Questions

1 If a solution contained only Ca(II), how could you determine the calcium concentration photometrically with EDTA, as above? Outline briefly how you would perform this titration.

2 Rough adherence to Beer's Law is necessary for a successful determination. As in conductimetry, dilution of the sample by adding titrant may produce sufficient deviation so as to require correction.

   a If V = initial volume of solution and v = volume of titrant added, by which factor $V/(V + v)$ or $(V + v)/V$ should you multiply the absorbance reading in order to correct the absorbance for dilution?

   b How may a photometric titration be carried out so that one can neglect the dilution caused by adding the titrant?

   c What percent error in absorbance (due to dilution) results at your Cu^{2+} end point? At your Ca^{2+} end point?

   d Would a plot of the corrected absorbance vs. volume of titrant be advisable in the titration just performed in the laboratory?

3 The following absorption data for Cu(II)–EDTA were obtained by use of a cuvette with identical thickness to the one you used:

Wavelength, nm	A	Wavelength, nm	A
550	0.013	725	0.210
575	0.026	750	0.191
600	0.045	755	0.175
625	0.070	800	0.145
650	0.127	850	0.103
675	0.164	900	0.070
700	0.195	950	0.040

    a What concentration was employed for this absorption spectrum?

    b The sensitivity of a photometric titration system can be altered simply by the selection of wavelength. Thus for the titration of more concentrated solutions of copper, a wavelength where the molar absorptivity is smaller should be employed. Assuming that an absorbance reading of no more than 0.7 (i.e., $\%T$ no less than 20) is a desirable limit during a titration, but that as large a change as possible also is desirable (e.g., 0.05 to 0.7 absorbance units), what wavelength smaller than 775 nm should be employed in the titration of a 10-mL sample of 1 M copper solution (diluted to 150 mL before photometric measurement)?†

4 **a** What is the wavelength relationship between first- and second-order spectra?

    b Orange light occurs at a wavelength of 600 nm. At what apparent wavelength will orange light occur because of second-order diffraction? At what apparent wavelength will orange light occur because of third-order diffraction?

    c Does a prism instrument suffer from *this* cause of "stray light"?

    d What is the purpose of the red filter? Would this red filter be suitable for use at 1300 nm? Why, or why not?

    e A substance absorbs strongly at 900 nm but not at all at 450 nm. If no red filter is employed, would you expect Beer's Law to be followed at 900 nm? Explain briefly.

5 What advantage(s) accrue from setting the 100 $\%T$ with a solution of copper rather than with distilled water?

6 In view of the fact that copper forms a more stable EDTA complex (log $K = 18.7$) than calcium (log $K = 10.7$), explain the order of titration found in this experiment.

SUPPLEMENTARY EXPERIMENTS

I. Photometric Titration of a Mixture of *p*-Nitrophenol and *m*-Nitrophenol

Chemicals

0.02 M *p*-nitrophenol
0.02 M *m*-nitrophenol
0.7 M sodium hydroxide [NaOH]
Unknown: Mixture of *p*-nitrophenol and *m*-nitrophenol

Procedure

A mixture of 50 mL of 0.02 M *p*-nitrophenol and 50 mL of 0.02 M *m*-nitrophenol are titrated with 0.07 M NaOH at a wavelength of 545 nm. Follow the general procedure outlined in the main experiment (pages 190–191). For further information refer to Ref. 8. Titrate the unknown mixture following the same procedure.

Treatment of Data

Plot the data obtained in the photometric titrations (absorbance vs. milliliters of 0.7 M NaOH). Determine the end points for *p*-nitrophenol and *m*-nitrophenol. Calculate the number of moles of *p*-nitrophenol and *m*-nitrophenol in the known and unknown solutions.

†See figure 1 in Ref. 7.

II. Photometric Titration of Iron

Chemicals

0.05 M iron(III) solution
Unknown iron(III) solution
0.100 M EDTA in 1.0 M acetic acid solution adjusted to pH 2 with HCl
1% solution of salicylic acid (add 4 mL of this to each 100 mL of solution to be titrated)

Procedure

Titrate the known and unknown iron(III) solutions with 0.100 M EDTA at a wavelength of 525 nm. Follow the general procedure outlined in the main experiment (pages 190–191).

Treatment of Data

Plot the data obtained in the photometric titrations (absorbance vs. milliliters of 0.100 M EDTA). Calculate the number of moles of iron(II) in the known and unknown solutions.

EXPERIMENT 6-4

The pK_a of an Indicator

Purpose

The use of spectrophotometry for the evaluation of the pK_a of an acid-base indicator is demonstrated in this experiment.

References

1 R. W. Ramette, "Chemical Equilibrium and Analysis," Addison-Wesley, Reading, Mass., 1981, chap. 13.
2 H. A. Laitinen and W. E. Harris, "Chemical Analysis," 2nd ed., McGraw-Hill, New York, 1975, pp. 48–51.
3 H. H. Bauer, G. D. Christian, and J. E. O'Reilly, "Instrumental Analysis," Allyn and Bacon, New York, 1978, pp. 180–182.
4 G. W. Ewing, "Instrumental Methods of Chemical Analysis," 4th ed., McGraw-Hill, New York, 1975, pp. 67–70.
5 W. B. Fortune in "Analytical Absorption Spectroscopy," M. G. Mellon (ed.), Wiley, New York, 1950, pp. 128–133.

Apparatus

Spectrophotometer
Cuvettes (2)
Pipets, 1 mL, 5 mL, and 10 mL
Volumetric flasks (4), 25 mL
Beakers (3), 100 mL

Chemicals

Bromothymol blue, 0.1% in ethanol (20%)
0.10 M sodium phosphate [Na_2HPO_4], dibasic
0.10 M potassium phosphate [KH_2PO_4], monobasic
Hydrochloric acid [HCl] (concd)
4 M sodium hydroxide [NaOH]

Theory

Bromothymol blue (3′,3″-dibromothymolsulfone-phthalein) is one member of a large class of acid-base indicators that are known as the "sulfone-phthaleins."

The possible forms of the indicator and their colors are

The equilibrium between the yellow acid form (HIn^-) and the royal-blue basic form (In^{2-}) can be represented as

$$HIn^- = H^+ + In^{2-} \tag{1}$$

and the equilibrium constant for the acid dissociation is

$$K_a = \frac{a_{H^+}a_{In^{2-}}}{a_{HIn^-}} \tag{2}$$

where a refers to activity.

Converting to logarithmic form and rearranging terms yields

$$pH = pK_a + \log \frac{a_{In^{2-}}}{a_{HIn^-}} \tag{3}$$

Since activity is directly related to concentration by activity coefficient f_i, this expression can be expressed as

$$pH = pK_a + \log \frac{[In^{2-}]}{[HIn^-]} + \log \frac{f_{In^{2-}}}{f_{HIn^-}} \tag{4}$$

The ratio of the activity coefficients may be incorporated into the equilibrium constant to give

$$pH = pK_a' + \log \frac{[In^{2-}]}{[HIn^-]} \tag{5}$$

where pK_a' is an *apparent indicator constant* (see Ref. 2). Thus a plot of pH vs. log ($[In^{2-}]/[HIn^-]$) should give a straight line with an intercept (at which $[In^{2-}] = [HIn^-]$) equal to pK_a'.

In this experiment, the indicator bromothymol blue is dissolved in a series of buffer solutions of known pH. The ratio $[In^{2-}]/[HIn^-]$ is then measured spectrophotometrically and pK'_a is calculated from a plot of Eq. (5).

Procedures

note   Bromothymol blue is unstable in acid solution over prolonged periods of time. Therefore, obtain all absorbance measurements with a solution on the same afternoon that you prepare the solution.

A. Absorption Spectra of Bromothymol Blue at Various pH Values

Obtain three complete absorption spectra of bromothymol blue at three pH values; use conditions of approximately pH 1, pH 7, and pH 13.

1   pH $\cong$ 1.   Carefully pipet 1 mL of bromothymol blue stock solution into a clean 25-mL volumetric flask. Add a few milliliters of distilled water, then 4 drops of concentrated HCl, and finally dilute to the mark with distilled water. Invert several times to effect mixing, and add about 5 mL of this solution to a cuvette that has been rinsed with this solution. Measure the %T (and calculate the absorbance) of this solution from 365 to about 575 nm, at 20-nm intervals. At maxima and minima in the curve, the intervals should be decreased to 10 nm. Plot absorbance vs. λ (nm), and indicate the color of the solution.

2   pH $\cong$ 6.9.   Pipet 1 mL of indicator into a 25-mL volumetric flask and add 5 mL each of 0.10 M Na_2HPO_4 and KH_2PO_4 from a pipet. Dilute to the mark and obtain the spectrum as above. Plot on the same graph as above [A vs. λ (nm)], and indicate the color of the solution.

3   pH $\cong$ 13.   To 1 mL of indicator in a 25-mL volumetric flask, add 12 drops of 4 M NaOH. Dilute to the mark and obtain the spectrum as above. Plot your values on the same graph as above, and indicate the color of the solution. The three curves should intersect each other at a single point, called an *isosbestic point*.

B. Absorbance of Solutions (Differing in pH) at Selected Wavelengths

Refer to the graph you have made and select two wavelengths at which further absorbance measurements will be made. You should select a wavelength to the left of the isosbestic point and one to the right of this point. Choose wavelengths where the acid and base forms of the indicator show a maximum difference in their absorbance.

Measurements will be made on solutions with seven different pH values other than the three solutions studied thus far.

	mL Indicator	mL $H_2PO_4^-$	mL HPO_4^{2-}	pH
1.	1.0	5.0	0.0	~4.5
2.	1.0	5.0	1.0	6.2
3.	1.0	10.0	5.0	?
4.	1.0	5.0	10.0	?
5.	1.0	1.0	5.0	?
6.	1.0	1.0	10.0	?
7.	1.0	0.0	5.0	~9.1

Pipet the above quantities for a given pH into a 25-mL volumetric flask and dilute to the mark. Measure the absorbance of each solution at the two selected wavelengths. (Remember to readjust the instrument to 0 and 100 %T whenever you change the wavelength setting.) Calculate the pH of the above solutions given the fact that the second ionization constant for H_3PO_4 is 1.3×10^{-7} ($pK_2 = 6.9$). Alternatively, pH values can be measured by a pH meter.

Treatment of Data

Combining the absorbance values at the two selected wavelengths (obtained in Procedure A) with the data obtained in Procedure B, plot absorbance (vertically) vs. pH (horizontally) for each of the two wavelengths studied. Connect the points with a smooth curve. The midpoint of each curve corresponds to equal concentrations of the acid and the base forms of the indicator. From each graph determine the pK_a' of the indicator. The two values thus obtained for the pK_a' may differ slightly.

Draw two horizontal lines across each of your A vs. pH plots; one corresponding to the absorbance of the pH $\cong 1$ solution of the indicator, and the other corresponding to the absorbance of the pH $\cong 13$ solution. The first line (I) gives the absorbance of the indicator when present entirely in the acid form. The second line (II) gives the absorbance of the indicator when present entirely in the base form. Any deviation of the actual absorbance from these two lines is a measure of the extent to which one form of the indicator has been converted to the other form.

Consider the absorbance reading obtained for the pH 6.2 solution (Fig. 6-11). By subtracting the absorbance at pH 6.2 from the absorbance at pH 1 (line I of the curve), a measure of the amount of In^{2-} in the solution at pH 6.2 can be obtained. By subtracting the absorbance at pH 13 (line II of the curve) from the absorbance at pH 6.2 a comparable measure of the amount of HIn^- in the pH 6.2 solution can be obtained. The ratio of In^{2-} to HIn^- may be obtained by dividing the results of the two preceding calculations. (The ratio also can be determined by measuring with a ruler the relative lengths of the two arrows indicated in Fig. 6-11.)

Determine the In^{2-}/HIn^- ratio that corresponds to each of the points plotted on *one* of your two absorbance vs. pH graphs, and plot log In^{2-}/HIn^- (vertically) vs. pH on a graph. That point where the line crosses the vertical axis at zero corresponds to equal concentrations of the basic and acidic forms of the indicator. From the pH at which the line crosses the vertical axis on your plot, determine the pK_a' of the indicator. Report your three values for the pK_a'.

Questions

1   a   What is meant by an isosbestic point?
      b   If the molar absorptivities of HIn^- and of In^{2-} were determined at a λ corresponding to an isosbestic point, how would they compare in value?
      c   What factors could contribute to the inability of an operator to obtain an isosbestic point in a study of a system of this nature?

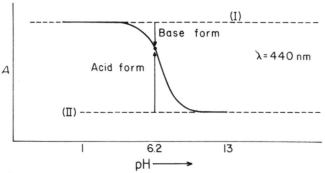

Figure 6-11 / Absorption of an acid-base indicator as a function of pH.

    d Could this experiment be applied to an indicator for which no isosbestic point was obtained in the available spectral range? Explain briefly.

2 If you were given a series of bromothymol-blue solutions of varying pH and were asked to determine the total amount of the indicator present in each solution, what wavelength(s) would you use for the analysis? Why? (A pH meter is not available.)

3 Suppose the indicator under study were a dibasic acid which dissociated according to the general equation

$$H_2In \rightarrow 2H^+ + In^{2-}$$

with both hydrogens coming off simultaneously in a single step (a rare case).

    a If $\log\,(In^{2-}/H_2In)$ were plotted against pH, what would the slope of the resulting line be?

    b When the line crosses the zero axis, does $pH = pK_a$? (Show your reasoning.)

4 If the indicator dissociated according to the general equation (not a rare case)

$$(HIn)_2 \rightarrow 2H^+ + 2In^-$$

what would you plot in order to get a straight line, and what would the slope of that line be? (Show your reasoning.)

5 (This one is to think about.) Reference books list the range of bromothymol blue as an indicator from pH 6.0 to pH 7.8. Can you account for the difference between the true pK_a and the "apparent" pK_a' deduced from this range?

SUPPLEMENTARY EXPERIMENTS

I. The pK_a' of Other Indicators

The general procedure and treatment of data outlined in the main experiment can be used to determine the values of pK_a' for numerous indicators other than bromothymol blue. Indicators that can be easily adapted to this experiment are listed below.

    Phenol red, 610 nm, pH 4.5 to 9, using a 0.04% solution of phenol red
    Bromcresol purple, 580 nm, pH 5 to 7.6, using a 0.04% solution
    Methyl red, 535 nm, pH 3.4 to 7.0, using a 0.04% solution
    Bromcresol green 580 nm, pH 3.1 to 7.2, using a 0.04% solution
    Any sulfonphthalein dye, 0.04% solution [0.1 g of indicator dissolved in water (and 1 to 3 mL of 0.1 M NaOH if necessary) and diluted to 250 mL]

II. The pK_a' of Methyl Orange: Simultaneous Analysis of Two Components in a Solution

Purpose

An acid-base indicator can be used to illustrate the simultaneous analysis of two components in a solution (see Experiment 6-1). In this experiment the concentrations of the two indicator forms are calculated at various pH's.

Chemicals

Methyl orange, 5.0×10^{-4} % solution

0.1 M hydrochloric acid [HCl]

0.1 M sodium hydroxide [NaOH]

Buffers: pH 3.40, 3.60, 3.80, 4.00

Procedure

Prepare six solutions that contain 5.0×10^{-4} % methyl orange with their acidities adjusted as follows: 0.1 M HCl; 0.1 M NaOH; pH 3.40; pH 3.60; pH 3.80; pH 4.00. Prepare the solutions so that each contains *exactly* the same concentration of methyl orange. Add the proper amount of methyl orange to each buffer mixture before diluting to the prescribed volume. Measure the absorbance of each solution at 10-nm intervals from 350 to 600 nm; use distilled water as the reference.

Treatment of Data

Plot all of the absorption spectra on the same graph. Assume that the indicator exists in its acid form exclusively in 0.1 M HCl and in its basic form in 0.1 M NaOH. Select the two wavelengths most suitable for analysis and determine the concentration of both acid and alkaline forms of the indicator in each solution of intermediate pH. Calculate the indicator constant from the data for each of the intermediate pH values. Compare the average value which you obtain for the indicator constant with published values.

EXPERIMENT 6-5

Spectrophotometric Determination of Formulas and Stability Constants of Complex Ions

---

Purpose

The formula and stability constant of a complex formed between copper(II) and the ligand iminodiacetic acid is determined spectrophotometrically by the method of continuous variations.

References

1 A. E. Martell and M. Calvin, "Chemistry of the Metal Chelate Compounds," Prentice-Hall, Englewood Cliffs, N.J., 1952.

2 G. P. Hildebrand and C. N. Reilley, *Anal. Chem., 29,* 258 (1957).

3 P. Job, *Ann. Chim., 10*(9), 113 (1928); *11*(6), 97 (1936).

4 R. T. Foley and R. C. Anderson, *J. Am. Chem. Soc., 70,* 1195 (1948).

5 H. H. Bauer, G. D. Christian, and J. E. O'Reilly, "Instrumental Analysis," Allyn and Bacon, Boston, 1978, pp. 178–180.

6 D. A. Skoog and D. M. West, "Fundamentals of Analytical Chemistry," 4th ed., Saunders, Philadelphia, 1982, pp. 552–554.

Apparatus

Spectrophotometer with red-sensitive phototube

Pyrex cells (2)

Pipets, 2 mL, 5 mL, 10 mL, 20 mL, and 25 mL

Volumetric flasks, two 250 mL and ten 25 mL

Beakers, 50 mL and 125 mL

Magnetic stirrer

pH meter

Chemicals

0.500 M cupric nitrate stock solution [Cu(NO$_3$)$_2$]

0.500 M iminodiacetic acid stock solution

Theory

Methods for
Determining the
Formula of a Complex

The *method of continuous variations* is a simple and widely used method for the spectrophotometric determination of complex formulas. However, it is reliable only for solution conditions where a single complex species is formed. The method also can be used for stability-constant determinations with systems that contain only one complex species.

1 If the sum of the total analytical concentrations C of complexing agent C_x and metal ion C_m is held constant and only their ratio is varied, then

$$C_x + C_m = C$$

2 A wavelength of light is selected where the complex absorbs strongly and the ligand and metal ions do not.

3 A plot of the mole fraction of ligand in the mixture X, where

$$X = \frac{C_x}{C}$$

$$\frac{C_x}{C} + \frac{C_m}{C} = \frac{C}{C}$$

and

$$1 - X = \frac{C_m}{C}$$

vs. absorbance gives the triangular-shaped curve shown in Fig. 6-12. The legs of the triangle are extrapolated until they cross. The mole fraction at the point of intersection gives the formula of the complex, because here the ligand and metal are in proper relative concentrations to give maximum complex formation for the complex MX_n,

$$n = \frac{C_x}{C_m} = \frac{X}{1 - X}$$

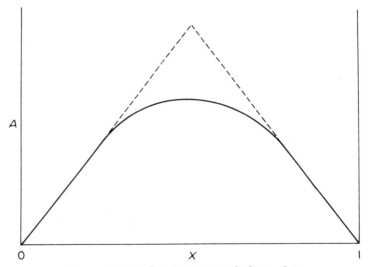

Figure 6-12 / Continuous-variations plot.

4 To determine whether there is more than one complex formed in solution (which would lead to erroneous results), the method is usually carried out at several different wavelengths and at several different values of C.

5 In the vicinity of maximum absorbance the actual curve may be observed to deviate somewhat from the extrapolated intersecting lines. Under certain conditions, the stability constant of the complex can be determined from the amount of deviation. For an explanation of this calculation and its limitations see the section below on the determination of stability constants.

The *mole-ratio method* is similar to the method of continuous variations. The difference lies in the fact that the total analytical concentration of metal (or ligand) is held constant rather than the sum of the ligand and metal concentrations. At a wavelength where only the complex absorbs, the absorbance vs. ligand-concentration curve is exactly the same as the curve obtained for a photometric titration where only the product absorbs.

1 The absorbance of solutions of constant total metal-ion concentration and varying ligand concentrations is measured.

2 A plot of absorbance vs. the ratio of ligand to metal concentration gives a curve like the one shown in Fig. 6-13 if only the complex absorbs.

3 The straight-line portions are extrapolated to where they cross. The ratio at this point is the ratio of ligand to metal ion in the complex.

4 The difference between the extrapolated values and the actual values of the absorbance can be used for stability-constant calculations. See the discussion below for this calculation.

5 If the dissociation constant of the complex is too large, then a smooth continuous curve is obtained which cannot be accurately extrapolated to a good value for the ligand/metal ratio. In this case the slope-ratio method is best.

The *slope-ratio method* is most valuable for weak complexes, because the absorbance measurements used with this method involve only solutions that contain a large excess of the metal and solutions that contain a large excess of the ligand. However, the method is applicable *only when one complex species is formed* and when *Beer's Law is followed,* and these are important limiting restrictions.

1 The absorbance of solutions containing a large excess of metal ion and various

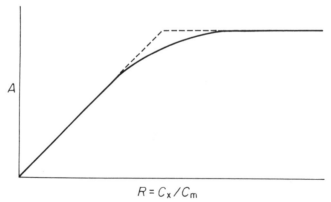

$$R = C_x / C_m$$

Figure 6-13 / Mole-ratio plot.

ligand concentrations is measured at a wavelength where only the complex $M_m X_n$ absorbs.

2   A plot of absorbance vs. total ligand concentration gives a straight line (if Beer's Law is obeyed) whose slope S_x is

$$S_x = \epsilon_c \frac{b}{n}$$

where ϵ_c = molar absorptivity of complex and b = cell length.

3   Similar measurements are made on solutions with a large excess of ligand and various metal-ion concentrations. Another plot of the absorbance vs. metal-ion concentration gives a line with the slope S_m equal to

$$S_m = \epsilon_c \frac{b}{m}$$

4   The formula can be determined from

$$\frac{S_m}{S_x} = \frac{n}{m} \qquad \text{for the complex } M_m X_n$$

Methods for Determining the Stability Constant of a Complex

The extrapolated values ($A_{\text{extp.}}$) near the "equivalence point" on the mole-ratio and continuous-variations plots correspond to the total absorbance of the complex if the complex formation was complete. Actually, the complex is slightly dissociated in this region, and the absorbance read is somewhat lower.

With a 1:1 complex (for which this method is generally applicable, because only one complex species forms), the ratio of the true absorbance to the extrapolated absorbance is the mole fraction of complex actually formed:

$$\frac{A}{A_{\text{extp.}}} = \frac{[\text{MX}]}{C}$$

where C is the total analytical concentration of the metal or ligand (whichever is the limiting concentration at the point in question). Then

$$[\text{MX}] = \left(\frac{A}{A_{\text{extp.}}}\right) C$$

$$[\text{M}] = C_m - [\text{MX}] = C_m - \left(\frac{A}{A_{\text{extp.}}}\right) C$$

$$[\text{X}] = C_x - [\text{MX}] = C_x - \left(\frac{A}{A_{\text{extp.}}}\right) C$$

$$K = \frac{(A/A_{\text{extp.}})C}{[C_m - (A/A_{\text{extp.}})C][C_x - (A/A_{\text{extp.}})C]}$$
$$= \frac{[\text{MX}]}{[\text{M}][\text{X}]}$$

where C_m and C_x are the total analytical concentrations of metal and ligand, respectively, and K is the stability constant.

Bjerrum's method is applicable when the ligand is either an acid or a base with a known ionization constant, or in general whenever there are two different equilibria which can compete for the metal or ligand and when one of these equilibrium constants is known. With an acid ligand protons compete with the metal ions for the ligand ion; if a large excess of metal ion is added and the complex concentration is determined as a function of pH, the stability constants for the complex can be computed by the use of the known ionization constants for the ligand. One metal ion can compete with another for the ligand ion and, by use of the stability constant of one metal complex, the stability constant for the other metal complex can be determined.

1 The absorbance of the complex in the presence of a hundredfold excess of metal ion is measured in the pH range where a proton can successfully compete with the metal ion for the complex.

2 A plot of absorbance vs. pH, where the complex is the only light-absorbing species, gives the curve shown in Fig. 6-14 where A_1 is the absorbance of the metal complex less the limiting absorbance of the solution at low pH and A_2 is the absorbance of the metal when completely complexed (at higher pH values).

3 The ratio of metal complex to acid ligand concentration is

$$\frac{[\text{MX}]}{[\text{H}_a\text{X}]} = \frac{A_1}{A_2 - A_1}$$

and for the competing equilibrium

$$\text{M}^{a+} + \text{H}_a\text{X} \rightleftharpoons \text{MX} + a\text{H}^+$$

$$\begin{aligned}
\text{p}K' &= -\log\frac{[\text{MX}][\text{H}^+]^a}{[\text{M}^{a+}][\text{H}_a\text{X}]} \\
&= -\log\frac{[\text{MX}]}{[\text{H}_a\text{X}]} - \log[\text{H}^+]^a + \log[\text{M}^{a+}] \\
&= -\log\frac{A_1}{A_2 - A_1} + a\text{pH} + \log[\text{M}^{a+}]
\end{aligned}$$

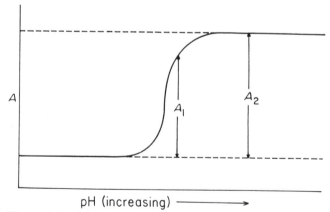

Figure 6-14 / Absorbance of a metal complex as a function of pH.

4 A plot of log ($[MX]/[H_aX]$) vs. pH gives a straight line whose slope is a, because K' and $[M^{a+}]$ are both constant. The value of pK' is determined from the preceding equation by inserting the value of pH at which log ($[MX]/[H_aX]$) is equal to zero and using the proper values of a and log $[M^{a+}]$.

The advantage of the log-plot method for determining the pH value lies in the fact that an average value, determined from several pH and absorbance values is obtained. Any association of the acid or complex forms will cause a deviation of the log plot from a straight line, and consequently this will serve as a warning that such behavior exists.

The *matching-absorbance method* is applicable only for determining stability constants.

1 The absorbances of two solutions that have different metal-to-ligand ratios are measured at a wavelength where only the complex absorbs.
2 The solution of higher absorbance is diluted until the absorbance of the two solutions is the same.
3 From the formal concentrations of metal and ligand in solution 1, $C_{m,1}$ and $C_{x,1}$, and the formal concentrations of metal and ligand in solution 2, $C_{m,2}$ and $C_{x,2}$, after dilution, the stability constant of the metal-ligand complex can be calculated.
4 For a 1:1 complex

$$[M]_1 = C_{m,1} - [MX]_1$$

$$[X]_1 = C_{x,1} - [MX]_1$$

$$[M]_2 = C_{m,2} - [MX]_2$$

$$[X]_2 = C_{x,2} - [MX]_2$$

When the absorbances of the two solutions are equal, then $[MX]_1 = [MX]_2 = c$, where c is some unknown concentration.
5 The stability constant K can then be calculated from the two stability-constant expressions which contain c and K as unknowns by eliminating c.

$$K = \frac{c}{(C_{m,1} - c)(C_{x,1} - c)} = \frac{c}{(C_{m,2} - c)(C_{x,2} - c)}$$

and c is

$$c = \frac{(C_{m,1}C_{x,1}) - (C_{m,2}C_{x,2})}{(C_{m,1} + C_{x,1}) - (C_{m,2} + C_{x,2})}$$

Procedure

Pipet 10 mL of stock iminodiacetic acid (IDA) into a 250-mL volumetric flask and dilute to the mark with distilled water. Also pipet 10 mL of stock $Cu(NO_3)_2$ solution into another 250-mL volumetric flask and dilute to the mark.

Pipet 2, 5, 7, 10, and 12 mL of the 0.02 M Cu^{2+} solution into five 25-mL volumetric flasks and dilute to the mark with the 0.02 M IDA solution. Also pipet 2, 5, 7, 10, and 12 mL of the 0.02 M IDA into five other 25-mL volumetric flasks and dilute to the mark with 0.02 M Cu^{2+}. Label these 10 flasks with the milliliters of the IDA that they contain.

Adjust each of the 10 solutions to pH 2.0 by use of a pH meter. After standardizing the pH meter with the standard buffer solution provided, place the contents of one of the 25-mL flasks into a clean, dry 50-mL beaker, drop in a stirring bar, and place the beaker on the stirrer. Lower the electrodes carefully into the beaker without allowing them to contact the side of the beaker. Add concentrated HNO_3 or 50% NaOH, as necessary, by filling a dropper with the proper solution and touching the tip of the dropper to the solution *without squeezing the bulb;* in this manner an extremely small amount of acid or base can be added by diffusion through the tip of the dropper. When the pH has been adjusted to *exactly* 2.0, return the solution to the volumetric flask. Repeat for each of the remaining solutions.

Measure the absorbance of these 10 solutions at 650 and 700 nm.

If time permits repeat the above, but use 20-mL aliquots of the stock solutions to make 0.04 M Cu^{2+} and 0.04 M IDA solutions for the preparation of the 10 solutions to be measured.

Treatment of Data

For the method of continuous variations, plot the absorbance of your Cu–IDA solutions vs. the mole fraction of IDA (for example, the solution containing 20 mL of the IDA has a mole fraction of $\frac{20}{25}$ or 0.80). Draw a straight line from the absorbance of pure IDA to that of pure Cu. For each point on the graph, subtract the absorbance due to free Cu (as represented by the straight line) and replot the absorbance values thus obtained against the mole fraction of IDA. Extrapolate the two straight sides of the curve until they cross. From the value of the fraction X at the intersection, calculate n.

From the ratio of the true absorbance to the extrapolated absorbance, $A/A_{\text{extp.}}$, calculate the overall formation constant K' for four points between $X = 0.4$ and 0.6.

$$K' = \frac{[H^+]^2[CuNH(CH_2COO)_2]}{[Cu^{2+}][NH(CH_2COOH)_2]}$$

$$= [H^+]^2 \frac{(A/A_{\text{extp.}})C}{[C_m - (A/A_{\text{extp.}})C][C_x - (A/A_{\text{extp.}})C]}$$

where C_m and C_x are the total analytical concentrations of metal and ligand, respectively, and C is the total analytical concentration of the metal or ligand, whichever is in limiting concentration. Given that $pK_{1,2} = pK_1 + pK_2 = 11.67$ for the diprotic ionization constant of IDA, calculate the absolute stability constant K

of the Cu–IDA complex where

$$K = \frac{[\text{CuIDA}]}{[\text{Cu}^{2+}][\text{IDA}]}$$

The literature value is $\log K = 10.3$.

Questions

1 Considering your experimental results, were both of the stock Cu and IDA solutions of exactly equal concentration, or, if not, which was more concentrated?

2 Why is the method of continuous variations more useful for weak complexes than the mole-ratio method?

3 By the method of continuous variations a metal M and a ligand X are found to form a 1:3 complex MX_3. Derive a simple formula for the calculation of the stability constant K at the point of maximum absorbance (assuming that the complex is the only light-absorbing species) in terms of the actual maximum absorbance A, the extrapolated maximum absorbance $A_{\text{extp.}}$, and the limiting concentration of the complex C. Note that in this case the *total analytical* concentration of metal in the solution $C_m = C$ and of the ligand $C_x = 3C$; but the *actual* concentration of metal (M) and of ligand (X) in the solution (due to dissociation of the complex formed) would actually be much less than this.

4 In Bjerrum's method, why is it necessary to add a large excess of metal ion?

SUPPLEMENTARY EXPERIMENTS

I. Method of Continuous Variations: Determination of the Formula for an Iron(III)–Sulfosalicylate Complex

Apparatus

Spectrophotometer
Cuvettes (2)
Burets (2), 10 mL
Volumetric flasks (10), 50 mL

Chemicals

$0.0100\ M$ sulfosalicylic acid in $0.1\ M$ $HClO_4$
$0.0100\ M$ ferric nitrate in $0.1\ M$ $HClO_4$
$0.1\ M$ perchloric acid

Procedure

Prepare a series of solutions by mixing the indicated amounts of $0.01\ M$ sulfosalicylic acid (solution A) and $0.01\ M$ ferric solution (solution B) and diluting to 50 mL with $0.1\ M$ perchloric acid:

1 mL solution A + 9 mL solution B
3 mL solution A + 7 mL solution B
5 mL solution A + 5 mL solution B
7 mL solution A + 3 mL solution B
9 mL solution A + 1 mL solution B

note The total number of moles added (ferric ion plus sulfosalicylic acid) is constant for each of the five solutions.

Allow the five solutions to stand for at least 30 min before any absorbance measurements are made.

While the series of solutions are standing, prepare five solutions which contain the amounts of solution B used in the other five solutions, but which do not contain any sulfosalicylic acid. Dilute each solution to 50 mL with 0.1 M perchloric acid.

Determine the absorption spectrum for the solution that contains 50 mol % iron over the spectral range from 350 to 625 nm; use distilled water as a blank. Next, measure the absorbance for each of the 10 solutions at the wavelength that corresponds to maximum absorption for the absorption spectrum; use distilled water as a blank.

Treatment of Data

Plot the absorbance vs. the wavelength for the 50 mol % iron solution. Next, plot the absorbance vs. concentration of iron for the five solutions which do not contain any sulfosalicylic acid. On a separate graph plot the absorbance vs. mol % iron for the series of solutions that contain iron and sulfosalicylic acid. From this graph make a preliminary conclusion concerning the formula of the complex and then correct the absorbance values for free iron(III) present; this can be accomplished by use of the absorbance–iron concentration plot. Plot the corrected absorbances on the same graph and make a final conclusion concerning the formula for the iron–sulfosalicylic acid complex.

Questions

1 Justify algebraically that the maximum absorbance will occur at the mole fraction corresponding to the formula of the complex when the method of continuous variations is applied. What conditions are necessary for the application of this method if valid results are to be obtained?

2 Suggest other types of absorbance plots for establishing the formula of a metal complex and indicate the advantages and limitations.

3 Discuss how it would be possible to use a continuous-variations plot to evaluate the stability constant of a complex. What are the limitations for using this method of evaluating a constant? Is this method applicable to the iron–sulfosalicylic acid complex? If it is, calculate the value of the stability constant. If it is *not* applicable, explain why.

II. Mole-Ratio Method: Determination of the Formula for a Fe(II)–2,2′-Bipyridine Complex in Water

Apparatus

Spectrophotometer
Cuvettes (2)
Pipets, $\frac{1}{2}$ mL, 1 mL, 2 mL, 3 mL, 4 mL, and 5 mL
Volumetric flasks, ten 10 mL, two 100 mL

Chemicals

1.16×10^{-3} M ferrous ammonium sulfate $[(NH_4)_2SO_4 \cdot FeSO_4 \cdot 6H_2O]$, 0.0455 g per 100 mL
1.16×10^{-3} M 2,2′-bipyridine, 0.0183 g per 100 mL

Procedure

Pipet 1-mL aliquots of the Fe(II) solution into each of ten 10-mL volumetric flasks. To the flasks, add successively 1-, 1$\frac{1}{2}$-, 2-, 3-, 4-, 5-, 6-, 7-, 8-, and 9-mL aliquots of the bipyridine solution. Fill each flask to the mark with distilled water. This gives solutions with 2,2′-bipyridine–to–Fe(II) mole ratios of 1:1, 1$\frac{1}{2}$:1, 2:1, 3:1, ..., 9:1.

Measure the absorbance of these 10 solutions at 522 nm (or as close to this wavelength as possible on the spectrophotometer used). Use distilled water as the blank.

Treatment of Data

Plot the absorbance vs. concentration of 2,2′-bipyridine for the 10 solutions measured. From this graph make a conclusion concerning the formula for the Fe(II)–2,2′-bipyridine complex.

III. Slope-Ratio Method: Determination of the Formula of the Fe(II)–1,10-phenanthroline Complex

Apparatus

Spectrophotometer
Cuvettes (2)
Pipets, 1 mL, 2 mL, 3 mL, 4 mL, 5 mL, 10 mL, and 30 mL
Volumetric flasks (10), 50 mL

Chemicals

5×10^{-4} M ferrous ammonium sulfate
5×10^{-4} M 1,10-phenanthroline
Buffer (pH 5.0), acetic acid–sodium acetate (0.1 M total acetate)
Hydroxylamine hydrochloride (5%)

Procedure

Pipet 30 mL of the Fe(II) solution into each of five 50-mL volumetric flasks. Add 10 mL of the pH 5.0 buffer to each flask, followed by 1 mL of 5% hydroxylamine hydrochloride to each. Add 1-, 2-, 3-, 4-, and 5-mL aliquots of the 1,10-phenanthroline solution, respectively, to each of these flasks. Dilute each solution to the mark and allow to stand 10 min. Read the absorbances at 510 nm.

Pipet 30 mL of 1,10-phenanthroline solution into each of five 50-mL volumetric flasks. Add 10 mL of the pH 5.0 buffer to each, as well as 1 mL of hydroxylamine hydrochloride. Add 1-, 2-, 3-, 4-, and 5-mL aliquots of the Fe(II) solution, respectively, to the flasks. Allow to stand 10 min after dilution to the mark. Read the absorbances at 510 nm.

Treatment of Data

Plot both sets of data on the same sheet of graph paper. Absorbance should be plotted vs. the concentration of the variable component. Determine the slopes of the two lines. The ratio of the slopes for the two straight lines equals the combining ratio of the two components.

EXPERIMENT 6-6

Qualitative and Quantitative Analysis of Aromatic Hydrocarbon Mixtures

Purpose

This experiment illustrates how ultraviolet spectra can be used to characterize and analyze conjugated systems.

References

1 H. H. Willard, L. L. Merritt, Jr., J. A. Dean, and F. A. Settle, Jr., "Instrumental Methods of Analysis," 6th ed., Van Nostrand, Princeton, N.J., 1981, chaps. 2 and 3.
2 G. W. Ewing, "Instrumental Methods of Chemical Analysis," 4th ed., McGraw-Hill, New York, 1975, chaps. 2 and 3.

3   H. H. Bauer, G. D. Christian, and J. E. O'Reilly, "Instrumental Analysis," Allyn and Bacon, Boston, 1978, chaps. 6 and 7.

4   E. D. Olsen, "Modern Optical Methods of Analysis," McGraw-Hill, New York, 1975, chap. 2.

5   D. A. Skoog and D. M. West, "Principles of Instrumental Analysis," 2nd ed., Saunders, Philadelphia, 1980, chaps. 6 and 7.

6   C. K. Mann, T. J. Vickers, and W. M. Gulick, "Instrumental Analysis," Harper and Row, New York, 1974, chaps. 11 and 16.

7   E. J. Meehan in "Treatise on Analytical Chemistry," P. J. Elving, E. J. Meehan, and I. M. Kolthoff (eds.), 2nd ed., part I, vol. 7, Interscience, New York, 1981, chaps. 1 and 2.

Apparatus

Automatic-recording spectrophotometer with quartz cells
Pipets, 1 mL, 2 mL, and 4 mL
Volumetric flasks, four 25 mL, one 100 mL
Medicine droppers, long type (furnished with solutions)

Chemicals

Benzene
Cyclohexane
p-xylene, 1.2 g/L in cyclohexane
o-xylene, 1.2 g/L in cyclohexane
Diphenyl, 0.052 g/L in cyclohexane
Unknown: Containing p-xylene in cyclohexane (between 0.05 and 0.20 g/L)

Theory

Ultraviolet spectrophotometry (200 to 400 nm) can be applied to the qualitative and quantitative analysis of many organic compounds. Recording spectrophotometers make measurements in this region especially convenient. Unlike the infrared region, the ultraviolet spectra of most molecules in solution are relatively simple, and usually have only one or two maxima compared with the possibility of 20 or more in the infrared.

When molecules absorb in the visible or ultraviolet, valence or bonding electrons are raised to higher energy levels, with accompanying vibrational and rotational excitations. The close spacing of the vibrational-rotational levels in relation to electron levels causes each electron transition to result in a large number of slightly different possible energy levels. This, along with solvent-solute interaction, tends to yield broad absorption bands in the ultraviolet.

Molecules of loosely bound electron clouds afford absorption of energy in the ultraviolet. Therefore ultraviolet spectra are diagnostic of unsaturation in absorbing molecules (chromophoric groups), because with few exceptions only molecules with multiple bonds have excited states of sufficiently low energy to give rise to absorption in the near ultraviolet.

For this reason, saturated hydrocarbons, alcohols, and ethers are transparent in the ultraviolet as are monofunctional olefins, acetylenes, carboxylic acids, esters, amides, and oximes which have absorption maxima at wavelengths just shorter than the near ultraviolet. Such unsaturated functional groups as aldehydes, ketones, aliphatic nitro compounds, and nitrate esters have maxima in the near ultraviolet, but their intensities are so low that only under special circumstances can they be useful. Conjugated double bonds, benzenoid, quinoid, azo, diazo, nitroso, and nitrile ester functions, however, have relatively high molar absorptivities in the ultraviolet.

Organic analysis in the ultraviolet would be severely limited but for the fact that olefinic, acetylenic, and carboxyl functions (in general, any double or triple bond group) will give strong absorption in the ultraviolet when conjugated with one another. (Most groups which do not absorb in the near ultraviolet do absorb at shorter wavelengths. Addition of chromophoric groups that enlarge the resonance system or attachment of electron-donating side groups usually causes the wavelength of absorption to shift into the near ultraviolet region.)

Therefore, ultraviolet spectra may be valuable for qualitative studies of conjugated systems. Solutions used are usually so dilute that Beer's Law holds and quantitative work also is possible.

Part I. Quantitative Analysis by Base-Line Method

A problem which often confronts the analytical chemist is the determination of a known component in the presence of one which is unknown but absorbs radiation in the same spectral region. Such problems can sometimes be solved by the application of the base-line method, as will be illustrated in this part of the experiment. The principle underlying this method is illustrated in Fig. 6-15. In (a) there is a portion of a spectrum of compound X that contains an absorption peak. Because the concentration of X is known, a value of A_x can be calculated. A value can also be calculated for the height of the peak itself by drawing a base line as

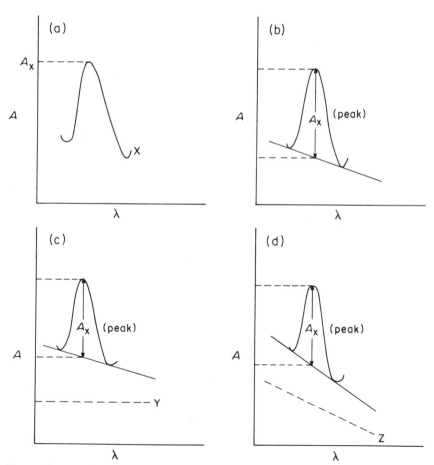

Figure 6-15 / Base-line technique for quantitative absorption spectroscopy. (*Note:* The data must be plotted in absorbance units, not %*T*.)

shown in (b). This value of A_x (peak) is also proportional to concentration and can be used for quantitative work. If compound X is mixed with another absorbing compound Y which has the simple spectrum indicated by the dashed line Y in (c), the value of A_x (peak) is not changed. The whole peak is, however, shifted to a higher value of absorbance. If compound X is measured in the presence of another compound Z [spectrum shown by dashed line in (d)] the same method can be used. In this case, the shape of the peak is changed, but the height of the peak itself A_x (peak) is unchanged.

This method can only be applied to a case in which there is a relatively sharp peak in the spectrum of the known component and the spectrum of the unknown component is essentially linear in the same wavelength region.

Procedure

Mix three solutions of *p*-xylene (0.05, 0.10, and 0.20 g/L) by diluting 1, 2, and 4 mL of *p*-xylene stock solution to 25 mL with cyclohexane. Also, prepare a solution that contains 0.1 g/L *p*-xylene and 0.05 g/L *o*-xylene and a solution that contains 0.1 g/L *p*-xylene and 2.5 mg/L diphenyl. These are prepared by diluting 2 mL of *p*-xylene stock solution plus 1 mL of the stock solution of the other compound to a volume of 25 mL with cyclohexane. Next prepare a solution by diluting 1 drop of benzene to 100 mL with cyclohexane.

Record the spectra of these solutions and of your unknown solution in a 1-cm cell from 285 nm to 245 nm. Use cyclohexane in both cells when balancing the instrument. Follow the set of operating instructions for instrument adjustments.

Treatment of Data

Using the base-line method, make a Beer's Law plot of concentration of *p*-xylene vs. absorbance (peak) for the peaks occurring at 274.4 and 265.8 nm. Calculate the height of the same peaks for the two solutions which had biphenyl and *o*-xylene as impurities and mark these points on your Beer's Law plots. Determine the concentration of your unknown *p*-xylene.

Part II. Qualitative Analysis of Cyclohexane and Use of Reference Cell to Compensate for Absorbance of Solvent

Procedure

Balance the instrument with an empty 2-cm (or 1-cm) cell in each light beam. Place cyclohexane in the sample cell and record the spectrum from 285 to 245 nm.

Next place cyclohexane in both cells and run the spectrum under the same conditions as above, leaving the balance control in the same position as it was for cyclohexane vs. air.

Next dry the sample cell and carefully place 1 drop of benzene in the bottom of the cell and stopper (do not get benzene on the windows). Record a spectrum from 245 to 285 nm.

Treatment of Data

Compare the spectrum of cyclohexane vs. air with the spectrum of benzene in solution and report, if benzene is the impurity present in the cyclohexane, as milliliters impurity per liter of cyclohexane. (Assume that 1 drop of benzene is approximately 0.05 mL.)

Questions

1 Why does a spectrum of benzene in the vapor phase show more fine structure than does benzene in the solvent?
2 Estimate the detection limit of this method for determining *p*-xylene. What is the limiting factor?
3 Calculate the molar absorptivity for *p*-xylene from the Beer's Law plots.

EXPERIMENT 6-7

Analysis of Pharmaceutical Mixtures

Purpose

This experiment demonstrates some of the problems that are encountered in the determination of the components of a pharmaceutical mixture which is commonly used for the relief of pain. One of the components must be extracted from the others. The remaining two ingredients are determined simultaneously in the same solution in the ultraviolet region of the spectrum.

References

1 H. H. Willard, L. L. Merritt, Jr., J. A. Dean, and F. A. Settle, Jr., "Instrumental Methods of Analysis," 6th ed., Van Nostrand, Princeton, N.J., 1981, chaps. 2 and 3.

2 G. W. Ewing, "Instrumental Methods of Chemical Analysis," 4th ed., McGraw-Hill, New York, 1975, chaps. 2 and 3.

3 H. H. Bauer, G. D. Christian, and J. E. O'Reilly, "Instrumental Analysis," Allyn and Bacon, Boston, 1978, chaps. 6 and 7.

4 E. D. Olsen, "Modern Optical Methods of Analysis," McGraw-Hill, New York, 1975, chap. 2.

5 D. A. Skoog and D. M. West, "Principles of Instrumental Analysis," 2nd ed., Saunders, Philadelphia, 1980, chaps. 6 and 7.

6 C. K. Mann, T. J. Vickers, and W. M. Gulick, "Instrumental Analysis," Harper and Row, New York, 1974, chaps. 11 and 16.

7 E. J. Meehan in "Treatise on Analytical Chemistry," P. J. Elving, E. J. Meehan, and I. M. Kolthoff (eds.), 2nd ed., part I, vol. 7, Interscience, New York, 1981, chaps. 1 and 2.

Apparatus

Ultraviolet spectrophotometer
Cells (2), silica (quartz)
Pipets (2), 20 mL
Volumetric flasks, two 100 mL, one 250 mL, and one 500 mL
Beaker, 150 mL
Separatory funnel, 250 mL
pH test paper
Dropping pipets for each solution

Chemicals

APC tablet or Anacin
Chloroform, reagent grade
Aspirin (acetylsalicylic acid), 100 mg/L of chloroform
Caffeine, 10 mg/L of chloroform
Phenacetin, 10 mg/L of chloroform
Sodium bicarbonate [$NaHCO_3$] (4%)
1 M sulfuric acid [H_2SO_4]

Theory

APC tablets are a mixture of aspirin, phenacetin, and caffeine in an inert binder. Each of these substances absorbs in the ultraviolet with principal absorption maxima at 277 nm for aspirin, 275 nm for caffeine, and 250 nm for phenacetin.

The sample is partitioned between chloroform and 4% aqueous sodium bicarbonate, with the aspirin only going into the aqueous layer. The phenacetin and caffeine are analyzed simultaneously in chloroform. The aspirin solution is acidified, extracted back into chloroform, and determined spectrophotometrically. A discussion of the theory for determining two substances simultaneously is given in Experiment 6-1.

Procedure

Weigh accurately one APC tablet and crush to a fine powder in a 150-mL beaker. Add 80 mL of chloroform and stir until the material has dissolved; then transfer quantitatively to a 250-mL separatory funnel. If there are any particles which did not dissolve, also transfer these to the separatory funnel. Extract the chloroform

safety caution

Perform extractions with chloroform in a hood.

solution with two 40-mL portions of chilled 4% sodium bicarbonate and then with one 20-mL portion of water. Wash the combined aqueous extracts with three 25-mL portions of chloroform and add the chloroform wash solutions to the chloroform solution. Filter the combined chloroform solution through paper previously wetted with chloroform into a 250-mL volumetric flask and dilute to the mark with chloroform. Pipet a 2-mL portion into a 100-mL volumetric flask and dilute to the mark with chloroform.

Acidify the bicarbonate solution, which is still in the separatory funnel, with 1 M sulfuric acid, making sure to add the acid slowly and in small portions. Do not mix vigorously until the carbon dioxide evolution has almost ceased. Extract the acidified solution with eight 50-mL portions of chloroform, and then filter the chloroform extracts through a chloroform-wet filter paper into a 500-mL volumetric flask. Dilute to volume with chloroform and pipet a 20-mL portion into a 100-mL volumetric flask and dilute to volume with chloroform.

Record the absorption spectra, using chloroform as the blank, for the three standard solutions and the two solutions obtained from the APC tablet.

Treatment of Data

Assume that Beer's Law is obeyed by all three components and determine the milligrams of aspirin, caffeine, and phenacetin per APC tablet. To do this, select the best wavelength for each component to make absorbance measurements. This can be determined from the absorption spectra of the standard solutions. The validity of your choice can be confirmed for phenacetin and caffeine by plotting the ratio of the absorbance for phenacetin relative to the absorbance of caffeine vs. wavelength. The graph should be included as part of your results.

Questions

1 Comment on the possibility that all three components of APC tablets can be determined simultaneously. Use the spectra of the three standard solutions as a reference for your answer.

2 Would it be more desirable to use spectra-grade chloroform for the blank solution? Justify your answer.

3 Would benzene be an equally satisfactory solvent for the APC tablet analysis? Why?

4 Explain the function of the bicarbonate.

EXPERIMENT 6-8

Simultaneous Determination of Vitamin C and Vitamin E†

Purpose

This experiment demonstrates how a two-component system [ascorbic acid (vitamin C) and α-tocopherol (vitamin E)] can be determined simultaneously in the ultraviolet region.

†From Janice M. Griewahn (Beebe), Master's Thesis, University of Pittsburgh, February 1950.

References

1 H. H. Willard, L. L. Merritt, Jr., J. A. Dean, and F. A. Settle, Jr., "Instrumental Methods of Analysis," 6th ed., Van Nostrand, Princeton, N.J., 1981, chaps. 2 and 3.

2 G. W. Ewing, "Instrumental Methods of Chemical Analysis," 4th ed., McGraw-Hill, New York, 1975, chaps. 2 and 3.

3 H. H. Bauer, G. D. Christian, and J. E. O'Reilly, "Instrumental Analysis," Allyn and Bacon, Boston, 1978, chaps. 6 and 7.

4 E. D. Olsen, "Modern Optical Methods of Analysis," McGraw-Hill, New York, 1975, chap. 2.

5 D. A. Skoog and D. M. West, "Principles of Instrumental Analysis," 2nd ed., Saunders, Philadelphia, 1980, chaps. 6 and 7.

6 C. K. Mann, T. J. Vickers, and W. M. Gulick, "Instrumental Analysis," Harper and Row, New York, 1974, chaps. 11 and 16.

7 E. J. Meehan in "Treatise on Analytical Chemistry," P. J. Elving, E. J. Meehan, and I. M. Kolthoff (eds.), 2nd ed., part I, vol. 7, Interscience, New York, 1981, chaps. 1, 2, and 3.

Apparatus

Recording ultraviolet spectrophotometer
Cells (2), silica
Volumetric flasks, eight 25 mL and two 1 L (for standards)
Pipets, 2 mL, 3 mL, 4 mL, and 5 mL

Chemicals

Ascorbic acid, Certified Reagent Grade stock solution, 0.0132 g/L in absolute ethanol (7.50×10^{-5} M)

α-Tocopherol, Certified Reagent Grade stock solution, 0.0488 g/L in absolute ethanol

Absolute ethanol

Unknown: Solution that contains ascorbic acid and α-tocopherol in absolute ethanol

Theory

The theory for the determination of two components in the same solution simultaneously is discussed in Experiment 6-1.

Ascorbic acid is referred to as a "water-soluble" vitamin, whereas α-tocopherol is known as a "fat-soluble" vitamin. However, they both will dissolve in absolute ethanol to the extent that they can be determined in the ultraviolet region. Because ascorbic acid solutions slowly oxidize to dehydroascorbic acid, fresh solutions must be prepared daily. α-Tocopherol is somewhat more stable. Both ascorbic acid and α-tocopherol behave as anti-oxidants; that is, they prevent fats from becoming rancid for a period of time. The effect of the two together is more than additive—they are "synergistic" in their behavior as anti-oxidants. For this reason, they are a useful combination for preserving various foods.

Procedure

Prepare standard solutions of ascorbic acid by diluting 2-, 3-, 4-, and 5-mL aliquots of the stock solution, respectively, in 25 mL with absolute ethanol in volumetric flasks. Prepare standards of α-tocopherol in the same manner—2-, 3-, 4-, and 5-mL aliquots of the stock solution to 25 mL.

Follow instructions for the instrument and obtain absorption spectra from 320 to 220 nm using absolute ethanol in the reference beam. Superimpose all four ascorbic acid spectra on the same graph. Also, superimpose all four α-tocopherol spectra on the graph. Obtain an absorption spectrum of the unknown.

Treatment of Data

Plot standard curves for ascorbic acid and α-tocopherol. Choose the wavelength of maximum absorption for each of the substances (about 246 nm for ascorbic acid and 292 nm for α-tocopherol), and prepare Beer's Law plots for each component by plotting absorbance vs. concentration.

Calculate the molar absorptivities of each solution at each wavelength or determine the slopes of the curves on the Beer's Law plots (see Experiment 6-1, Part IV). Calculate the concentration of the ascorbic acid and of α-tocopherol in your unknown.

Identify the isosbestic point on your graph and determine its wavelength and molar absorptivity.

Questions

1 Draw structures for ascorbic acid and α-tocopherol. Explain why one is a *water-soluble* vitamin and the other a *fat-soluble* vitamin.

2 Discuss the possible use of this procedure for determining vitamin C in orange juice, cranberry juice, spinach, and vitamin C tablets.

3 Is the method used in this experiment more sensitive for the determination of ascorbic acid or α-tocopherol? Explain.

7

Absorption Spectroscopy of Electronic Transitions

Absorption of light in the visible region (700 to 400 nm) and the near ultraviolet region (400 to 200 nm) of the electromagnetic spectrum results in electronic transitions for both organic and inorganic compounds.

Absorption of Organic Compounds

Absorption of light by the electrons of a single (sigma, σ) bond occurs in the vacuum-ultraviolet region (below 180 nm) where atmospheric components also absorb strongly. Obtaining spectra for these transitions is experimentally difficult. In general, spectra are limited to absorption by various functional groups, or chromophores, which contain lower energy valence electrons. In the near ultraviolet region, most of the observed absorption bands are due to transitions of the pi (π) electrons. Pi orbitals are formed by parallel overlap of p orbitals. A double bond contains a sigma orbital and a pi orbital; a triple bond contains a sigma orbital and two pi orbitals. Nonbonded or unshared electrons, which occur with nitrogen and oxygen, are known as *n* electrons. Interactions between *pi* and *n* electrons affect absorptions in the ultraviolet and visible regions.

Electron types:

$$
\begin{array}{c}
\text{H} \\
\sigma \Big\{ \quad \text{C} = \overset{..}{\text{O}} \colon \quad \nearrow n \text{ (nonbonded)} \\
\text{H} \quad \Big\{ \pi + \sigma
\end{array}
$$

The relative energies of the four transitions are illustrated:

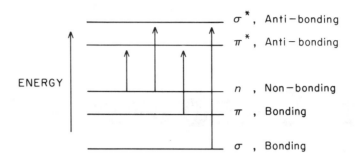

Chromophores usually contain either unshared electrons which are localized on atoms such as oxygen, sulfur, nitrogen, or halogens (nonbonding electrons), or bonding electrons such as those found in double and triple bonds. The commonly encountered transitions are $n \rightarrow \sigma^*$, $n \rightarrow \pi^*$, and $\pi \rightarrow \pi^*$. Transitions that involve excitation of electrons in single bonds are known as $\sigma \rightarrow \sigma^*$ transitions and are rarely observed because of matrix interference and instrument limitations.

Ketones ($R-\overset{\overset{\textstyle O}{\|}}{C}-R'$) exhibit $\pi \rightarrow \pi^*$ and $n \rightarrow \pi^*$ transitions, which can be distinguished by their molar absorptivities. For $n \rightarrow \pi^*$ transitions, ϵ ranges from 10 to 100 M^{-1} cm^{-1}, whereas for $\pi \rightarrow \pi^*$ transitions, ϵ ranges from 1000 to 100,000 M^{-1} cm^{-1}. As solvent polarity increases, $n \rightarrow \pi^*$ transitions tend to shift to shorter wavelengths. This type of shift is referred to as a hypsochromic, or blue, shift. For many $\pi \rightarrow \pi^*$ transitions, an increase in polarity causes a shift to a longer wavelength, and this shift is known as a bathochromic, or red, shift.

The absorbance of molecules that contain more than one like-functional group often can be predicted, if the spectrum of the monofunctional groups is known. The principle of additivity says that the energy at which a ketone carbonyl absorbs should not vary with the number of carbonyls per molecule, but the intensity should. For example, the absorbance due to the $n \rightarrow \pi^*$ transition for 2,5-hexadione is approximately twice that for acetone, but the wavelength remains constant. This increase in intensity is known as a hyperchromic shift. A decrease in intensity is referred to as a hypochromic shift.

The principle of additivity fails for conjugated systems because delocalization of electrons lowers the π^* state and a bathochromic shift for the chromophore results. The $\pi \rightarrow \pi^*$ transition that is observed for the methyl vinyl ketone spectrum occurs in the vacuum-ultraviolet region for nonconjugated systems.

Because conjugation alters spectral properties, the absorption spectra of aromatic systems are quite different from those of nonaromatic systems; benzene exhibits typical aromatic features. The band at 256 nm, known as the B band, is the weakest of the three $\pi \rightarrow \pi^*$ transitions. Another band known as the E_2 band occurs at 204 nm ($\epsilon = 7900$) and an even stronger band E occurs at 184 nm ($\epsilon = 60,000$). The series of sharp peaks in the 256-nm band are the vibrational transitions which are superimposed on the electronic transition. The effect of solvation on the benzene spectrum can be observed by comparing the solution and the gas-phase spectra.

Functional groups which don't absorb in the ultraviolet region can also affect the intensity and energy of chromophore transitions. Auxochromes have at least one pair of nonbonding electrons which interact with the aromatic ring. This interaction stabilizes the π^* state which causes a bathochromic shift. Some auxochromes include the $-NH_2$ group of aniline and the $-OH$ group of phenol. The spectrum of sodium phenolate shows an increased bathochromic shift and increased peak intensity due to the additional electron pair present on the oxygen.

Absorption of Inorganic Compounds

In contrast to organic compounds and most inorganic compounds, the ions of lanthanide and actinide elements absorb ultraviolet and visible light in narrow bands. These bands represent $f \rightarrow f^*$ transitions. The f electrons are highly shielded from external influences by occupied orbitals of higher principal quantum number.

In contrast to lanthanide and actinide elements, the spectra of transition-metal ions are extremely subject to environmental effects. This is due to the involvement of d orbitals in the coordination bonding of transition-metal ions. In solution, these ions form complexes either with specific ligands or complexing agents, or with solvent molecules. Complex formation removes the degeneracy of the five d orbitals, such that electronic transitions from the lower energy d orbitals to the higher energy d orbitals are observed upon absorption of the proper frequency of light. The light required for these transitions ranges from the near infrared to the ultraviolet region.

Transition metals tend to form six coordinate complexes with octahedral symmetry. Four coordinate complexes with tetrahedral or square-planar symmetry also occur. The d orbitals lose their degeneracy upon complex formation to different degrees. This depends on the symmetry of the resultant complex.

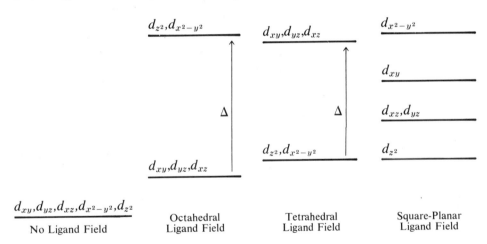

The amount of splitting between d orbitals depends on the ligand field strength. This means that the energy difference between d orbitals depends on the nature of the ligand, the oxidation state of the metal, and the symmetry. Ligand field strength has been determined to follow the order: $I^- < Br^- < SCN^- < OH^- < NH_3 < o\text{-phen} < NO_2^- < CN^-$. For example, CN^- causes d orbitals to "split" more than I^-; hence, more energy is required for a transition for a cyano complex than for an iodo complex.

In addition to ligand field strength, symmetry also influences the positions of $d \rightarrow d^*$ bands. Cobalt(II) can exist in octahedral or tetrahedral symmetry. The pink aqueous solution contains $Co(H_2O)_6^{2+}$, whereas the blue methanol solution is made up of tetrahedral $Co(MeOH)_4^{2+}$. In general, $d \rightarrow d^*$ transitions exhibit low intensity with ϵ values that range from 10 to 200 M^{-1} cm^{-1}.

In addition to $d \rightarrow d^*$ transitions, many transition metal complexes exhibit intense absorption bands ($\epsilon > 10,000$) due to charge transfer. With this type of transition, one component of the complex acts as an electron donor, while the other component of the complex acts as an electron acceptor. Upon absorption of a photon, the electron from the donating component is transferred via an internal oxidation-reduction process, to the accepting component. The spectra of tris(1,10-o-phenanthroline)iron(II) and of triiodide illustrate charge-transfer transitions. Some organic compounds also exhibit charge-transfer transitions. Examples include quinone:hydroquinone and iodine:amine complexes.

EXPERIMENT 7-1
Electronic Transitions in Organic Molecules

Purpose   In this experiment, the major types of electronic transitions observed in organic compounds will be studied, as well as the relative sensitivity of these systems to environmental perturbations.

References
1 A. B. P. Lever, "Inorganic Electronic Spectroscopy," Elsevier, Amsterdam, 1968.
2 D. A. Skoog and D. M. West, "Principles of Instrumental Analysis," 2nd ed., Saunders, Philadelphia, 1980, chap. 7.
3 H. H. Bauer, G. D. Christian, and J. E. O'Reilley, "Instrumental Analysis," Allyn and Bacon, Boston, 1978, chap. 7.
4 R. M. Silverstein, G. C. Bassler, and T. C. Morrill, "Spectrometric Identification of Organic Compounds," 4th ed., Wiley, New York, 1981, chap. 6.
5 R. C. Pecsok, L. D. Shields, T. Cairns, and I. G. McWilliam, "Modern Methods of Chemical Analysis," 2nd ed., Wiley, New York, 1976, chap. 12.
6 G. D. Olsen, "Modern Optical Methods of Analysis," McGraw-Hill, New York, 1975, chap. 2.

Apparatus   Recording spectrophotometer (with spectral range from 200 nm to 650 nm)
Silica cuvettes (2)

Chemicals 2.0×10^{-3} M iodomethane, in hexane
2.0×10^{-3} M iodomethane, in methanol
1.0×10^{-2} M acetone, in water
1.0×10^{-2} M 2,5-hexanedione, in water
2.5×10^{-2} M methyl vinyl ketone, in hexane
2.5×10^{-2} M methyl ketone, in water
6.6×10^{-5} M methyl vinyl ketone, in hexane
6.6×10^{-5} M methyl vinyl ketone, in water
Benzene, spectrograde
4.0×10^{-3} M benzene, in methanol
3.1×10^{-4} M phenol, in methanol
3.1×10^{-4} M sodium phenolate, in methanol

Theory   The general theory for this experiment has been discussed in the chapter introduction.

The most frequently encountered transitions are

Transition	Spectral Region	Example	Molar Absorptivity
$\sigma \to \sigma^*$	Vacuum ultraviolet (below 180 nm)	Methane, 125 nm	
$n \to \sigma^*$	Far or near ultraviolet	Water, 167 nm	1,480
		Iodomethane, 258 nm	258
$\pi \to \pi^*$	Ultraviolet	$C_6H_{13}CH{=}CH_2$, 177 nm	13,000
$n \to \pi^*$	Near ultraviolet and visible	Acetamide, 214 nm	60
$\pi \to \pi^*$	Ultraviolet	Benzene, 256 nm	200

Chromophores are the absorbing groups in a molecule. They are functional groups which absorb between 185 and 1000 nm when they are bonded to a nonabsorbing saturated residue which has no unshared, nonbonding valence electrons. An example would be a nitrite group introduced on *n*-octane, which exhibits a strong absorption at 230 nm.

Chromophores can be classified according to the type of bands they exhibit: K bands are due to $\pi \to \pi^*$ transitions; R bands to $n \to \pi^*$ transitions; and B bands to forbidden $\pi \to \pi^*$ transitions of aromatic molecules (see Ref. 4 for a detailed explanation of these bands).

Auxochromes (functional groups with nonbonding electron pairs) do not themselves absorb radiation, but can increase the intensity of absorption or shift the wavelength of absorption when attached to a chromophore.

Tables of absorption bands for various chromophores can be found in Refs. 2, 3, 4, and 5.

When two or more chromophores occur in a molecule, their relative positions determine the effect produced on the absorption of the molecule. A few general rules may be stated:

1 Two chromophores in the same molecule, when separated by more than one carbon atom, give an absorption equal to the summation of each of the two chromophores.
2 Two chromophores in the same molecule, when adjacent to each other, give an absorption displaced to a longer wavelength and the intensity of the absorption is increased.
3 Two chromophores attached to the same carbon atom result in intermediate absorption between the two extremes.

Position and intensity of absorption vary with the solvent used. These effects are due to (*a*) the nature of the solvent, (*b*) the nature of the absorption band, and (*c*) the nature of the solute. Polar solvents tend to shift $n \to \pi^*$ transitions to a shorter wavelength and $\pi \to \pi^*$ transitions to a longer wavelength.

Procedure

Obtain spectra from 650 nm to 200 nm for all of the solutions that are listed under Chemicals. Handle cells carefully and fill the reference cell with the solvent used in the sample.

The benzene vapor spectrum can be obtained by putting a drop of benzene in the bottom of the sample cell and capping it. No reference cell is needed.

safety caution

Methyl vinyl ketone is a lachrymator and should be discarded in a waste container in the hood.

Treatment of Data   (Read the chapter introduction on electronic transitions.)

Tabulate all of the wavelengths of absorption maxima and the molar absorptivities for each of the 12 solutions. Include the type of transition for each band.

Questions

1 How large a shift did you observe for the absorption maximum of iodomethane in water vs. hexane, in nanometers?

2 Why are water, methanol, and hexane commonly used as solvents in ultraviolet and visible spectroscopy?

3 In methyl vinyl ketone, which band arises from a $\pi \rightarrow \pi^*$ transition, and which is the result of an $n \rightarrow \pi^*$ transition? Explain your reasoning.

4 Predict the spectrum for 0.1 M 2,5,8-nonanetrione and for 0.1 M 2-ene-5,8-nonanedione.

5 How does conjugation affect the spectrum of methyl vinyl ketone? Predict the spectra for 1-pentene, 1,5-hexadiene, and for 1,3,5-hexatriene. [*Hint:* 1-butene has a $\pi \rightarrow \pi^*$ ($\epsilon = 10,000$) transition at 180 nm.] Is there an $n \rightarrow \pi^*$ transition?

6 What difference is observed between the gas-phase and the solution-phase spectra for benzene? Why are they different?

7 How would the auxochrome $-NH_2$ affect the B band in aniline and how would protonation to give the anilinium cation alter this effect?

EXPERIMENT 7-2
Electronic Transitions for Transition-Metal Complexes

Purpose   In this experiment, the electronic transitions for transition-metal complexes will be determined by absorption spectrophotometry.

References

1 A. B. P. Lever, "Inorganic Electronic Spectroscopy," Elsevier, Amsterdam, 1968.
2 D. A. Skoog and D. M. West, "Principles of Instrumental Analysis," 2nd ed., Saunders, Philadelphia, 1980, chap. 7.
3 C. K. Mann, T. J. Vickers, and W. M. Gulick, "Instrumental Analysis," Harper and Row, New York, 1974, chap. 16.
4 H. H. Jaffe and M. Orchin, "Theory and Applications of Ultraviolet Spectroscopy," Wiley, New York, 1962.
5 E. D. Olsen, "Modern Optical Methods of Analysis," McGraw-Hill, New York, 1975, chap. 2.

Apparatus   Recording spectrophotometer (with spectral range from 200 to 700 nm)
Silica cuvettes (2)

Chemicals 7.5×10^{-2} M praseodymium chloride [$PrCl_3$], in water
1.0×10^{-1} M nickel sulfate [$NiSO_4$], in water
1.0×10^{-1} M nickel sulfate, in 12 M hydrochloric acid
1.0×10^{-1} M tris(ethylenediamine)nickel(II) sulfate [$Ni(en)_3SO_4$], in water
1.0×10^{-1} M cobalt(II) sulfate [$CoSO_4$], in water
1.0×10^{-1} M cobalt(II) sulfate, in anhydrous methyl alcohol
8.0×10^{-5} M tris(1,10-phenanthroline)iron(II), in water
3.0×10^{-5} M potassium triiodide [KI_3], in water

Theory

The general theory for this experiment is discussed in the chapter introduction.

Absorption of ultraviolet or visible radiation by metal complexes may be due to one or more transitions: (*a*) excitation of the ligand, (*b*) excitation of the metal, or (*c*) charge transfer. Ligands, most of which are organic chelating agents, undergo $n \rightarrow \pi^*$ and $\pi \rightarrow \pi^*$ transitions. Complexation of a ligand with a metal ion resembles protonation of the molecule; slight changes in wavelength and absorption intensity result from this complexation in the majority of cases. Low molar absorptivity ($\epsilon = 1$ to 200) is exhibited by transition-metal ion excitation in a complex ($d \rightarrow d^*$ transitions).

Charge-transfer transitions in metal chelates often result in intense colors due to movement of an electron from the metal ion to the ligand or the ligand to the metal ion. Frequently, the ligand is oxidized and the metal is reduced, but a ligand with a high electron affinity that is complexed with a metal in a lower oxidation state may form a complex in which the metal is oxidized. The 1,10-phenanthroline complex of iron(II) is an example.

Procedure

Obtain spectra from 650 nm to 200 nm for all of the solutions listed under Chemicals. Fill the reference cell with the solvent used in the sample.

Treatment of Data

(Read the chapter introduction.)

Tabulate all of the wavelengths of absorption maxima and the molar absorptivities for each of the eight solutions. Include in the table the type of transition for each band.

Questions

1   On the basis of the series of nickel spectra, determine the relative ligand field strength of H_2O, Cl^-, and ethylenediamine.
2   What feature of charge-transfer spectra makes them important for analytical purposes?
3   In the $Fe(1,10\text{-phen})_3^{2+}$ system, does iron act as an electron acceptor or an electron donor?
4   In the triiodide complex, what is the electron acceptor and what is the electron donor?

8
Infrared Spectroscopy

The Infrared Region

Wavelengths that are longer than those for the visible region are referred to as the near infrared. The overtone region begins at about 12,500 cm^{-1} (0.8 μm). The fundamental region, which is the area generally used, extends from 4000 cm^{-1} (2.5 μm) to 650 cm^{-1} (15 μm). The fundamental region is further divided into the group frequency region [4000 cm^{-1} (2.5 μm) to 1300 cm^{-1} (8 μm)] and the fingerprint region [1300 cm^{-1} (8 μm) to 650 cm^{-1} (15 μm)]. In the group-frequency region the position of the absorption peaks is more or less dependent only on the functional group that absorbs and not on the complete molecular structure. The positions of the peaks in the fingerprint region, however, are dependent on the complete molecular structure and are thus more difficult to identify and correlate. The far infrared extends from the fundamental region [650 cm^{-1} (15 μm)] to 25 cm^{-1} (400 μm), and beyond this lies the microwave region.

Origin of Infrared Spectra

In a molecule, the atoms move together in straight-line translation and rotate and vibrate periodically about their centers of mass. These latter motions arise from two opposing forces, the repulsive forces between nuclei with their completed electron shells and the bonding forces due to valence electrons. To describe these complex motions in an orderly fashion, three coordinates (for example, the x, y, z cartesian coordinates) must be specified for each nucleus. Thus there are $3n$ coordinates in all when there are n atoms in the molecule. Three of these coordinates describe the translation of a molecule as a rigid unit. Another three coordinates describe the rotation of a nonlinear molecule (while only two coordinates are necessary to describe the rotation of a linear molecule). Therefore, the normal vibrations of a nonlinear and a linear molecule can be described by $3n - 6$ and $3n - 5$ coordinates, respectively.

Each normal vibration is associated with a characteristic frequency. Toluene, for example, with 15 atoms has 39 normal vibrations, and 39 fundamental frequencies can be found spectroscopically.

The correlation between the expected number and the observed number of absorption bands does not hold in every case. The number may be increased by bands which are not fundamentals, namely, combination tones, overtones, and difference tones. A combination tone is the sum of two or more frequencies, such as $\nu_1 + \nu_2$. An overtone is a multiple of a given frequency, as 2ν (first overtone), 3ν (second overtone), etc. A difference tone is the difference between two frequencies, such as $\nu_1 - \nu_2$.

Conversely, the number of observed bands may be diminished by the following effects:

1 The general requirement for infrared activity of a vibration is that the vibration must produce a periodic change in the dipole moment. If the molecule is highly symmetrical it is probable that selection rules will forbid the appearance of some of the frequencies in the infrared spectrum.
2 High symmetry also often results in certain pairs or triads of the fundamental frequencies being exactly identical. They are then said to be degenerate and are observed as only one band.
3 Some vibrations may happen to have frequencies so nearly alike that they are not separated by the spectrometer.
4 Some fundamental bands may be so weak that they are not observed or are overlooked.
5 Some of the fundamentals may occur at such low wave numbers that they fall outside the range of the usual infrared spectrophotometers.

Appearance of Spectral Bands

The energy of a vibrational transition [corresponding to the absorption of radiation of wavelengths 2 to 20 μm (500 to 5000 cm^{-1})] is on the order of 0.6 to 0.06 electron volt (eV) per molecule; in contrast, rotational transitions require energies of only 0.01 eV or less. A frequency of pure vibrational transition (where only a vibrational quantum number changes) differs, therefore, very little in energy from the transitions where both vibrational and rotational quantum numbers change. Hence, every pure vibrational transition is observed spectroscopically in the immediate neighborhood of a collection of less probable vibrational-rotational transitions and give the effect of widening each vibration line into a band. For this reason, we observe in the near infrared spectrum of a molecule a number of bands, the vibrational-rotational bands, rather than single sharp lines of pure vibrational transitions.

Types of Vibration

The fundamental vibrations of a molecule are

1 *Stretching or valency vibration.* The stretching vibration in which the distance between two atoms varies with time are of two types, symmetrical and unsymmetrical. In the symmetrical stretching vibration of CO_2, for example, the two oxygen atoms move away from the central carbon atom along the molecular axis and, after reaching maximum displacement, move back toward the carbon atom. In the unsymmetrical stretching vibration the carbon atom approaches one oxygen atom while receding from the other.
2 *Bending or deformation vibrations.* In the deformation vibration, the angle between two atoms varies with time.
3 *Vibrations involving a whole structural group.*

These can be further classified as

1   Wagging, where the structural unit swings back and forth in the plane of the molecule
2   Rocking, where the structural unit swings back and forth out of the plane of the molecule
3   Twisting, where the structural unit rotates about the bond which joins it to the rest of the molecule
4   Scissoring or bending, where, for example, the two hydrogens of a methylene group move toward each other

Group Frequencies

In many of the normal modes of vibrations the main participants in the vibration will be two atoms held together by a chemical bond. Such modes of vibration have frequencies which depend primarily on the weight of the two vibrating atoms and on the force constant of the bond joining them. The frequencies are only slightly affected by the other atoms attached to the two atoms concerned; hence these vibrational modes are characteristic of the groups in the molecule and are thus very useful in identifying a compound. These frequencies are known as *group frequencies*.

As a first approximation, the frequency of vibration is given by the equation for a harmonic oscillator

$$\nu = \frac{1}{2\pi}\sqrt{\frac{k}{\mu}} = \frac{c}{\lambda} \qquad \text{Hz} \tag{1}$$

When divided by c the relation becomes

$$\bar{\nu} = \frac{1}{\lambda} = \frac{1}{2\pi c}\sqrt{\frac{k}{\mu}} \qquad \text{cm}^{-1} \tag{2}$$

where $\bar{\nu}$ is the frequency in cm^{-1}, c is the velocity of light, μ is the reduced mass ($1/\mu = 1/m_1 + 1/m_2$ where m_1 and m_2 are the masses of the two oscillating objects), and k is the force constant (equivalent to the stress-strain ratio for a perfect spring in classical mechanics). The value of k is proportional to the strength of the bond when it is in its equilibrium position. In practice vibrations are not truly harmonic (equivalent to a deviation from Hooke's Law), so that additional terms predicted by quantum theory must be used to obtain a satisfactory agreement between theory and experiment. These higher-order terms are ignored here, but they significantly influence the observed spectra.

Upon substituting the proper universal constants into Eq. (2), Eq. (3) is obtained:

$$\bar{\nu} = 1307\sqrt{\frac{k}{\mu}} \tag{3}$$

With this equation, the absorption frequency of a given bond can be estimated. For the C—H bond in methane, k and μ have values close to 5×10^5 dynes/cm and 1 dyne/cm, respectively. Thus,

$$\bar{\nu}_{\text{C–H}} = 1307\sqrt{\tfrac{5}{1}} = 2900 \text{ cm}^{-1} \tag{4}$$

For the C—O in methyl alcohol, the force constant is still close to 5×10^5 dynes/cm, but μ jumps to about 6.85.

$$\nu_{\text{C}-\text{O}} = 1307 \sqrt{\frac{5}{6.85}} = 1110 \text{ cm}^{-1} \tag{5}$$

For C=O in acetone,

$$\nu_{\text{C}=\text{O}} = 1307 \sqrt{\frac{12}{6.8}} = 1730 \text{ cm}^{-1} \tag{6}$$

and for C≡N in HCN,

$$\nu_{\text{C}\equiv\text{N}} = 1307 \sqrt{\frac{15}{6.5}} = 2000 \text{ cm}^{-1} \tag{7}$$

In the infrared spectra of these four compounds there are strong absorption bands at 2915, 1034, 1744, and 2089 cm^{-1}, in reasonable agreement with the predicted values. Moreover, a series of molecules that contain one or *more* of these linkages have characteristic bands at approximately the same calculated frequencies.

However, the observed vibrational transitions are due to the vibrations of the whole molecule. While the approximations discussed above are useful and sufficiently good for many purposes, accurate calculations of the normal modes of vibration must consider the molecule as a whole.

Factors Influencing Group Frequencies

Resonance

The effect of resonance is to decrease the bond order of the double bonds and their vibrational frequencies by about 30 cm^{-1} (bathochromic shift) and simultaneously to increase the bond order and frequency of the interspersed single bonds (hypsochromic shift).

Hydrogen Bonding

The spectra of alcohols are characterized by an absorption band at 3300 cm^{-1} (3 μm), which corresponds to the O—H stretching vibration. The exact position of this band depends to a large extent upon the degree of hydrogen bonding or association to which the hydroxyl group is subject. Upon association, the energy and force constant of the O—H bond decreases, and the absorption band is therefore shifted to a lower frequency by about 200 cm^{-1}.

Effects of Ring Strain

For cyclic ketones, the carbonyl frequency increases as the size of the ring diminishes. In the series, cyclohexanone, cyclopentanone, and cyclobutanone, the C=O frequencies increase from 1710 to 1740 to 1775 cm^{-1}, respectively. The same trend is found in lactones and lactams.

Qualitative Analysis for Specific Chemical Substances

This type of analysis is based on the fact that the infrared spectrum is one of the most specific molecular properties known. The vibrational frequencies of a molecule depend on the number, weight, and geometrical arrangement of the atoms and the force constant of each interatomic bond. A change in any one of these factors will alter the infrared spectrum of a molecule. Thus if two compounds have the same infrared spectrum, they can be considered to be identical.

Some limitation on this specificity must be acknowledged, however. For example, detection of differences in the spectra of $CH_3(CH_2)_x CH_3$ if x were varied from 15 to 17 is unlikely. The added methylene groups do not add any bands that are not already present in large numbers, nor do they change the structural geometry of the molecule.

Qualitative Analysis for Groups of Atoms

In the vibrational spectra of thousands of molecules, some of the vibrational frequencies are essentially those of small groups of atoms within the molecule. These frequencies are characteristic of the groups of atoms regardless of the composition of the rest of the molecule. Carbonyl, hydroxyl, sulfhydryl, cyano, and nitro groups, for example, can be associated with characteristic absorption frequencies.

Infrared spectra, then can be used to give information on the functional groups in a molecule, as well as the molecular structure as a whole. As is demonstrated below, the functional groups of a molecule are not pinpointed at a definite frequency in the spectrum. At best they can be narrowed down to certain ranges of frequencies.

In the neighborhood of 3300 cm^{-1} the appearance of absorption bands is characteristic of the hydrogen stretching vibration. The location of absorption peaks in this region is fairly constant and not too dependent on molecular configuration as a whole. Thus absorptions here are characteristic of such hydrogen bonds as occur in molecules containing C—H, O—H, and N—H groups. The occurrence of hydrogen bonding in general causes a broadening of the absorption bands coupled with a shift towards lower frequencies.

The intermediate frequency region extends approximately from 2500 cm^{-1} to 1600 cm^{-1}. This region is often termed the unsaturated region and it is a transitional region. That is to say, the location of an absorption band that arises from a given functional group is influenced to a greater degree by the nature of the molecule in which it is found. Nevertheless, within limits, an absorption at a given wavelength can be assigned to a particular atomic configuration. The bands due to C≡C and C≡N appear in the high-frequency section of this region (2500 to 2000 cm^{-1}) and next come the carbonyl frequencies between 1900 and 1650 cm^{-1}. By judicious application of accumulated empirical data it is possible to distinguish among carbonyl bands arising from aldehydes and ketones, acids, esters, and amides. For example, if an acid is suspected the 3300-cm^{-1} region can be examined for the presence of —OH to substantiate the carbonyl of an acid radical. Carbon-carbon bonds of the benzene ring give characteristic absorptions in the region from 1650 cm^{-1} to 1450 cm^{-1}, particularly the two medium peaks at about 1600 cm^{-1}. These absorptions occur in the low-frequency range of the double-bond region because of the decreased bond order of the conjugated-benzene system. In the region from 1600 cm^{-1} to 1300 cm^{-1} the CH bending vibrations of methyl and methylene groups as well as the OH bending vibrations of alcohols and acids are present. The main usefulness of this region and the fingerprint region below 1300 cm^{-1}, where the rest of the single-bond stretching and bending vibrations occur, is in the identification of chain structure in aliphatics and ring substitution in aromatics. These vibrations are all more sensitive to overall molecular structure as you go from high to low frequency.

Only in exceptional cases will it be possible to assign correctly all observed bands in the spectrum to particular modes of vibration or to the vibration of particular chemical groups. In general the ratio of useful bands to observed bands will be moderately small, although this increases significantly with experience. Above 1600 cm^{-1} it is frequently possible to explain all bands of at least moderate intensity; in fact, if there is a moderately strong band in this region which cannot be explained, there is reason to suspect that some important feature of the sample has been overlooked. This is not true for the region below 1300 cm^{-1}. The location of many bands in this region depends significantly on the particular member of a homologous series which is being examined. For this reason this low-frequency region is called the fingerprint region.

Quantitative Analysis   The fundamental rule that correlates component concentration with the absorption intensity of the component is the well-known Beer's Law $P = P_0 e^{-\epsilon b C}$. Theoretically, the application of Beer's Law to the analysis of any mixture is quite straightforward, especially when a band unique to a single component can be found. Otherwise a procedure similar to that given in Experiment 6-7 is followed.

Direct application of Beer's Law is difficult in infrared work because of instrumental limitations. The source and detector have limited output. The instrument thus uses a large slit width and a wide spectral bandwidth in comparison to the sharp peaks found in most infrared spectra. For this reason the instrument cannot measure the actual height of the peak chosen for analysis but measures an average or integrated height across the peak in the range of wavelengths passed by the instrument. Another instrumental factor that produces deviation from Beer's Law, especially in the infrared, is the large amount of stray radiation.

A calibration curve must be prepared and checked frequently for direct quantitative work in the infrared. A base-line technique similar to that described in Experiment 6-6 is generally used. An internal standard method is also frequently employed. For a more detailed discussion, see the references given at the beginning of Experiment 8-1.

EXPERIMENT 8-1
Infrared Spectrophotometry

Purpose   This experiment demonstrates the various methods of sample preparation for analysis in the infrared region. It illustrates how to identify functional groups from infrared spectra and determine the principal structure of an unknown component from these groups.

References

1 L. Bellamy, "The Infrared Spectra of Complex Molecules," 3rd ed., Wiley, New York, 1975.
2 M. G. Mellon (ed.), "Analytical Absorption Spectroscopy," Wiley, New York, 1950.
3 H. M. Randall, R. G. Fowler, N. Fuson, and J. R. Dangl, "Infrared Determination of Organic Structures," Van Nostrand, Princeton, N.J., 1949.
4 R. L. Shriner, R. C. Fuson, and D. Y. Curtin, "The Systematic Identification of Organic Compounds," 4th ed., Wiley, New York, 1956.
5 A. Weissberger (ed.), "Technique of Organic Chemistry," vol. IX: "Chemical Applications of Spectroscopy," Interscience, New York, 1956, pp. 247–580.
6 R. M. Silverstein, G. C. Bassler, and T. C. Morrill, "Spectrometric Identification of Organic Compounds," 4th ed., Wiley, New York, 1981, chap. 3.
7 Sadtler Standard Spectra, Sadtler Research Laboratories, Inc., Philadelphia.
8 A. L. Smith in "Treatise on Analytical Chemistry," P. J. Elving, E. J. Meehan, and I. M. Kolthoff (eds.), 2nd ed., part 1, vol. 7, Interscience, New York, 1981, chap. 5.

Apparatus

Double-beam infrared spectrophotometer and accessories
Volumetric flask, 50 mL
KBr pellet equipment (press, die, grinder, etc.)
Cell-filling equipment (syringe, dry nitrogen, Scott wipers, etc.)

Chemicals

Benzoic acid
Polystyrene film (0.0015 in thick)

Group I:	Benzophenone	Acetone
	Methyl ethyl ketone	Cyclohexanone
	Benzaldehyde	Acetaldehyde
	iso-Butyraldehyde	Acetophenone

| Group 2: | n-Butyl alcohol | Benzyl alcohol |
| | sec-Butyl alcohol | tert-Butyl alcohol |

Group 3:	Benzonitrile	p-Nitrotoluene
	Nitrobenzene	2,4,6-Trinitrotoluene
	o-Nitrotoluene	Benzene

Group 4:	Ethyl benzoate	Ethyl acetate
	Ethyl malonate	Propionic anhydride
	Acetic anhydride	Phthalic anhydride

| Group 5: | Acetamide | Dimethyl formamide |
| | Acetanilide | |

Unknown: Compound from among the 355 organic compounds for which complete infrared spectra are reproduced in Ref. 3
Chloroform
Carbon tetrachloride
Mineral oil, "Nujol"
Potassium bromide (200 mesh)

Sample Preparation

Solid Samples

The infrared spectra of solid materials may be obtained by a variety of methods of which the four most common are listed with some of their possible advantages and disadvantages.

Mull Technique

The earliest and probably the most common of the solid-sample methods is to suspend the finely ground sample in a mineral oil such as Nujol. The main problem with all solid methods is light scattering. By grinding the sample into a fine powder and suspending it in a liquid of about the same refractive index, the amount of scattering is minimized.

Generally the sample is ground with the mulling agent in a mortar and pestle. A mixture of about 10:1 (oil to sample) is used, and the mixture is ground to a fine paste. A vortex mixer makes a convenient device for rapidly preparing a good mull. A few drops of the paste are placed on one plate of the demountable cell and covered with the other; the sample thickness is varied by rotating and squeezing the plates. The proper concentration of solid and cell thickness can be judged by the translucency of the sample. A sample film that appears brownish in visible light indicates the fine particle size necessary to reduce scattering and window scratching. To disassemble a mull, one should not attempt to pull the windows apart but simply slide or shear them apart. Some practice is necessary to prepare a mull which shows low scattering losses and sharp absorption bands of the proper intensity. Once this skill is obtained, the mull technique is the fastest and easiest of the solid methods.

The disadvantage of the method is the lack of control of cell thickness and thus the necessity of an internal standard if quantitative results are desired. The most common mulling agent is mineral oil or Nujol. These materials have the disadvantage of absorption frequencies at 2900, 1450, and 1375 cm^{-1} (typical of the C—H bond). Alternative mulling agents which avoid this difficulty are hexachlorobutadiene and perfluorocarbon oil.

Pellet Method

The mull-suspension method is frequently replaced by the newer method of suspending the solid in a potassium bromide pellet prepared under great pressure. In this method, a few milligrams of the organic compound are mixed with approximately 0.5 to 1 g of potassium bromide (a substance which is highly transparent in the infrared region). A portion of this mixture is subjected to pressure of several tons in a hydraulic press. The resulting pellets are visually transparent and may be used for qualitative and quantitative infrared spectrophotometry. The laboratory equipment for preparing the KBr pellets and the use of this equipment will be demonstrated by the instructor. This method is rapid and easy but does suffer from some disadvantages. The main ones are the uncertainty in the appearance of the spectrum caused by the process of producing the pellet and the difficulty in avoiding some moisture pickup, which gives rise to interfering O—H bond frequencies that are usually less desirable than the interfering C—H bond absorptions encountered in the mulling technique.

Film Techniques

Solid films produced by a variety of methods can be used directly. Some of the methods used have been to cut the sample into sheets of suitable thickness with a microtome or to melt the sample and allow it to dry as a film or evaporate the sample from solution. These methods are usually difficult and involve the possibility of interference fringes which are recorded as absorption bands.

Solution Methods

Solids are commonly run as solutions of about 5% in some suitable solvent. This method is identical to that used for liquids.

Liquid Samples

Liquid samples can be run either as the pure liquid if a cell of suitable thickness is available (approximately 0.02 mm) or as a solution in a longer cell (approximately 0.50 mm) if a suitable solvent can be found. Unfortunately there is no nonabsorbing solvent in the infrared. The best solvents for infrared use are nonpolar, nonhydrogen liquids, such as CS_2 or CCl_4. A more polar solvent such as $CHCl_3$ is useful except that the C—H stretch region may be distorted.

The cells used for liquids are quite fragile and must be handled carefully, particularly during the filling and emptying operations. Because of the thin spacing the cells can normally be filled by capillary action using a syringe to introduce the sample. A clean dry syringe can be used to empty the cells prior to rinsing them with solvent. After the rinsing operation, dry nitrogen should be passed through the cells until all of the solvent has been evaporated.

Preliminary Report

The following questions should be answered in the notebook prior to starting the laboratory work of this experiment.

1 Complete the following table:

	Colorimeter	Visible Spectro-photometer	Ultraviolet Spectro-photometer	Infrared Spectro-photometer
Source	Incandescent lamp or sunlight			
Monochro-mator	None or colored filters			
Sample holder	Glass or clear plastic			
Sample thickness	1–10 cm			
Detector	Eye or photo-cell			
Recorder	Brain or galva-nometer			
Wavelength range	400–800 nm			

2 Calculate what a wavelength of 5 μm in the infrared region would correspond to if it were expressed in terms of (a) wave numbers, (b) frequency, (c) angstrom units, and (d) nanometers.

3 Express 3333 wave numbers in terms of micrometers.

4 If an organic compound were an ester, in what wavelength region would infrared absorption be expected?

5 If an organic compound exhibited strong absorption at both 6.2 and 6.7 μm, what functional group or type of structure would be suspected?

Procedure

caution The optical parts and absorption cells employed in the infrared are made of NaCl and are etched by moisture. Following a few simple rules will help to preserve them:

Do not touch the cell window surfaces. If it is necessary to handle the windows as in preparing mulls, handle them by the edges only and preferably with gloves.

Do not breathe on the cell windows.

Do not remove the cells from the constant-temperature room.

Avoid hygroscopic liquids and solvents.

If it is necessary to use hygroscopic materials, do so only in the special cell provided.

Do not use any material which has any moisture content.

The following spectra are to be recorded:

	Sample	**Reference**
Chart 1:	Carbon tetrachloride	NaCl
	Group 1 compound	NaCl or solvent
Chart 2:	Benzoic acid in KBr (2 mg benzoic acid per 100 mg KBr)	KBr or air
	0.10 M benzoic acid in CCl_4 (6.11 g in 50 mL CCl_4)	CCl_4
Chart 3:	Group 2 compound	NaCl or solvent
	Group 3 compound	NaCl or solvent
Chart 4:	Group 4 compound	NaCl or solvent
	Group 5 compound	NaCl or solvent

After the above samples have been run, infrared spectra are obtained for (*a*) a strip of polystyrene 0.0015 in thick vs. air (for the purpose of obtaining accurate wavelength calibration data), and (*b*) an unknown sample vs. a rock salt window or solvent blank. The unknown may be run as a pure liquid, a solution, a solid in KBr, or a Nujol mull.

The spectrum of the unknown is immediately examined to see if the absorption is too great (much absorption in the vicinity of 0% transmission) or too little (most of the spectrum in the 100% transmission region). If the spectrum is unsatisfactory, steps should be taken to increase the absorption (increase the path length) or decrease the absorption (decrease the path length or dilute the sample with a suitable solvent).

The cells should be cleaned with chloroform and dried with a slow flow of nitrogen. Return the cells to the desiccator after drying.

Treatment of Data

From some standard source, such as Refs. 1 and 3, and Fig. 8-1 obtain frequencies of the functional groups and mark these on the appropriate charts of the known samples.

Make notations in your notebook regarding the differences in the spectra obtained for the known samples, especially for chart 2.

For the unknown compound tabulate all the wavelengths of significant absorption. Select the wavelengths of maximum absorption in the 2- to 7-μm region. With the aid of the infrared charts in Fig. 8-1 and the references, attempt to deduce the principal structural or functional groupings. Then, using the classification scheme of Randall et al. (Ref. 3) narrow the number of possible compounds down to a few whose spectra are completely given. Finally, tabulate point by point the absorption maxima of the unknown compound with the same data for the compound which is believed to be identical with the unknown. If this comparison agrees in sufficient detail, then the compounds are identical.

It should be noted that in the normal procedure for identification of organic compounds, infrared spectra are used as a *supplement* to other information—namely, physical measurements such as melting and boiling points and refractive indices; chemical information, such as elemental quantitative analysis for C, H, N, etc.; and the standard qualitative chemical-analysis scheme of Shriner, Fuson, and Curtin (Ref. 4).

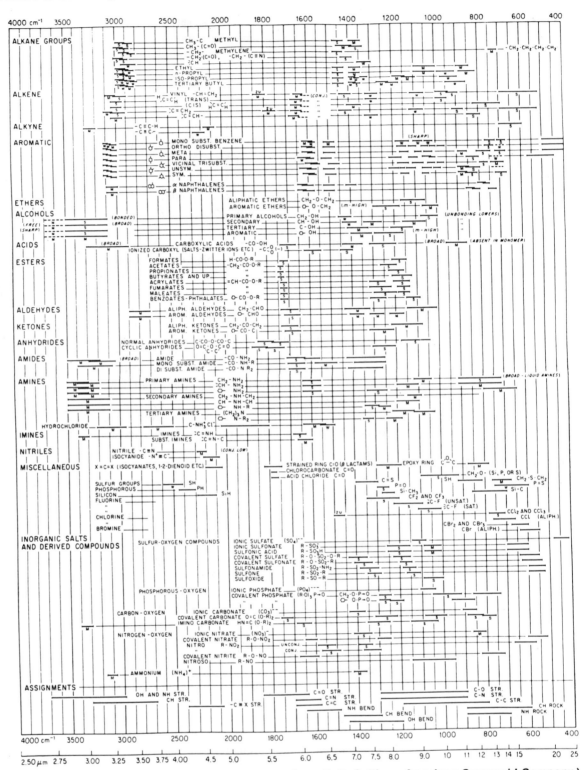

Figure 8-1a / Correlation chart. (*By permission from N. B. Colthup, American Cyanamid Company.*)

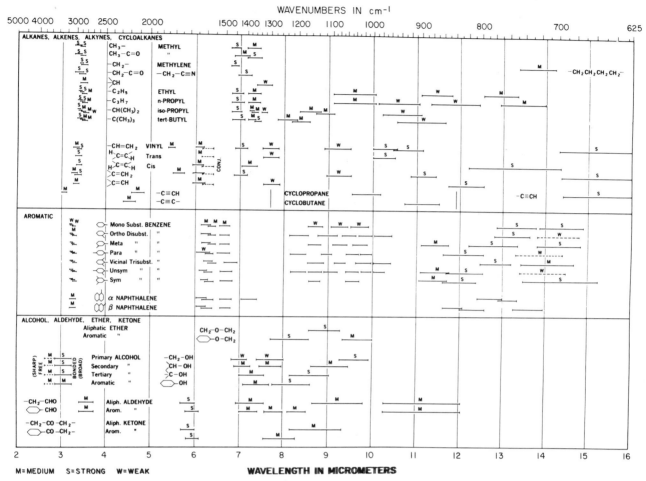

Figure 8-1b / Correlation chart. (From L. E. Kuentzel, Wyandotte Chemicals Corporation.)

Questions

1 Discuss various methods which are possible for the infrared study of compounds that are only soluble in water. Be specific concerning techniques, sample preparation, and limitations.

2 What are the specific advantages and disadvantages of a double-beam infrared spectrophotometer over a single-beam instrument?

3 Discuss the particular advantages of the following materials as prism materials for infrared spectrophotometers: NaCl, KBr, CaF$_2$, LiF, CaBr$_2$, and quartz. Compare specifically such qualities as useful wavelength range and resolution.

4 Why is it possible to separate the fundamental region of the near infrared into the characteristic and fingerprint region? Discuss this separation in terms of bond strengths, relative weight of the vibrating atoms, and interactions between group vibrations and the rest of the molecule.

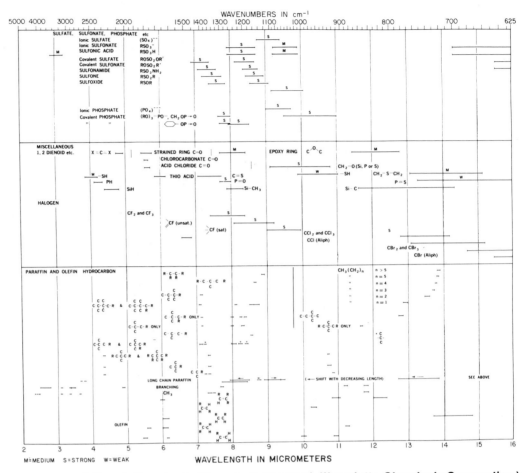

Figure 8-1c / Correlation chart. (From L. E. Kuentzel, Wyandotte Chemicals Corporation.)

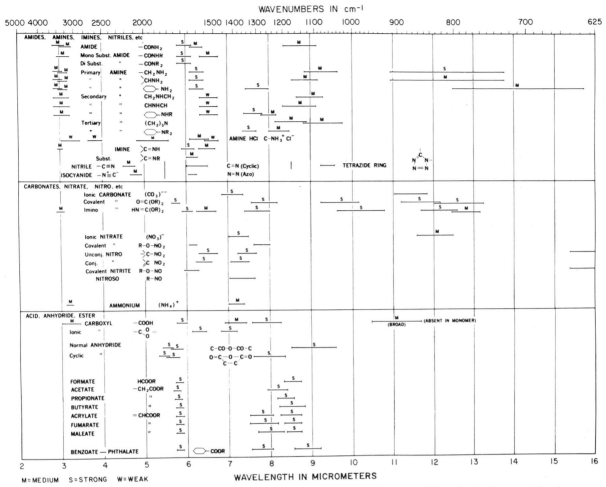

Figure 8-1d / Correlation chart. (From L. E. Kuentzel, Wyandotte Chemicals Corporation.)

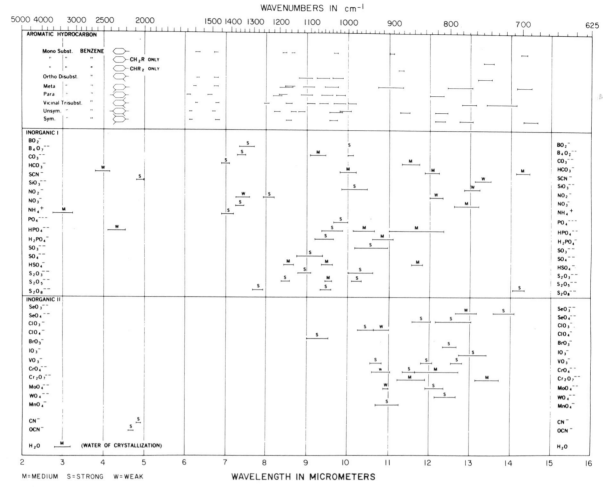

Figure 8-1e / Correlation chart. (*From L. E. Kuentzel, Wyandotte Chemicals Corporation.*)

EXPERIMENT 8-2

Spectra of Aldehydes and Ketones†

Purpose

Comparison of the carbonyl absorption frequencies for selected aldehydes and ketones will be made to illustrate the effect of substituents and conjugation. Assignments of the main bands for individual aldehydes and ketones also will be made.

References

1 R. L. Pecsok, K. Chapman, and W. H. Ponder (eds.), "Modern Chemical Technology," vol. IV, rev. ed., American Chemical Society, Washington, D.C., 1972.

†From a similar experiment in R. L. Pecsok, K. Chapman, and W. H. Ponder (eds.), "Modern Chemical Technology," vol. IV, rev. ed., American Chemical Society, Washington, D.C., 1972, p. 764.

2 R. M. Silverstein, G. C. Bassler, and T. C. Morrill, "Spectrometric Identification of Organic Compounds," 4th ed., Wiley, New York, 1981, chap. 3.

3 E. G. Brame, Jr., and J. G. Graselli (eds.), "Infrared and Raman Spectroscopy," Dekker, New York, 1977.

4 N. B. Colthup, L. H. Daly, and S. E. Wiberley, "Introduction to Infrared and Raman Spectroscopy," 2nd ed., Academic, New York, 1974.

5 A. L. Smith in "Treatise on Analytical Chemistry," P. J. Elving, E. J. Meehan, and I. M. Kolthoff (eds.), 2nd ed., part 1, vol. 7, Interscience, New York, 1981, chap. 5.

Apparatus

Infrared spectrophotometer and accessories
Pellet press

Chemicals

Pure potassium bromide for discs
Benzaldehyde
Cinnamaldehyde
n-Butyraldehyde
Benzophenone
Cyclohexanone
Acetophenone

Theory

Aldehydes and ketones exhibit a strong C=O stretching band in the region from 1870 to 1540 cm^{-1} (5.35 to 6.5 μm). Because the position is relatively constant and the band is of high intensity, it is easily recognized in infrared spectra. The actual position of the C=O band is affected by several factors: the physical state, neighboring substituents, conjugation, hydrogen bonding, and ring strain.

Aliphatic aldehydes absorb in the region from 1740 to 1720 cm^{-1} (5.75 to 5.82 μm). Electronegative substitution on the α-carbon increases the frequency of absorption of the C=O band. For example, acetaldehyde absorbs at 1730 cm^{-1}, whereas trichloroacetaldehyde absorbs at 1768 cm^{-1}. Conjugation of a double bond with the carbonyl group reduces the frequency of the carbonyl absorption. Aromatic aldehydes absorb at lower frequencies. Internal hydrogen bonding also shifts the absorption to lower frequencies.

The carbonyl group of ketones absorbs at slightly lower frequencies than that of a corresponding aldehyde. A saturated aliphatic ketone has a carbonyl absorption frequency of about 1715 cm^{-1} (5.83 μm); conjugation with a double bond causes an absorption shift to lower frequency. Intermolecular hydrogen bonding between a ketone and a solvent such as methyl alcohol also decreases the carbonyl frequency.

Procedure

Determine the infrared spectra of benzaldehyde, cinnamaldehyde, *n*-butyraldehyde, benzophenone, cyclohexanone, and acetophenone. For liquids, use thin films of about 0.015 to 0.025 mm thickness of the pure liquid; for solids, prepare KBr pellets.

Treatment of Data

Determine the carbonyl absorption frequency for each compound and write the structure of the compound on each spectrum.

On the spectrum of benzaldehyde, make assignments to the main bands found in the region around 3000 cm^{-1} and between 675 and 750 cm^{-1}. Indicate the bonds or groups of bonds in the molecule responsible for these bands.

On the spectrum of cyclohexane, make similar assignments in the regions around 2900 cm^{-1} and 1460^{-1}.

Compare the carbonyl frequencies of the aldehydes. Comment on the effect of conjugation and aromaticity on the frequency of absorption of the carbonyl group for cinnamaldehyde and benzaldehyde as compared with *n*-butyraldehyde.

Make a similar comparison of the carbonyl frequencies of the ketones with regard to conjugation and aromaticity.

Questions

1 Explain how a shift in the carbonyl frequency also can occur by replacement of an alkyl group with a chlorine atom.
2 At what frequency would you expect the overtone for the C=O stretch of acetophenone?

EXPERIMENT 8-3

Determination of Monomers and Polymers†

Purpose

In this experiment organic films will be prepared and subsequently identified by their fingerprint spectra. Some absorption bands are used to identify a functional group and others are specific for a particular compound.

References

1 R. L. Pecsok, K. Chapman, and W. H. Ponder (eds.), "Modern Chemical Technology," vol. IV, rev. ed., American Chemical Society, Washington, D.C., 1972.
2 Sadtler Research Laboratories, "Infrared Spectra of Commercial Products: Monomers and Polymers," Philadelphia.

Apparatus

Recording infrared spectrophotometer and accessories
Glass tube, 6 to 8 mm diameter by 20 to 30 cm long, bent to 90° to 120° angle
Test tubes (2)
Cork, one-hole (to fit above test tube)
Beaker, 400 mL
Bunsen burner
Ring stand
Cardboard or heavy paper to mount films

Chemicals

Old sodium chloride or potassium bromide crystals (large enough to fit the sample beam)
Spray cans of lacquers, resins, varnishes, and acrylic clear plastic
Plexiglas or Lucite (approximately 1 g)
Benzoyl peroxide, 10 mg
NaOH solution (to dissolve benzoyl peroxide for disposal)

Theory

Infrared spectra can be used as "fingerprints" to identify specific polymeric films. Certain polymers, when depolymerized to their monomers (whose spectra are different) can be repolymerized to a nearly identical material. For example, polymethylmethacrylate can be depolymerized to methylmethacrylate, its simplest repeating structural unit.

---

†From a similar experiment in R. L. Pecsok, K. Chapman, and W. H. Ponder (eds.), "Modern Chemical Technology," vol. IV, rev. ed., American Chemical Society, Washington, D.C., 1972, pp. 636–639.

The monomer can be converted back to the polymer with benzoyl peroxide. Infrared spectra of the original polymer, the repolymerized sample, and the film from an acrylic spray are almost identical.

Procedure

In the event that the films provided by the instructor are unmounted and without a holder, prepare a holder from a piece of cardboard or heavy paper. For a card that is approximately 2 in × 4 in, cut a window about $\frac{3}{4}$ in × $1\frac{1}{2}$ in starting $\frac{3}{4}$ in from one end, as indicated in Fig. 8-2. Cut a piece of film slightly larger than the window and tape it to the card. Obtain a spectrum for each film prepared, from 4000 to 650 cm^{-1}. Identification of each film should be made by comparison with the fingerprint spectra provided by the instructor. Some of the typical films are as shown in Table 8-1.

Prepare a film from one or more polymeric preparations: lacquer, varnish, resins, acrylic plastic protective coating, etc. (use spray cans). Spray a thin coat of any of these commercial products on one side of an old sodium chloride or potassium bromide crystal (one that will be large enough to fit into the sample beam of the spectrophotometer); allow to dry and record the spectrum in the same manner as for the previous films. The next part of the experiment should be started while the films are drying. These films can be cleaned off the crystal by use of methylene chloride as solvent. Acrylics (polymers of methylmethacrylate) such as Lucite and Plexiglas yield spectra similar to the one in Fig. 8-3. The formula for polymethylmethacrylate is given in Table 8-1. The R—C—O—CH$_3$ segments
‖
O

of the polymer give rise to large peaks at 1735 and 1150 cm^{-1} (see Fig. 8-3).

The polymer can be readily depolymerized to its simplest repeating structural unit (methylmethacrylate) by first placing 1 g of Plexiglas or Lucite in a test tube. Next, a cork stopper that contains a bent (90° angle) 8-mm-diameter glass tube (20 to 30 cm long) is inserted in the test tube. The test tube is clamped onto a ring stand, as shown in Fig. 8-4, and a second test tube is placed over the long end of the glass tube to act as receiver of the liquid methylmethacrylate (bp 100 to 101°C). The tube that contains the polymer should be heated gently with a flame until all the solid has been sublimed. To avoid condensation of the vapor, keep moving the flame along the test tube, but keep it away from the cork stop-

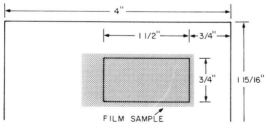

Figure 8-2 / Holder for polymer film samples.

Table 8-1 / Structures of Some Common Polymers

Name	Polymers	Structure
	Polyethylene	$\left[CH_2\text{—}CH_2\text{—}CH_2\text{—}CH_2\text{—}CH_2\text{—}CH_2\right]_n$
	Polypropylene	$\left[CH_2\text{—}CH(CH_3)\text{—}CH_2\text{—}CH(CH_3)\text{—}CH_2\text{—}CH(CH_3)\right]_n$
Teflon	Polytetrafluoroethylene	$\left[CF_2\text{—}CF_2\text{—}CF_2\text{—}CF_2\text{—}CF_2\text{—}CF_2\right]_n$
PVC	Polyvinyl chloride	$\left[CH_2\text{—}CH(Cl)\text{—}CH_2\text{—}CH(Cl)\text{—}CH_2\text{—}CH(Cl)\right]_n$
	Polystyrene	$\left[CH_2\text{—}CH(\phi)\text{—}CH_2\text{—}CH(\phi)\text{—}CH_2\text{—}CH(\phi)\right]_n$
Lucite Plexiglas	Polymethylmethacrylate	$\left[CH_2\text{—}C(CH_3)(CO\text{-}O\text{-}CH_3)\text{—}CH_2\text{—}C(CH_3)(CO\text{-}O\text{-}CH_3)\text{—}CH_2\text{—}C(CH_3)(CO\text{-}O\text{-}CH_3)\right]_n$
Nylon	Several polyamide types (6-6 nylon shown)	$\left[NH\text{—}(CH_2)_6\text{—}NH\text{—}CO\text{—}(CH_2)_4\text{—}C(O)\right]_n$
Dacron	Polyethyleneglycol-terephthalate	$\left[O\text{-}C(O)\text{—}C_6H_4\text{—}C(O)\text{-}O\text{-}CH_2\text{-}CH_2\right]_n$
Lexan	Polybisphenol A carbonate	$\left[O\text{—}C_6H_4\text{—}C(CH_3)_2\text{—}C_6H_4\text{—}O\text{-}C(O)\right]_n$

per and the open receiver tube. Continue the distillation until you collect about 1 mL of the liquid in the receiver.

Place 2 drops of the methylmethacrylate between two sodium chloride plates and record a spectrum.

To repolymerize the methylmethacrylate (i.e., convert the monomer back to the polymer), add about 10 mg of benzoyl peroxide to the receiver tube of methylmethacrylate.

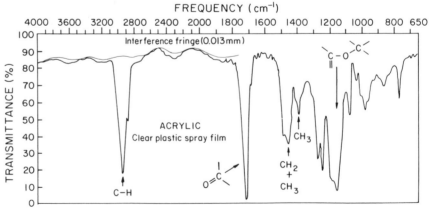

Figure 8-3 / IR spectrum of an acrylic.

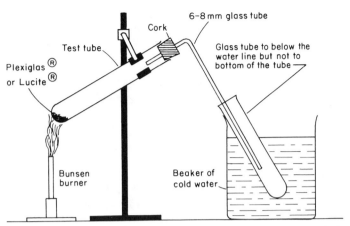

Figure 8-4 / Setup for depolymerizing a polymethyl-methacrylate.

safety caution

Treat benzoyl peroxide with care. It decomposes with heat and can decompose violently if heated above its melting point, 103.5°C. Benzoyl peroxide should be kept in cardboard containers, never in bottles or other rigid containers, and it should never be powdered by grinding. Avoid twisting the cardboard cover when capping or uncapping the container. Excess benzoyl peroxide can be safely dissolved in NaOH solution and washed down the drain. In case of a benzoyl peroxide fire, extinguish it with water.

Loosely stopper the receiver tube and heat in a boiling water bath. After 20 to 25 min the liquid will become viscous and after 5 to 10 min more, it will become a clear, slightly yellow (from the benzoyl peroxide) solid. While the heating is being done, prepare a film of the original sample by dissolving a small piece in 5 to 10 drops of methylene chloride. Spread a few drops of the solution on a salt plate and allow to dry for 5 min. Record its spectrum as well as one for a sample of the repolymerized solid (prepared in the same way as the original polymer).

Treatment of Data

Compare the film spectra which you ran with the fingerprint spectra provided by the instructor for identification. Identify the major peaks.

Compare the spectra of the starting polymer and the repolymerized solid with each other and with the spectrum of the acrylic spray. Indicate any differences.

Compare the spectra of the liquid monomer and polymer, indicating any new bands and any shifts in previous bands. Is the structure of the liquid different from the solid polymer? Explain.

Question

1 Would UV absorption spectroscopy be useful for identifying any of the polymers shown in Table 8-1?

9
Atomic Absorption and Atomic Emission Spectroscopy

Atomic spectroscopy is used for the qualitative and quantitative identification and determination of trace levels of metals in all types of materials and solutions. In atomic absorption spectroscopy measurement is made of the radiation absorbed by the nonexcited atoms in the vapor state. In emission spectroscopy, measurement is made of energy emitted when atoms in the excited state return to the ground state. Flame emission spectroscopy is a special area of emission spectroscopy in which a flame is used to excite the atoms. Figure 9-1 illustrates the absorption and emission processes.

Flame absorption and flame emission techniques usually involve introduction of a sample solution in aerosol form into a flame. Solvent evaporation and vaporization of the salt occur prior to dissociation of the salt into free gaseous atoms. At the temperature of an air-acetylene flame (approximately 2300°C) atoms of many elements exist largely in the ground state. When a beam of radiant energy that consists of the emission spectrum for the element that is to be determined is passed through this flame, some of the ground state atoms absorb energy of characteristic wavelengths (resonance lines) and are raised to a higher energy state.

For example, at 283.3 nm,

$$Pb^0 \xrightarrow{h\nu} Pb^*$$

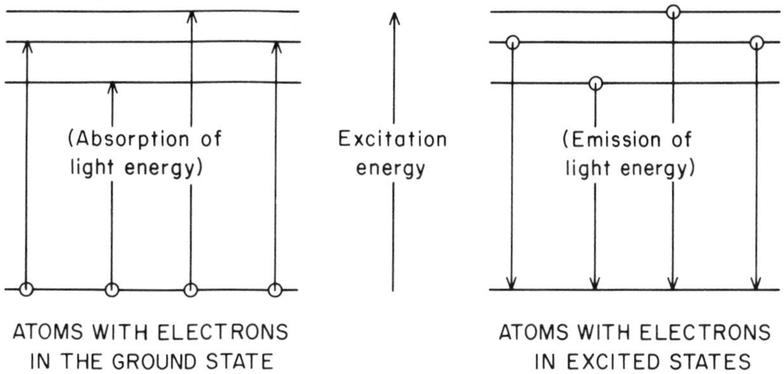

Figure 9-1 / Relationship between emission and atomic absorption spectroscopy. [*From "Modern Chemical Technology," rev. ed., R. L. Pecsok, K. Chapman, and W. H. Ponder (eds.), copyright © 1972 by American Chemical Society, Washington, D.C. Reprinted with permission.*]

The amount of radiant energy *absorbed* as a function of concentration of an element in the flame is the basis of *atomic absorption spectroscopy*.

For a few elements, such as the alkali metals Na and K, an air-acetylene flame is hot enough not only to produce ground state atoms, but to raise some of the atoms to an excited electronic state. The radiant energy *emitted* when the atoms return to the ground state is proportional to concentration and is the basis of *flame emission spectroscopy*.

For example, at 589 nm,

$$Na \xrightarrow[\text{flame}]{\text{Energy from}} Na^*$$

$$Na^* \rightarrow Na + h\upsilon$$

The emitted radiant energy from flame emission or the external lamp radiant energy not removed by atomic absorption is dispersed by a monochromator and detected by a photomultiplier.

At the much higher energies of the electric arc, spark, and plasma sources, significant fractions of the atoms for many elements are excited. However, this comes at the expense of an extremely complex emission spectrum with many lines. Complex and expensive instrumentation is necessary to adequately resolve and distinguish the many lines which result from the emission spectrum of a metal alloy. Because the electrons can only exist in discrete energy levels, which are different for each element, the wavelengths of the emitted light are characteristic of the element. Not only does this characteristic emission of light provide a means of qualitative analysis but quantitative analysis is also possible with appropriate calibration curves. The unique wavelengths of light for each element make this physical property an extremely selective and sensitive means of analysis. With adequate equipment, emission spectroscopy is one of the most general and versatile instrumental methods of analysis, and it is one of the most sensitive.

EXPERIMENT 9-1

Atomic Absorption Determinations†

Purpose

The factors which affect the free atom concentrations in atomic absorption spectroscopy and, thereby, the analytical results (flame temperature, aspiration rate, solvent, chemical form of the element, height of observation in the flame, type of burner and nebulizer, fuel-to-oxidant ratio of the flame, and the degree of ionization of the element in the flame) are examined. Copper is determined by the absorbance vs. concentration method and by the standard addition method.

References

1 J. F. Dean and T. Rains (eds.), "Flame Emission and Atomic Absorption Spectrometry," vols. I and II, Dekker, New York, 1969 and 1971.
2 D. A. Skoog and D. M. West, "Principles of Instrumental Analysis," 2nd ed., Saunders, Philadelphia, 1980, chap. 11.
3 H. H. Willard, L. L. Merritt, Jr., J. A. Dean, and F. A. Settle, Jr., "Instrumental Methods of Analysis," 6th ed., Van Nostrand, Princeton, N.J., 1981, chap. 5.
4 H. H. Bauer, G. D. Christian, and J. E. O'Reilly, "Instrumental Analysis," Allyn and Bacon, Boston, 1978, chap. 10.
5 G. W. Ewing, "Instrumental Methods of Chemical Analysis," 4th ed., McGraw-Hill, New York, 1975, chap. 7.
6 E. D. Olsen, "Modern Optical Methods of Analysis," McGraw-Hill, New York, 1975, chap. 5.
7 C. K. Mann, T. J. Vickers, and W. M. Gulick, "Instrumental Analysis," Harper & Row, New York, 1974, chap. 13.
8 R. L. Pecsok, K. Chapman, and W. H. Ponder (eds.), "Modern Chemical Technology," vol. V, rev. ed., American Chemical Society, Washington, D.C., 1972.
9 M. D. Amos and J. B. Willis, *Spectrochimica Acta, 22,* 1325 (1966).
10 J. W. Robinson in "Treatise on Analytical Chemistry," P. J. Elving, E. J. Meehan, and I. M. Kolthoff (eds.), 2nd ed., part 1, vol. 7, Interscience, New York, 1981, chap. 8.

Apparatus

Recording atomic absorption spectrophotometer and accessories
Hollow cathode lamps, Cu, Zn, Ca
Compressed air, nitrous oxide, and acetylene with regulator valves
Pipets, two 10 mL, 20 mL, 25 mL, 30 mL, and 50 mL
Volumetric flasks, four 100 mL and 1000 mL
Beakers (5), 100 mL
Separatory funnels (4), 250 mL

Chemicals

Cupric sulfate [$CuSO_4 \cdot 5H_2O$] stock solution, 0.4 g/L (100 μg/mL as Cu)
Magnesium chloride [$MgCl_2$], 5.0 μg/mL Mg
Stannous chloride [$SnCl_2$], 5.0 μg/mL Sn
Manganous chloride [$MnCl_2$], 5.0 μg/mL Mn
Ethanol or propanol (20%)
Methyl isobutylketone [MIBK]

---

†In part, from a similar experiment in R. L. Pecsok, K. Chapman, and W. H. Ponder (eds.), "Modern Chemical Technology," vol. V, rev. ed., American Chemical Society, Washington, D.C., 1972, pp. 947–959.

Ammonium pyrolidine dithiocarbamate [APDC] (1% w/v 20% ethanol)
Calcium, 1000 μg/mL stock solution
14.7 M H_3PO_4 (85% by wt.)
Potassium chloride [KCl]

Theory

To analyze a sample by atomic absorption, it is heated to a high temperature —most commonly in a flame. After evaporation of the solvent, the flame dissociates chemical bonds and releases free metal atoms which absorb light that is characteristic of the individual elements. The band of wavelengths at which each element absorbs is narrow and almost unique. The unexcited atoms absorb light, which raises the valence electron(s) to an excited state; as a result of this absorption the intensity of the original light is reduced. The amount of light absorbed is proportional to the concentration of the element present. Within the flame there are many more atoms in the ground state than in the excited state. For example, for calcium in a 2000 K flame, there are 1.2×10^7 atoms in the ground state for every atom in the excited state; for zinc there are 7.3×10^{15} ground state atoms for every excited atom. Some 65 elements can be determined with good precision and sensitivity by atomic absorption spectroscopy, in contrast to about 10 by the conventional flame photometric method.

The source of light for the atomic absorption spectrophotometer must produce a narrow band of adequate intensity and stability for prolonged periods of time. An ordinary monochromator is incapable of yielding a band of radiation as narrow as the peak width of an atomic absorption line (approximately 0.02 to 0.05 Å). In order for Beer's Law to be followed, a source of radiation which emits a line of the same wavelength as that used for an absorption analysis is essential. The radiation emitted must also be for excited state atomic lines of the element. To obtain these characteristics, hollow-cathode lamps are employed which have cathodes that are constructed from the element of interest. The lamps are filled with argon or neon at low pressure, which ionizes when voltage is impressed across electrodes. The inert gas bombards the cathode, heating it and causing it to sputter; a metal vapor fills the area. Collision of metal atoms with gas atoms causes them to become excited and to emit light. Figure 9-2 illustrates geometric features of a hollow-cathode tube.

Tubes that contain several metals are available, which permit the determination of these elements simultaneously without changing tubes.

A critical part of atomic absorption spectroscopy is the introduction of the sample into the flame and its vaporization and atomization to ground state atoms. Any method which results in a reproducible cloud of ground state atoms in the light path can be used in atomic absorption. In a typical flame atomizer all or part of a solution is sprayed as a fine mist and spread throughout the flame. Fuel and

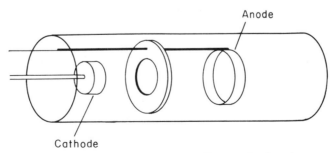

Figure 9-2 / Schematic of a hollow cathode tube.

oxidant are needed to produce a flame in which to burn a sample, the choice being dependent upon the type of burner used. There are two basic types of burner systems: (1) total consumption burner, in which the fuel and the oxidizing gas come through separate passages and meet at the base of the flame, and (2) the premix, or laminar flow, burner, in which the sample mist, fuel, and oxidant are mixed and then forced to the burner opening. The advantages and disadvantages of each type of burner are discussed in the references at the beginning of this experiment. Oxygen and acetylene are commonly used in total consumption burners; air-acetylene or nitrous oxide–acetylene are used in the premix burners.

An alternative source of free atoms is a furnace atomizer in which resistive heating causes vaporization and atomization of the molecules in the sample. Furnace atomizers, which have conversion efficiencies that are much higher than do flame atomizers, use a few microliters of sample in a horizontal graphite tube or a carbon rod. After a drying step, the sample is ashed to destroy organic material. Finally, it is atomized thermally by resistive heating of the tube or rod to give a transient cloud of atomic vapor above the atomizer. The area of the atomic absorption peak (or its height) is related to the quantity of metal atomized.

The optics of an atomic absorption spectrophotometer are similar to those in any type of spectrophotometer. The monochromator is adjusted to focus the wavelength of an atomic line for the element onto a photomultiplier tube in the *single-beam dc system*. A *double-beam ac system* has a chopper (rotating sector mirror) with a dual function: It divides the light into pulses which pass alternately through the flame and through a reference path. The sample and reference beams are recombined and the ratio of the intensity of the two types of pulses is measured electronically. See Figs. 9-3 and 9-4 for single- and double-beam systems, respectively.

Quantitative determinations in atomic absorption spectrophotometry rely, like any other spectrophotometric method, on the applicability of Beer's Law. Some instruments indicate $\%T$; others indicate A. Absorbance is directly proportional to concentration; however, percent transmittance vs. concentration is a logarithmic relationship. The relationships between A, $\%T$, and percent absorption (100 $\%T$) are shown in Fig. 9-5.

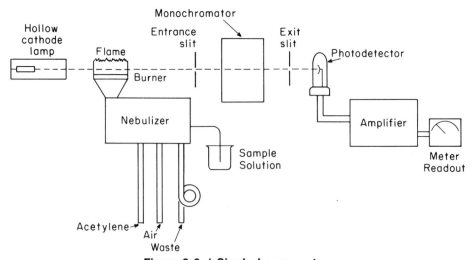

Figure 9-3 / Single-beam system.

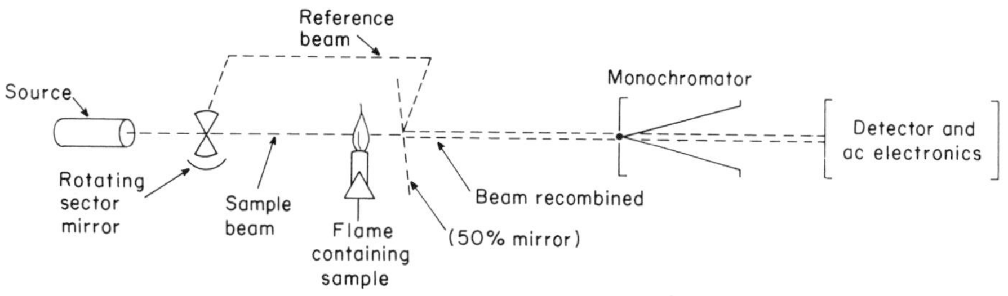

Figure 9-4 / Double-beam ac system.

Figure 9-6 provides a schematic representation of atomic absorption spectrophotometry.

Quantitative Performance of the Instrument

Sensitivity and detection limit are two terms used to describe the quantitative performance of an instrument. *Sensitivity* (for atomic absorption only) is defined as the concentration of the test element in aqueous solution which will produce a percent absorption of 1%, or an absorbance of 0.0044. It is expressed in terms of $\mu g/mL$ per 1% absorption or $\mu g/g$ per 1% absorption most frequently. The *detection limit* is the concentration which gives a signal twice the noise level of the background. Figures 9-7 and 9-8 illustrate sensitivity and detection limit, which are determined at the primary resonance lines of elements for optimum operating

Absorbance	% Transmittance	% Absorption
0	100	0
0.045	90	10
0.097	80	20
0.155	70	30
0.229	60	40
0.301	50	50
0.398	40	60
0.523	30	70
0.699	20	80
1.00	10	90
∞	0	100

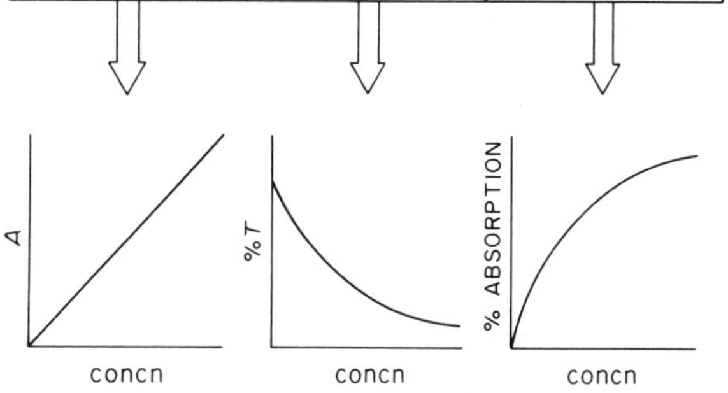

Figure 9-5 / Relationship between *A*, %*T*, % absorption, and concentration.

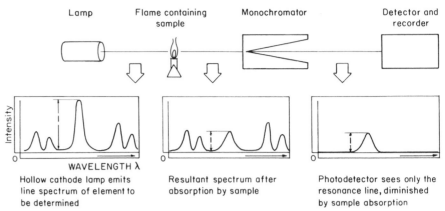

Figure 9-6 / Schematic of atomic absorption spectroscopy.

conditions. If a signal of twice the background is less than the amount required for a 1% absorption of an element at its primary resonance line, the presence of the element at the concentration for a given sensitivity can easily be measured. However, if the detection limit and the sensitivity are almost the same, a large percentage of the signal will be due to noise and the error in measuring that particular element at that concentration will be great.

Calibration of an Atomic Absorption Spectrophotometer

The strongest absorption line usually is used for analysis. For example, to determine Pb from lead poisoning by analysis of blood, the absorption line at 217 nm is used. A certain portion of the light that is emitted by a lead hollow-cathode tube will be absorbed; the amount depends upon the concentration of lead in the sample. Theoretically, a straight line should result when absorbance vs. concentration is plotted. However, the line often is curved. For example, the 217-nm line for lead often is broadened by self-absorption from an overdriven hollow-cathode lamp (made necessary by the optical losses at this wavelength), which results in a deviation from Beer's Law. Because of this problem, one manufacturer uses the less sensitive 283.3-nm line of Pb. Calibration curves must be prepared for each type of sample system. Deviations from Beer's Law-type plots are due to pressure broadening, source characteristics, and flame properties.

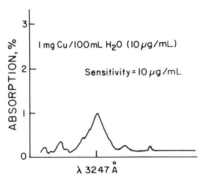

Figure 9-7 / The sensitivity of an atomic absorption spectrophotometer.

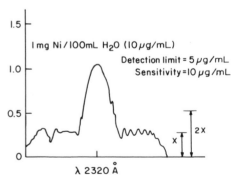

Figure 9-8 / The detection limit of an atomic absorption spectrophotometer.

Interferences result from:

1 Molecular emission by metal oxides and refractory compounds.
2 Light scattering or absorption by solid particles or unvaporized solvent droplets.
3 Ionization interference because many atoms ionize in hot flames. Because the measurement assumes that only un-ionized atoms are present, the formation of ions yields low absorbance values.
4 Refractory compound formation:
   a Anions in the sample may react with the test element to form a refractory compound and cause reduction in absorbance values.
   b The sample element may react with oxygen and hydroxyl ions in the flame, which results in a decrease in the number of atoms.

Background absorption must be corrected with the system that is a part of the instrument.

Factors Affecting the Quantity of the Material Reaching the Flame

Several factors affect the quantity of material that reaches the flame.

1 *The rate of aspiration of the material to the flame:* This usually is affected by gas pressure and, sometimes, by viscosity.
2 *Viscosity of the material to be analyzed:* A high concentration of dissolved solids will affect the rate of flow and cause it to move slowly through the aspirator.
3 *Choice of solvent:* Some organic solvents will increase the absorbance by a factor of two to four. Organic solvents also improve the efficiency of the aspirator because their surface tensions are lower than that of water.
4 *Surface tension of the solvent:* A lower surface tension allows smaller droplets to be formed. Smaller droplets permit more of the sample to reach the flame in the premix burner and more of the sample to be burned in the total consumption burner.

Temperature of the Flame

The percentage of atoms in the ground state is controlled by the temperature of the flame. Occasionally interfering elements prevent the formation of ground state atoms. Sometimes the element to be determined forms compounds of stable complexes which do not dissociate into atoms at the same temperature as the calibration standard. For example, phosphorus causes fewer calcium ions to be converted to free atoms than would be formed in its absence.

Calibration Standards

Ideally, the calibration standard is in a similar matrix to that of the sample solution. If samples of sera are to be analyzed for Ca^{2+}, then the standard solutions should be made to simulate serum.

Most elements produce a signal corresponding to 1% absorption at a concentration of less than 1 μg/mL. Therefore, a solution of about 1 μg/mL of any of these elements can readily be measured by atomic absorption. At concentrations above 50 μg/mL, the linearity between absorbance and concentration decreases. Thus, the calibration standards should be kept within an optimum range for each element.

In the event standards cannot be prepared identical to the unknowns, the technique of standard additions should be used. For example, in an analysis for copper in urine, a known amount of copper is added to the sample so that it contains 10 μg/mL of additional copper. To another portion of sample, copper is added to give an additional 20 μg/mL, etc. After reading the absorbances of these solu-

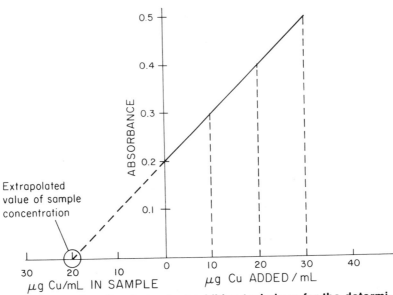

Figure 9-9 / Results of standard addition technique for the determination of Cu.

tions, a plot similar to Fig. 9-9 is made. The line is extrapolated to the value for the copper concentration in the original urine sample.

Procedures

A. Instrument Parameters

Prepare the instrument according to the instruction manual. Place the copper hollow-cathode tube into position. Set all voltage and current indicators to 0 and turn on the instrument. Set the proper current, slit width, and wavelength for a copper determination. Allow the tube to warm up for the specified time, align it and the burner, and make sure the latter is clean. Turn on the gases (*air first*, then acetylene). Never use acetylene when less than 15 psi remains in the tank because of the traces of acetone present. (Copper usually is measured at 324.7 nm with an air-acetylene flame.)

Burner Height

Adjust the gas pressures (air and acetylene) until the flame is a pale yellow color. Then turn up the acetylene pressure to give a strong yellow glow. This is fuel rich and is yellow due to unburned carbon particles in the flame. In a lean flame, there is excess oxygen present and the flame is blue. Aspirate a 5-μg/mL solution of copper and record its absorbance at 324.7 nm. Adjust the wavelength setting to obtain the maximum absorbance reading. The monochromator is now set exactly at the copper line. Raise the burner with the height adjustment knob so that the light beam just passes over the tip of the burner (base of flame). Zero the instrument with distilled water and measure the absorbance of the 5-μg/mL copper solution. Lower the burner in increments of about six steps, recording the absorbance at each height.

Plot absorbance vs. height of observation in the flame. Select the optimum height from the graph.

Fuel/Air Ratio

With the air pressure constant, adjust the acetylene pressure in increments from a fuel-rich (yellow) to a fuel-lean (blue) flame. Record the absorbance of the 5-μg/mL copper solution at each step and determine the acetylene pressure for maximum sensitivity for copper.

With the optimum acetylene pressure constant, adjust the air pressure in increments from a fuel-rich to a fuel-lean flame. Determine the optimum air and fuel settings by use of the 5-μg/mL copper solution. Indicate whether the flame is rich, stoichiometric, or lean.

Treatment of Data

Show the optimization of burner height and fuel/air ratio by means of the following graphs: (a) absorbance vs. height of observation in the flame, (b) absorbance vs. acetylene pressure, and (c) absorbance vs. air pressure.

B. Quantitative Analysis of Copper

Ordinary Calibration Method

From the stock solution of 0.40 g of $CuSO_4 \cdot 5H_2O$ per liter (which is 100 μg/mL in Cu), prepare solutions that contain 1, 2.5, 5, and 10 μg/mL Cu. This can be done by pipetting 10 mL of the stock solution (100 μg/mL) into a 100-mL volumetric flask and diluting to the mark; this solution will be 10 μg/mL Cu. From the solution that contains 10 μg/mL Cu, the remaining solutions can be made by pipetting 10-, 25-, and 50-mL aliquots into separate 100-mL volumetric flasks. Store all solutions in plastic bottles to avoid contamination from glass containers.

Put some water into a small, clean beaker and permit it to aspirate into the flame. Adjust the recorder base line to zero (if a recorder is being used). Transfer the four solutions of Cu to small, clean beakers. Aspirate the first solution into the flame until an absorption plateau is reached (Fig. 9-10). Aspirate the remaining solutions into the flame, each followed by water. A typical recording of the response from these solutions appears in Fig. 9-10b. Aspirate the unknown into the flame last.

Record the absorbance for each solution.

Calibration by Standard Addition Technique

To a 10.0-mL aliquot of the unknown (which is between 1 and 10 μg/mL), add 0.2 mL of the 100-μg/mL stock solution of Cu in a clean, dry beaker. This will increase the Cu concentration of the unknown by 2 μg/mL. Repeat, with 10.0-mL aliquots of the unknown and the addition of 0.4, 0.6, 0.8, and 1.0 mL of the Cu stock solution, respectively, to increase the unknown samples by 4, 6, 8, and 10 μg/mL. Zero the instrument with distilled water and aspirate the unknown solutions with the added copper successively, aspirating water between each sample.

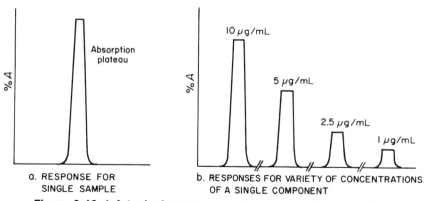

Figure 9-10 / A typical response pattern for atomic absorption.

Treatment of Data

Ordinary Calibration Method

Construct a calibration curve by plotting absorbance vs. concentration in micrograms per milliliter. (If necessary, convert % absorption to absorbance.) Determine the concentration of the unknown from the calibration curve. Is the plot linear? If it is not, explain causes of the nonlinearity.

Standard Addition Technique

Make volume corrections and plot the data in a manner similar to Fig. 9-9. Extrapolate the line to the value of the unknown sample concentration.

C. Sensitivity. Comparison of the Sensitivities of Different Elements

Compare sensitivities for the determination of Cu^{2+}, Mg^{2+}, Sn^{2+}, and Mn^{2+} using 5.0-μg/mL solutions for each of the ions.

Optimize the instrument according to the instruction manual. Use at least three of the solutions and, for each, set the most sensitive resonance wavelength and the appropriate slit width. Determine the absorbance readings.

Treatment of Data

Calculate the analytical sensitivity for each element, i.e., the concentration which will give an absorbance of 0.0044.

D. Effect of Solvents

Miscible Organic Solvents

Prepare solutions of 1, 2.5, and 5 μg/mL Cu^{2+} in the following three solvent systems by appropriate dilution of the 100-μg/mL Cu^{2+} stock solution: (1) water, (2) 20% ethanol or propanol in water, and (3) 50% ethanol or propanol in water.

Use a wavelength of 324.7 nm and a slit width that gives a 0.2-nm bandpass, and optimize the instrument with an air-acetylene flame and a 5-μg/mL Cu standard. Aspirate and read the aqueous solutions; use water as a reference. Aspirate the 20% alcohol solutions and again note the readings. Readjust the flame to remove the yellow color, and read the 20% alcohol solutions again. Aspirate the 50% alcohol solutions, read, readjust the flame, reread.

Solvent Extraction

This is a method to increase the concentration of the sample species.

Prepare aqueous solutions of 0, 0.05, 0.10, and 0.20 μg/mL Cu^{2+}.

Adjust the pH of each of the Cu solutions to between pH 5 and pH 7 by transferring 100 mL of each to a separatory funnel and adding dilute ammonia or hydrochloric acid. Mix 5 mL of 1% APDC with each Cu solution, add 10 mL MIBK to each, shake 2 min and allow the layers to separate. Transfer the lower, organic layer to a small beaker. Use a wavelength of 324.7 nm and a slit width of 0.5 mm. Aspirate each of the aqueous solutions and plot a calibration curve. Rinse the spray chamber with pure MIBK, fill the constant level liquid trap with MIBK, and aspirate each organic extract, making adjustments of air and acetylene flow for a maximum signal. Measure the absorbance of each copper solution.

Treatment of Data

Miscible Organic Solvents

Use one sheet of graph paper to plot the calibration curves for all the solutions. Which solvent system exhibits the best sensitivity for copper?

Solvent Extraction

Plot the calibration curve on the same graph with the aqueous solutions (Procedure B).

E. Detection Limits

Prepare aqueous solutions of 0.005, 0.015, and 5 μg/mL Cu^{2+}.

Using a 324.7-nm wavelength and a 1.0-nm slit width, adjust the instrument with an air-acetylene flame and the 5-μg/mL Cu solution. Allow the instrument to stabilize for 5 min. Aspirate water for 2 min to clean the chamber. Set the instrument for the longest possible integration time and the maximum scale expansion. Aspirate water, take a reading; aspirate the 0.005-μg/mL Cu solution and water alternately for at least 10 more readings of each until there are 11 readings for water and 10 for the Cu solution. Repeat the water and Cu aspirations as before, using the 0.015-μg/mL solution. Use a chart recorder and obtain a trace for each peak.

Treatment of Data

Average the absorbance readings for the first two blank readings (peaks 1 and 3). Subtract this value from the corresponding reading for the Cu solution (peak 2).

Repeat the above procedure for each reading from each Cu solution using the blank readings that bracket the copper reading; continue until 10 values are obtained for peak heights.

Calculate the standard deviation for the 10 values.

Calculate the detection limit by the expression:

$$\frac{2 \times \text{Concentration} \times \text{standard deviation}}{\text{Mean}} = \text{detection limit}$$

An estimation of the limit of detection can be made by assuming that it is twice the noise level. First, estimate the "peak-to-peak" height of the noise. Then measure the average peak height of the blank solution and the 0.015-μg/mL solution of Cu. Subtract the blank from the 0.015-μg/mL height to obtain the true height. If linear responses are assumed, at what concentration will the signal be twice the noise level?

F. Interferences

Ionization Interference in Calcium Determinations (see Ref. 9)

Ions of an element absorb light at a different wavelength than its neutral atoms. Addition of an excess of a more easily ionized element will produce a larger number of free electrons in the flame and shift the ionization equilibrium so that the element of interest will be insignificantly ionized. In this experiment the effect of potassium on the calcium signal at two concentrations will be observed.

Prepare (*a*) calcium solutions (as $CaCl_2$) of 8 μg/mL and 200 μg/mL, each containing 0, 1, 10, 100, or 1000 μg/mL of potassium (as KCl) and (*b*) calcium solutions of 8 μg/mL and 200 μg/mL without potassium.

Use a wavelength of 422.7 nm and a slit width of 0.2 mm; a calcium hollow-cathode tube is used to obtain the resonance wavelength. The air-acetylene flame will suffice, but a nitrous oxide–acetylene flame provides a more noticeable effect.

For the 8-μg/mL solutions, adjust the aspiration to a fairly fast rate. Zero the instrument with distilled water. Record the absorbance for each sample. With the 200-μg/mL samples, adjust the aspiration to a much slower rate. Aspirate each sample and record the absorbances.

Phosphate Interferences

Atomic absorption signals for alkaline earth elements are depressed by the presence of anions such as phosphate which form relatively nonvolatile compounds in the flame (calcium pyrophosphate). This type of chemical interference can be minimized by solvents which cause smaller particles to form inside the nebulizer. The small particles are more easily vaporized in the flame. Hotter flames also produce more vaporization. In this experiment, the phosphate interference with calcium in 50% ethanol will be studied.

Prepare solutions of 200 μg/mL calcium (as $CaCl_2$) in 50 mL of 95% ethanol plus 0, 0.1, 0.3, 1.0, 3.0, or 10.0 mL of concentrated phosphoric acid (85% acid, 14.7 M) with dilution to 100 mL with water. Flames of nitrous oxide–acetylene (2990°K) and air-acetylene (approximately 2500°K) are used.

Use a wavelength of 422.7 nm and a slit width that gives a 0.2-mm bandpass. Aspirate the solutions and use an appropriate blank that contains phosphoric acid and ethanol to check zero. Use the nitrous oxide–acetylene flame first and then the air-acetylene flame to measure the absorbance by assuming a linear response.

Treatment of Data

Ionization Interference in Calcium Determination

Convert the % absorption to absorbance. Plot graphs (on the same sheet of paper) of absorbance vs. concentration of potassium added. From the graphs, determine the minimum amount of KCl required to overcome ionization effects.

Phosphate Interference

Plot the apparent calcium concentration vs. the logarithm of the mole ratio (Ca/PO_4) for both sets of data (air-acetylene and nitrous oxide–acetylene) on the same sheet of graph paper.

Questions

1 A premix burner does not introduce all the material into the flame, and the larger droplets are drained to waste. How is air prevented from backing up into the burner and possibly causing an explosion?

2 What other anions are likely to cause low responses for alkaline earth samples?

3 In atomic absorption spectroscopy why is the monochromator located after the sample compartment (the flame) rather than before as in the case of a UV-visible absorption spectrophotometer?

4 What is beam modulation and why is it used in atomic absorption?

5 Why are atomic absorption lines so sharp compared to the absorption spectrum of a molecule dissolved in solution?

6 How does the detection limit of atomic absorption compare with flame emission? Give some specific examples.

EXPERIMENT 9-2

Determination of Mercury in Air, Water, and Fish by Flameless Atomic Absorption Spectrophotometry†

Purpose

Due to its volatility, mercury atomizes inefficiently in conventional flames. This experiment illustrates how mercury can be determined by a flameless atomic absorption technique.

References

1 J. Y. Hwang, *American Laboratory*, International Scientific Communications, Inc., Fairfield, Conn., December 1970, p. 50.

†In part, from a similar experiment in G. D. Christian, "Analytical Chemistry," 3rd ed., Wiley, New York, 1980.

2 V. T. Lieu, A. Cannon, and W. E. Huddleston, *J. Chem. Ed., 51,* 752 (1974).

3 J. Y. Hwang, P. A. Ullucci, and S. B. Smith, Jr., *American Laboratory,* International Scientific Communications, Inc., Fairfield, Conn., August 1971.

4 P. J. Thistlethwaite and M. Trease, *J. Chem. Ed., 51,* 687 (1974).

5 A. L. Malenfant, S. B. Smith, Jr., and J. Y. Hwang, "Mercury Pollution Monitoring by Atomic Absorption Utilizing a Gas Cell Technique," presented at the 12th Annual Eastern Analytical Symposium, Nov. 19, 1970, published by Instrumentation Laboratory, Inc., Lexington, Mass.

6 J. Y. Hwang and G. P. Thomas, *American Laboratory,* International Scientific Communications, Inc., Fairfield, Conn., August 1974.

Apparatus

Atomic absorption spectrophotometer and accessories
Cell for determining mercury (see Fig. 9-11) with fritted glass scrubber
Flowmeter
Aspirator or air pump
Mercury hollow-cathode tube
Pipets, 0.5 mL, 1 mL, 2 mL, and four 5 mL
Graduated cylinder, 100 mL

Chemicals

Argon
Stannous chloride [$SnCl_2$] (10% w/v), 10 g $SnCl_2$, 1 mL 9 M sulfuric acid [H_2SO_4] plus water to dissolve; dilute to 100 mL (Prepare the day of use.)
Potassium permanganate [$KMnO_4$] (10%), 1 g in 10 mL water
1.5 M hydroxylamine, 10 g of hydroxylamine hydrochloride [$NH_2OH \cdot HCl$], dissolve in 100 mL water
Stock mercury solution, 0.1354 g of mercuric chloride [$HgCl_2$] in 100 mL 0.5 M H_2SO_4; dilute to 1 L (This is 100 μg/mL in Hg.)
Mercury working standard, dilute the stock solution 1:100 on day of use (This is now 1 μg/mL in Hg.)
6 M nitric acid [HNO_3]
9 M H_2SO_4
Water samples
Air samples (if other than from laboratory atmosphere)

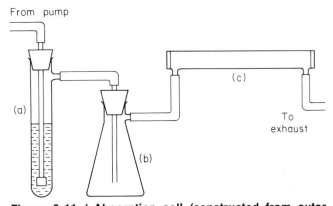

Figure 9-11 / Absorption cell (constructed from outer tube of 250-mm Liebig-type demountable condenser and quartz windows). [*Reprinted with permission from V. T. Lieu, A. Cannon, and W. E. Huddleston, J. Chem. Ed., 51, 752 (1974). Copyright © 1974, Division of Chemical Education, American Chemical Society.*]

Theory

Although flame emission methods exist for the determination of mercury, their detection limits are not sufficient for trace levels. A technique has been devised whereby mercury vapor is drawn into a cell through which light from a mercury lamp passes. For atmospheric samples, the mercury first is converted to mercuric ion in a solution trap by oxidation with potassium permanganate. The mercury salts are then reduced to elemental mercury by stannous chloride. The elemental mercury (ground state Hg) is volatile, and when argon is bubbled through, the mercury vapor is circulated through the cell. The absorbance is measured when it reaches a maximum value at 253.6 nm. Standards are run in a similar manner. With this technique, as little as 1 ng of mercury per milliliter can be detected (i.e., less than 1 ng/mL). Mercury is marked as a polluter of the environment, and as a source of contamination of fish, animals, and humans. In the natural environment mercury is found in soil, air, and water. The natural upper limit in soil in the northeast has been estimated to be 0.04 μg/mL $\pm$ 0.02 μg/mL. In air, the naturally occurring concentrations amount to a few nanograms per cubic meter. Variations between 2 ng/m^3 and 20 ng/m^3 have been observed. In water, in the northeast, the natural level is approximately 0.06 ng/mL. The concentration limit of mercury in fish has been set at 0.5 μg/mL by both the U.S. and Canadian governments. Fishing bans in some areas have been imposed after detecting concentrations as high as 7 μg/mL.

Procedures

A. Collection of Air Samples

To a fritted glass scrubber, add 5 mL of 6 M HNO$_3$, 5 mL of 9 M H$_2$SO$_4$, 2 drops of 10% KMnO$_4$, and 90 mL of deionized water. If the solution becomes colorless, add another drop of permanganate solution. Connect the inlet of the gas scrubber to a flowmeter and the outlet to an aspirator or air pump. Adjust the flow rate to between 10 and 15 L/h (or 0.2 L/min). Record the initial time of sampling, and continue the sampling for at least 3 h, occasionally checking the flow rate. Calculate the volume of air aspirated from the flow rate and the sampling time.

B. Determination of Mercury

Preparation of the Atomic Absorption Cell

Figure 9-11 shows a representation of a nonflame atomic absorption attachment. The argon flow rate should be constant and just sufficient to give smooth aspiration; a fritted glass bubbler can be used as the sample vessel. The absorption cell (c) should be wrapped with Nichrome wire and asbestos to warm it. The Nichrome wire is then attached to a Variac and warmed to about 70°C. The purpose of heating the cell is to keep moisture from condensing in the cell. An alternative approach is to replace the aerosol trap (b) with a granular CaCl$_2$ trap.

Preparation of a Calibration Curve

Prepare a calibration curve just before the samples are to be run. Add 90 mL of deionized water, 5 mL of 6 M HNO$_3$, and 2 drops of 10% KMnO$_4$ to a reaction flask. Add successively, in separate measurements, 0.5-, 1.0-, 2.0-, and 5.0-mL portions of the 1-μg/mL mercury solution, swirling 15 s each time. A magnetic stirrer may be used to expedite the oxidation-reduction reaction instead of swirling the flask. Because the permanganate oxidizes and dissolves inorganic and organic mercury compounds, the solution should remain colored; if it becomes colorless, add another drop of KMnO$_4$. Add 5 mL 9 M H$_2$SO$_4$ and 5 mL 1.5 M hydroxylamine hydrochloride, mix, and wait 30 s for the solution to clear. If the permanganate color remains, add another portion of hydroxylamine hydrochloride. Add 5 mL 10% SnCl$_2$ and immediately stopper the flask. Wait 10 s, then turn on the argon tank to aspirate the mercury into the cell. Record the absorbance on a strip chart recorder and note the maximum absorbance on the instrument meter. Run at least two blanks in the same manner, omitting the addition of mercury. The cell should be flushed with argon between runs until the absorbance drops back to near the original base line. Correct for the blank absorbance and plot a calibration curve of net maximum absorbance vs. micrograms of mercury.

Analysis of an Air Sample

Transfer the contents of the sampling scrubber to the sample flask. Add 5 mL 1.5 M hydroxylamine hydrochloride, mix, and wait 30 s for the solution to clear. Add 5 mL 10% $SnCl_2$, stopper, wait 10 s, and then aspirate with argon. Record the absorbance as above. Correct the reagent blank reading that has been obtained above. From the calibration curve determine the micrograms of mercury in the analyzed sample. From the volume of sample collected, calculate the concentration of mercury in the laboratory atmosphere in micrograms per liter.

Analysis of Water Samples

Tap water, river water, or other water sources also may be analyzed for mercury by this method. Likewise, unknowns may be prepared by adding mercuric chloride to deionized water. Tap water should contain in the neighborhood of 1 ng/mL or less mercury and so a 90-mL sample will contain 0.1 μg or less (near the detection limit). Analyze triplicate 90-mL water samples in the same manner that the standards were run. Correct for the reagent blank absorbance and from the calibration graph determine the quantity and the concentration of mercury in the sample.

Analysis of Mercury in Fish Samples

Treat 1.0 g of fish with 10 mL H_2SO_4 and digest at 50 to 60°C to wet-ash the sample. Add 3 mL hydrogen peroxide to completely decolorize the digestion mixture. Remove excess peroxide by the addition of 5% potassium permanganate until the solution retains a purple color, and dilute the sample to 25 mL. An appropriate aliquot is taken, depending upon the concentration of mercury in the fish. (If the concentration is lower than 0.5 ng/g, another calibration curve can be run for a lower range of concentrations.) Add 5 mL 1.5 M hydroxylamine hydrochloride, mix, and wait 30 s for the solution to clear. Add 5 mL 10% $SnCl_2$, stopper, wait 10 s, and then aspirate with argon. Record the maximum absorbance. An alternate method for determination of mercury in fish will be found in Ref. 6.

Treatment of Data

Plot a calibration curve of net maximum absorbance vs. micrograms of mercury. From the calibration curve, determine the concentration of mercury in the laboratory atmosphere in micrograms per liter.

From the calibration curve, determine the quantity and concentration of mercury in the water sample. Report the mean concentration from the three trials in nanograms per milliliter.

Again, using the calibration curve, determine the quantity of mercury in the fish sample in nanograms per gram.

Questions

1 Write equations for the reactions that occur during the oxidation and the reduction processes.

2 Would this technique work for the determination of any other metals?

EXPERIMENT 9-3

Determination of Calcium, Iron, and Copper in Food by Atomic Absorption†

Purpose

This experiment is designed to illustrate how trace amounts of several nutritionally important elements can be determined by atomic absorption.

References

1 R. L. Pecsok, K. Chapman, and W. H. Ponder (eds.), "Modern Chemical Technology," vol. V, Rev. ed., American Chemical Society, Washington, D.C., 1972.
2 J. W. Robinson in "Treatise on Analytical Chemistry," P. J. Elving, E. J. Meehan, and I. M. Kolthoff (eds.), 2nd ed., part 1, vol. 7, Interscience, New York, 1981, chap. 8.

Apparatus

Atomic absorption spectrophotometer and accessories
Hollow-cathode tubes, Cu, Fe, Ca
Volumetric flasks, thirteen 100 mL, two 50 mL
Erlenmeyer flask, 250 mL
Graduated cylinder, 100 mL
Pipets, two 1 mL, two 3 mL, three 5 mL, four 10 mL, one 20 mL
Funnel, 65 mm
Filter paper, Whatman No. 1 to fit funnel

Chemicals

8 M HCl, 1500 mL
Lanthanum oxide [La_2O_3], add 50 mL deionized water to 58.64 g La_2O_3, then add 250 mL HCl to dissolve the La_2O_3, and dilute to 1 L with deionized water
Copper stock solution (1000 μg/mL of Cu), dissolve 1 g Cu metal in 50 mL 1:1 HNO_3–H_2O, and dilute to 1 L with 1% (v/v) HNO_3
Calcium stock solution (500 μg/mL of Ca), 1.249 g primary standard calcium carbonate [$CaCO_3$] in 50 mL H_2O, add 10 mL HCl dropwise until the $CaCO_3$ is dissolved, and dilute to 1 L
Iron stock solution (1000 μg/mL Fe), dissolve 1.000 g Fe wire in 50 mL 1:1 HNO_3–H_2O and dilute to 1 L with deionized water
Food sample: cereal, milk, fruit, vegetables

Theory

Atomic absorption is a popular technique for the determination of metals in many types of samples. It is commonly used for the analysis of food. The food is first digested in acid to release the metals into a soluble form for determination.

Procedures

A. Preparation of the Sample to Be Analyzed

Place a 2 g sample in a 250-mL Erlenmeyer flask; add 25 mL 8 M HCl. Boil slowly on a hot plate for 5 min. Cool and add 10 mL deionized water. Filter through a Whatman No. 1 filter paper. Transfer the filtrate to a 50-mL volumetric flask, dilute to the mark with deionized water, and use this solution directly for determination of copper and iron. For a calcium determination, pipet 10 mL of this solution to another 50-mL volumetric flask, add 20 mL 8 M HCl and 10 mL of the lanthanum solution, and dilute to the mark. The lanthanum prevents the calcium from forming complexes with phosphates; such complexes hinder atomization in the flame.

†From a similar experiment in Ref. 1, pp. 962-965.

B. Determination of Calcium

Calibration Curve

Use the instruction manual for the instrument to set the parameters for the determination of calcium. Prepare a Ca^{2+} standard by pipetting 10 mL of the stock standard (500 μg/mL) into a 100-mL volumetric flask and diluting to the mark. This diluted solution contains 50 μg/mL Ca, and is used to make the standards for the calibration curve. Prepare these standards by pipetting 2-, 5-, 10-, and 20-mL aliquots of the 50-μg/mL solution into successive 100-mL volumetric flasks. Add 50 mL 8 M HCl and 20 mL of the lanthanum solution to each flask and fill to the mark with deionized water. These diluted solutions must be prepared fresh each day. Aspirate the diluted standards into the flame and record the absorbance.

Analysis of the Food Sample

Aspirate the sample solution that was prepared in Procedure A. If the absorbance is within the range of the standards of the calibration curve, record the value. If the absorbance is higher than that of the most concentrated standard, dilute the sample ten times and repeat the analysis. To make a tenfold dilution, pipet 10 mL into a 100-mL volumetric flask, add 45 mL 8 M HCl, and 18 mL of the lanthanum solution, and dilute with deionized water to the mark.

C. Determination of Iron

Calibration Curve

Use the instruction manual for the instrument to set the parameters for the determination of iron. Pipet 10 mL of the stock iron solution (1000 μg/mL Fe) into a 100-mL volumetric flask and dilute with deionized water to the mark. Pipet 1, 3, 5, and 10 mL of this diluted standard (100 μg/mL) into successive 100-mL volumetric flasks. Add 50 mL 8 M HCl to each and dilute with deionized water. Aspirate each of the calibration standards (1, 3, 5, and 10 μg/mL Fe) and record the absorbance.

Analysis of Food Sample

Aspirate the sample prepared in Procedure A into the AA spectrophotometer. If the absorbance is higher than that of the highest standard, make the appropriate dilution. For example, try a tenfold dilution by pipetting 10 mL of the sample in a 100-mL volumetric flask and adding 45 mL 8 M HCl; dilute to the mark.

D. Determination of Copper

Calibration Curve

After setting the instrument parameters for copper, make appropriate dilutions to give calibration solutions of 1, 3, 5, and 10 μg/mL Cu by use of the same procedure as for iron. Aspirate the diluted standards and record the absorbance.

Analysis of Food Sample

Aspirate the sample as prepared in Procedure A in the spectrophotometer. Adjust the concentration as for iron in the event the absorbance is too high for the standards.

Treatment of Data

Plot calibration curves for the calcium, iron, and copper solutions (absorbance vs. concentration).

Determine the amount of each of these elements in your sample. Compare your values, as % by weight, with those on the food package.

Questions
1 How does lanthanum aid in the determination of calcium?
2 Why are the diluted calibration solutions stable for only 1 or 2 days?

EXPERIMENT 9-4

Quantitative Determination of Alkali and Alkaline Earth Metals by Flame Photometry

---

Purpose

This experiment demonstrates the analytical uses of flame photometry and compares the direct-intensity method with the internal-standard method of analysis.

References

1 H. H. Willard, L. L. Merritt, Jr., J. A. Dean, and F. A. Settle, Jr., "Instrumental Methods of Analysis," 6th ed., Van Nostrand, Princeton, N.J., 1981, chap. 5.
2 J. F. Dean and T. Rains (eds.), "Flame Emission and Atomic Absorption Spectrometry," vols. I and II, Dekker, New York, 1969, 1971.
3 D. A. Skoog and D. M. West, "Principles of Instrumental Analysis," 2nd ed., Saunders, Philadelphia, 1980, chap. 11.
4 A. Syty in "Treatise on Analytical Chemistry," P. J. Elving, E. J. Meehan, and I. M. Kolthoff (eds.), 2nd ed., part 1, vol. 7, Interscience, New York, 1981, chap. 7.

Apparatus

Flame photometer
Fuel and oxygen
Pipets, 1 mL, 2 mL, 3 mL, 4 mL, 10 mL, and 25 mL
Volumetric flasks (5), 100 mL (one containing the unknown solution; dilute to mark)

Chemicals

Sodium chloride [NaCl], 2000 $\mu g/mL$ as Na
Lithium chloride [LiCl], 1000 $\mu g/mL$ as Li
Sodium chloride, 5, 10, 25, 50, 75, and 100 $\mu g/mL$ as Na
Sodium chloride, 0, 5, 10, 25, 50, 75, and 100 $\mu g/mL$ as Na—with each solution also containing 100 $\mu g/mL$ lithium ion
Unknown: Solution of sodium chloride in a 100-mL volumetric flask
Sodium chloride, 25 $\mu g/mL$ as Na in
   1 Distilled water
   2 Ethanol (10%)
   3 Ethanol (50%)
   4 Glycerine (50%)
Calcium ion, 200 $\mu g/mL$
   1 Using calcium chloride [$CaCl_2$]
   2 Using calcium nitrate [$Ca(NO_3)_2$]
   3 Using $CaCl_2$ in 0.1 M phosphoric acid [H_3PO_4]
   4 Using $CaCl_2$ in 0.1 M aluminum chloride [$AlCl_3$]

Theory

Flame photometry is one method in the general field of emission spectroscopy. The main distinction between this and the other emission techniques is the source of energy for excitation. The more general emission techniques use the arc and spark sources which produce much higher energy than the flame. The instrument used also is different in that the complicated spectra from the high-energy sources require a high resolution monochromator, while the simple spectra of the flame do not.

The flame is especially useful because of its simplicity for routine analyses which are difficult by other means. Thus the alkali metals with their low excitation energies are easily determined with this method, while they are difficult to do by other means.

Flame photometry has many disadvantages. Perhaps the most difficult of these is the control of the many variables. The intensity from a flame is dependent on the flame temperature, the rate of flow of liquid into the flame, the pressure and rate of flow of fuel gases, and any of many other variables which affect the character of the flame or atomizing of the sample. Thus the compounds in which the ion is found and the viscosity of the solution have a great effect. The experimental results are therefore only empirical.

To circumvent these difficulties two methods are used. An internal standard may be employed. In this technique a constant amount is added of a metal which is not already present in the sample and which has excitation characteristics similar to those of the metal being determined. The ratio of the intensities of the standard and sample are compared, and the logarithm of the intensity ratio is plotted against the logarithm of the concentration. Another technique is the method of standard addition. Here a known amount of the element being determined is added, and the increase in intensity is observed. After correcting for dilution, the observed increase in concentration is divided into the actual increase, and this correction factor is multiplied by the originally observed concentration of the sample.

The spectrum obtained with the flame is due to the following order of events. The solution of the metal salt in question is sprayed into the flame and the solvent evaporates leaving the finely powdered salt. The salt is vaporized, atomized, and a valence electron is raised to a higher-energy state. The energy emitted when this electron drops down into a vacant lower level is given off as radiant energy of a wavelength determined by the Planck-Einstein relationship

$$\Delta E = h\nu = \frac{hc}{\lambda}$$

where
h = Planck's constant
c = velocity of light
ν = frequency of light emitted
λ = wavelength

The change in energy corresponding to light of wavelength 500 nm is

$$\Delta E = \frac{6.6 \times 10^{-27} \text{ erg·s} \times 3 \times 10^{10} \text{ cm/s}}{500 \times 10^{-7} \text{ cm}} = 3.96 \times 10^{-12} \text{ ergs}$$

In the case of lithium there are two $1s$ electrons and one $2s$ valence electron. When a lithium atom is excited in a flame, the one $2s$ electron may absorb enough energy to be raised to the $2p$ orbital. This energy is absorbed from the flame as a change in the angular momentum of this electron. When the electron releases this energy and returns to the normal or ground $2s$ state, it emits the red light of 670.8 nm normally associated with the principal line of the lithium spectrum. Such a transition from an excited state to a normal state is called a ground state transition. In contrast the 610-nm line is not a ground state transition since the valence electron in this case falls from the 3*d* electronic state to the 2*p*, which is itself an excited state. Flame spectra are mainly composed of ground state transitions, while with higher-energy sources like the arc, the transitions are mainly between excited states.

The scope of elements determinable by flame photometry and also the possibility of interference from a given element can be deduced from Fig. 9-12.

Flame photometry is still largely an empirical method, a point which cannot be emphasized too strongly. A consequence of this situation is that reliable results can be obtained only after painstaking attention to details, with repeated checks of reproducibility and the effects of altering conditions. The experiments to be done here will illustrate the importance of controlling some of these conditions, and the analysis to be done will be under ideal conditions. However, the technique has a great value and can be used, with appropriate precautions and control tests, for many otherwise difficult or tedious analyses. Further exploration of its potentialities and limitations, particularly in regard to the effects of varying

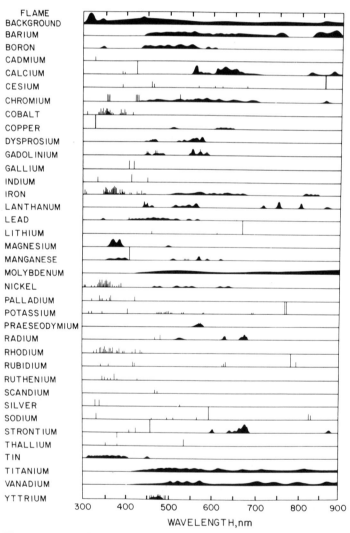

Figure 9-12 / Spectra for elements determinable by flame analysis. (*By permission from Beckman Instruments, Inc.*)

the composition of the solution, might well be made the subject of a further extended experiment. Studies of flame characteristics and flame reactions, while outside the scope of a course in instrumental analysis, are particularly pertinent and are currently under progress in many laboratories, especially in connection with the kinetics of fast reactions.

For direct-intensity measurements the following variables must be kept constant if reproducible results are to be obtained: oxygen pressure, fuel pressure, sample composition (volatile matter, ionic content, and viscosity), and the amount of sample introduced into the flame. Commercial flame photometers are designed to provide careful control of these variables, but extreme care is necessary to obtain accurate analyses by the direct-intensity method.

The internal-standard method provides an accurate way of analyzing samples in those cases where there is considerable inaccuracy in the direct-intensity method. A constant concentration of some element (the internal standard) is added to the standard samples and the unknown. The ratio of the intensity of the analysis line to the internal-standard line is measured. Variations in the constancy of the energy of the flame and of sample delivery to the flame, etc., are compensated and nullified by this technique. The restriction is that the internal standard and sample must have approximately the same volatility and excitation energy.

Other Applications

The flame photometer is especially useful in the determination of alkali and similar metals because of the scarcity of other good quantitative methods for these substances. Suggested applications are in the study of the coprecipitation of sodium ion with colloidal ferric hydroxide (other precipitates or ions might also be used), in the study of equilibrium constants involving ion-exchange resins, in the study of the quantitative aspects of ion-exchange column separations, in the determination of lithium in the presence of magnesium or other alkaline earth or alkali elements, in the determination of Ca(II) and Mg(II) in cement, and in the analysis of a high-detergency motor oil for barium. If you wish to try these or other experiments, consult with the instructor. Basic studies of the fundamentals of the technique (effect of foreign ions, etc.) might be particularly valuable and interesting.

Procedure

note Potassium ion can be used equally well in place of sodium ion if it is so desired.

The atomizer system should be carefully cleaned and checked for contamination. This can often best be accomplished by atomizing distilled water through the system with the flame operating.

Next the instrument should be turned on and allowed to warm up for 15 to 30 min with the flame on to allow approximate establishment of a thermal steady state. The instructor will supply additional operating instructions and precautions for the particular instrument used.

Determine a working curve for sodium ion using the direct-intensity method. Use the standard sodium ion solutions (0 to 100 μg/mL) and set the 0 %T with distilled water and the 100 %T with the 100-μg/mL solution. Without changing the slit settings determine the %T for the other concentrations. Before making any measurements make certain that the proper wavelength is set on the instrument for sodium ion (589 nm). (If potassium is to be studied use a wavelength of 767 nm.)

Pipet 25 mL of the unknown solution into a 100-mL volumetric flask, dilute to the mark, and mix thoroughly. Measure this diluted sample on the flame photometer at the same time the direct-intensity working curve is prepared using the same instrument settings.

Determine the data for a working curve of sodium ion using the internal-standard method. Use the standard sodium ion solutions (0 to 100 μg/mL) which contain 100 μg/mL lithium ion; set the 0 %T with the 0-μg/mL sodium ion solution and the 100 %T with the 100-μg/mL sodium ion solution. Determine the relative %T for sodium and lithium, using the slit settings found for the 100-μg/mL solution. Use a wavelength of 671 nm for lithium ion measurements.

Pipet 25 mL of the unknown solution into a 100-mL volumetric flask; next pipet 10 mL of the lithium stock solution (1000 μg/mL) into the 100-mL volumetric flask; dilute to the mark and mix thoroughly. Measure this diluted sample on the flame photometer at the same time the internal-standard working curve is prepared using the same instrument settings.

The importance of scrupulous cleanliness can be emphasized by (a) dipping two fingers into 20 mL distilled water and measuring the resulting sodium ion concentration on the flame photometer (use the direct-intensity method), (b) measuring the sodium ion content of tap water, and (c) measuring the sodium content of various reagents which are not supposed to contain sodium ion. A correction for the latter effect would normally be taken care of by carrying out a "blank" determination.

Determine the effect of the composition and viscosity of a solution by running, under identical conditions, four solutions containing equal concentrations of sodium (e.g., 25 μg/mL) but containing respectively 0, 10, and 50% ethanol, and 50% glycerine.

In order to illustrate the effect of the ionic composition of the solution (much more marked with some ions than with others), compare the readings obtained for Ca with the following four solutions (under constant conditions): each solution contains 200 μg/mL of Ca(II) (a) using $CaCl_2$, (B) using $Ca(NO_3)_2$, (c) using $CaCl_2$ in 0.1 M phosphoric acid, and (d) using $CaCl_2$ in 0.1 M $AlCl_3$. Use a wavelength of 423 nm for calcium.

Treatment of Data

Plot intensity vs. concentration of sodium ion from the direct-intensity data and note on the graph the reading for the diluted unknown sodium ion sample.

Plot the ratio of sodium ion intensity to lithium ion intensity ($I_{Na}:I_{Li}$) vs. sodium ion concentration from the internal-standard data and note on the graph the reading for the diluted unknown sodium ion sample by the internal-standard method.

Report the sodium ion concentration of the original unknown solution in micrograms per milliliter as determined by (a) the direct-intensity method and (b) the internal-standard method.

Plot the intensity reading vs. the viscosity for the 25-μg/mL sodium ion solutions containing various amounts of ethanol and glycerine. (The viscosities for the respective solutions can be found in the "International Critical Tables" and chemistry handbooks.)

Prepare a table of intensity readings for the 200-μg/mL calcium solutions which tabulates the effect of the ionic composition on the readings. What general conclusions can be made concerning ionic effects?

Questions

1 What advantages, if any, does flame photometry have, other than economic, over conventional emission spectroscopy?

2 Suggest a possible procedure for the determination of barium in a solution containing about 10 μg/mL barium in the presence of 1000 μg/mL sodium ion.

3 Describe the types of interferences that are encountered in flame emission spectrometry.

4 Explain why a typical standard curve for flame emission deviates from linearity at very low and very high concentrations of the metal ion.

SUPPLEMENTARY EXPERIMENT

Determination of Sodium and Potassium in Cement

Apparatus

Flame photometer, oxygen-fuel regulator
Volumetric flasks, three 1 L (for stock solutions), twelve 50 mL
Pipets (2 of each), 5 mL, 10 mL, 15 mL, 20 mL, and 25 mL

Chemicals

Stock Na solution (0.1 mg Na per milliliter), dissolve 0.254 g sodium chloride [NaCl] per liter H_2O

Stock K solution (0.1 mg K per milliliter), dissolve 0.190 g potassium chloride [KCl] per liter H_2O

Stock Ca solutions (5.0 mg Ca per milliliter), dissolve 12.4 g calcium carbonate [$CaCO_3$] in 400 mL 1:3 HCl and dilute to 1 L

Unknown cement samples: Pulverize the cement, dry and weigh a sample, and dissolve in 12 M hydrochloric acid [HCl]. Make appropriate dilutions to give readings within the calibration curves.

Procedure

Pipet 25-mL aliquots of the stock Ca solution (5.0 mg/mL) into each of six 50-mL volumetric flasks. Then pipet 5-, 10-, 15-, 20-, and 25-mL aliquots of stock Na solution into the first, second, third, fourth, and fifth flasks, respectively; dilute to the mark with distilled water. Fill the sixth flask with distilled water to use for a blank. Pipet aliquots of the stock Ca solution into the second set of six flasks, and make up solutions of potassium in the same manner as sodium. Keep all twelve solutions throughout the experiment. Adjust the flame photometer according to the instruction manual; use oxygen-hydrogen or oxygen-acetylene for the flame for Na. (Use a wavelength of 589 nm.) Set the instrument to read 100% for the most concentrated Na solution; determine the intensity of the other Na solutions as compared to this. For K, use a wavelength of 767 nm (this may require a change to a red-sensitive phototube if the instrument requires it). Set the instrument to 100% with the most concentrated K solution. Read the intensity of the K solutions as compared to this. Run an unknown cement sample.

Treatment of Data

Tabulate the data. Prepare calibration curves from the standards, and from these, determine the concentration of sodium and potassium in the unknown cement sample.

Questions

1 Explain the shape of the curves.
2 Why isn't the y-axis intercept at zero?

10

Fluorescence Spectroscopy

Fluorescence is the emission of light by a molecule which has absorbed radiant energy; the radiation is emitted at a longer wavelength than the incident absorbed energy. Figure 10-1 illustrates absorption transitions from the ground state to various vibrationally excited states of the upper electronic level on the left.

S_0 represents the singlet ground state (electrons are paired, of opposite spin, and there is no splitting of energy levels). S_1 refers to the lowest singlet excited state (electrons are of opposite spin, but one is excited to a higher energy level). Occasionally, excitation of an electron to a higher energy level can result in a triplet state T_1 (electrons are of the same spin, but one is excited to a higher energy level).

Electrons in the excited state (with a lifetime of about 10^{-8} s) will drop to the lowest vibrational energy level due to collisions via vibrational deactivation or "relaxation." Electrons can now return to the ground state by internal conversion, emitting heat through vibrational relaxation. They can also return to various vibrational levels of S_0 at a longer wavelength by fluorescence. The longer wavelength is a consequence of loss of energy during the vibrational deactivation. In certain cases, electrons cross over to the triplet state T_1 via intersystem crossing. When they return to the various vibrational levels of S_0 the light emitted is known as phosphorescence. This latter process has a lifetime of the order of 10^{-4} s or longer. Fluorescence can be quenched by the presence of certain ions or molecules; for example, bromide ions suppress the emission of quinine sulfate.

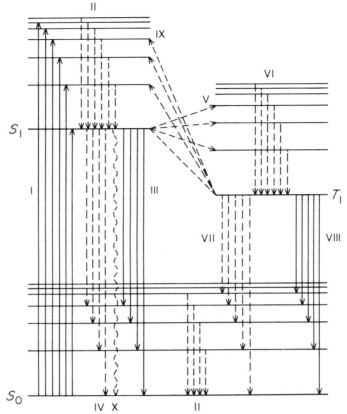

Figure 10-1 / Energy-level diagram of a typical organic molecule which includes only ground singlet, first excited singlet, and its corresponding triplet states. Solid lines indicate radiational transitions, and dashed lines indicate nonradiational transitions. Process I: absorption. Process II: vibrational deactivation. Process III: fluorescence. Process IV: quenching of excited singlet state. Process V: intersystem crossing to triplet state. Process VI: vibrational deactivation in the triplet system. Process VII: quenching of triplet state. Process VIII: phosphorescence. Process IX: intersystem crossing to excited singlet state. Process X: internal conversion. (*From "Fluorescence Assay in Biology and Medicine," vol. II, S. Udenfriend, copyright © 1969 by Academic Press, New York, p. 45. Reprinted with permission.*)

Structure and Fluorescence

Saturated molecules and molecules with only one double bond do not exhibit significant fluorescence. Molecules with at least one aromatic ring or multiple conjugated double bonds are prone to have fluorescence spectra in the visible or near ultraviolet regions. Substituents such as $-OH$, $-OCH_3$, and $-NH_2$ which are electron-donating groups enhance fluorescence. Compounds which are nonfluorescent often can be converted into fluorescent derivatives. Phosphorescence often is associated with solid materials (including those solidified by freezing).

Relation between Fluorescence Intensity and Concentration

In fluorometry, the amount of fluorescent light detected is related to the concentration of the fluorescing species:

$$F = P_0(1 - 10^{\epsilon bc})Q_f k$$

$$F \cong P_0(2.3 \, \epsilon bc)Q_f k \quad \text{(for dilute solutions, transmitting more than 80\% of } P_0\text{)}$$

where F = intensity of fluorescence
ϵ = molar absorptivity
b = path length
c = concentration
P_0 = power of incident radiation
Q_f = quantum efficiency = photons emitted/photons absorbed
k = photons measured/photons emitted

A plot of F vs. concentration is used as a calibration curve for quantitative analyses, and ideally should be linear if $c < 0.01\ M$. Extremely dilute samples can be analyzed because of the high sensitivity of the method. However, extraneous conditions often cause background emissions or suppression of the fluorescence for the sample species.

Instrumentation

In order to separate the emitted radiation from the incident radiation, fluorescence measurements are made at right angles to the incident beam. This is possible because fluorescence is emitted in all directions, but the incident radiation passes straight through the cell.

Figure 10-2 illustrates the features of a filter fluorometer. The excitation light source is a mercury arc lamp (or xenon arc—usually found in spectrofluorometers). The radiation passes through a range selector to control the amount of light. It then goes through a primary filter which allows a certain range of wavelengths (in the UV) to strike the sample. The sample cell may be made of glass, optical grade quartz, or synthetic silica. The fluorescent light generated by the

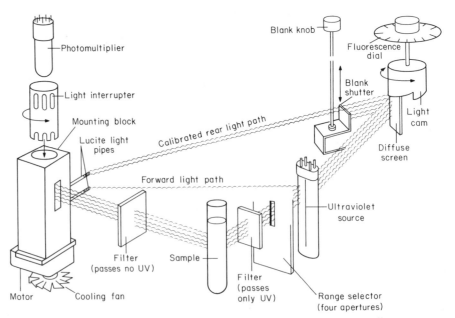

Figure 10-2 / Schematic diagram for the Turner Model 110 fluorometer optical system. (*Courtesy of Amsco Instrument Co.*)

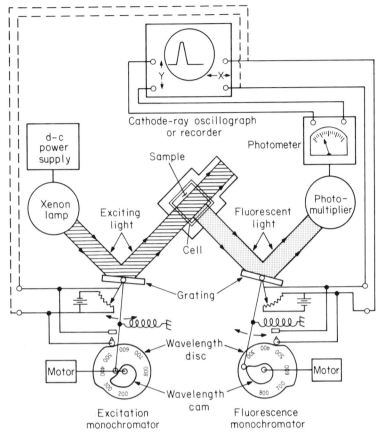

Figure 10-3 / Schematic diagram for the Aminco-Bowman spectrofluorometer. (*Courtesy of American Instrument Co.*)

the sample passes (at a right angle) through a secondary filter which is selected to pass the maximum fluorescent emission and reject any scattered light. The fluorescent light then strikes a photomultiplier tube which transforms the light signal into electric energy. In turn, this energy is amplified and displayed on a readout meter. A single detector with a light interrupter is used to measure the light differential between the fluorescent emission and a standard calibrated (rear light path) beam. Rotation of the fluorescence dial adjusts the calibration beam to be equal to the sample intensity by means of the diffuse screen. When balanced, the photomultiplier detects no difference in signal intensity. Hence, the instrument is operated as an optical-null system.

In contrast, a fluorescence spectrofluorometer has a pair of monochromators in place of the filters (Fig. 10-3). The primary light source usually is a xenon arc which yields continuous emission from 200 to 800 nm. Excitation and emission spectra can both be obtained from a spectrofluorometer. This is a decided advantage over a filter instrument. With a spectrofluorometer spectra are recorded automatically. Figure 10-4 illustrates an excitation (absorption) spectrum and the resultant emission (fluorescence) spectrum for anthracene.

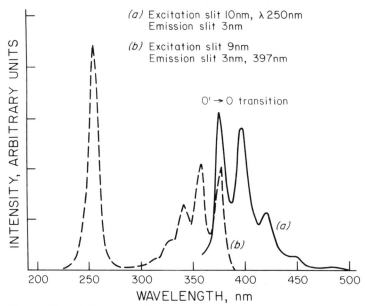

(a) Excitation slit 10nm, λ 250nm
   Emission slit 3nm

(b) Excitation slit 9nm
   Emission slit 3nm, 397nm

O' → O transition

Figure 10-4 / Spectra of anthracene. (a) Fluorescence emission; (b) excitation. Note that the mirror-image relationship does not extend to the absorption maximum at 250 nm. (From "Research Development," Smith, copyright © 1968 by Technical Publishing Co. Reprinted with permission.)

EXPERIMENT 10-1

Characterization of Quinine and Its Determination†

Purpose

Quinine is a strongly fluorescent compound in dilute acid solution with two excitation wavelengths (250 and 350 nm) and a fluorescence emission at 450 nm. The factors which affect the quantitative determination of quinine (such as concentration quenching and chemical quenching) will be studied as well as the interpretation of the emission spectra.

References

1 A. J. Pesce, C. G. Rosen, and T. L. Pasby (eds.), "Fluorescence Spectroscopy: An Introduction for Biology and Medicine," Dekker, New York, 1971.

2 H. H. Bauer, G. D. Christian, and J. E. O'Reilly (eds.), "Instrumental Analysis," Allyn & Bacon, Boston, 1978, chap. 9.

3 H. H. Willard, L. L. Merritt, Jr., J. A. Dean, and F. A. Settle, Jr., "Instrumental Methods of Analysis," 6th ed., Van Nostrand, New York, 1981, chap. 4.

4 G. W. Ewing, "Instrumental Methods of Chemical Analysis," 4th ed., McGraw-Hill, New York, 1975, chap. 4.

5 J. E. O'Reilly, *J. Chem. Ed., 53,* 191 (1976).

6 J. E. O'Reilly, *J. Chem. Ed., 52,* 610 (1975).

†Adapted, in part, from J. E. O'Reilly, *J. Chem. Ed., 52,* 610 (1975); *53,* 441 (1976); with permission from the Division of Chemical Education, American Chemical Society, copyright © 1975 and 1976.

7 M. W. Legenza and C. J. Marzzacco, *J. Chem. Ed., 54,* 183 (1977).

8 J. D. Winefordner, S. G. Schulman, and T. C. O'Haver, "Luminescence Spectrometry in Analytical Chemistry," Wiley, New York, 1972, p. 293.

9 S. Udenfriend, "Fluorescence Assay in Biology and Medicine," Academic Press, New York, vol. I, 1962; vol. II, 1969.

10 G. G. Guilbault, "Practical Fluorescence: Theory, Methods, Techniques," Dekker, New York, 1973.

11 C. E. White and R. F. Argauer, "Fluorescence Analysis. A Practical Approach," Dekker, New York, 1970.

12 E. D. Olsen, "Modern Optical Methods of Analysis," McGraw-Hill, New York, 1975, chap. 8.

13 W. R. Seitz in "Treatise on Analytical Chemistry," P. J. Elving, E. J. Meehan, and I. M. Kolthoff (eds.), 2nd ed., part 1, vol. 7, Interscience, New York, 1981, chap. 4.

Apparatus

Fluorometer (filter instrument or, preferably, a spectrofluorometer)
Cuvettes (2 matched)
Volumetric flasks, one 1000 mL; six 100 mL; twelve 25 mL
Pipets, six 10 mL; twelve 2 mL

Chemicals

Quinine stock solution, (100.0 μg/mL), 120.7 mg quinine sulfate dihydrate or 100.0 mg quinine, transfer to a 1-L volumetric flask, add 50 mL 1 M sulfuric acid [H_2SO_4], and dilute to mark with water (prepare fresh daily and protect from light)
Unknown: Quinine solutions (in 0.05 M H_2SO_4)
0.05 M H_2SO_4 (for dilutions)
Quinine stock solution (10.0 μg/mL), prepared from the 100.0-μg/mL solution
0.05 M sodium bromide [NaBr]
Buffers (pH 1, 2, 3, 4, 5, 6)
Quinine fluorescence spectra for Procedure E

QUININE

Theory

The relationship between concentration and fluorescence intensity has been discussed in the chapter text. For low concentrations, the fluorescence intensity is directly proportional to the concentration as well as to the intensity of the incident radiation. The equation $F = P_0(2.3\epsilon bc)Q_f k$ holds generally for concentrations up to a few micrograms per milliliter. Reduction in intensity of fluorescence can be due to specific effects of constituents of the solution itself. The term quenching is used to describe any such reduction in intensity. Types of quenching include *concentration quenching* (a decrease in the fluorescence-per-unit-concentration as the concentration is increased), also referred to as an inner filter effect, and *chemical quenching*. Concentration quenching results from excessive absorption of either primary or fluorescent radiation by the solution. Collisional quenching may be caused by nonradiative loss of energy from the excited molecules, and the quenching agent (such as oxygen) may facilitate conversion of the molecules from the excited singlet to a triplet level. Chemical quenching is due to actual changes in the chemical nature of the fluorescent substance (conversion of a weak acid to its anion with increasing pH). Aniline is an example: It fluoresces as the molecule between pH 5 and pH 13, below pH 5 it exists as the anilinium cation, and above pH 13 it exists as the anion; both do not fluoresce.

Excitation and Emission Spectra

The fluorescence (emission) spectrum of an organic molecule is roughly a mirror image of its excitation (absorption) spectrum. This is due to the fact that the vibrational levels in the ground and excited states have nearly the same spacing (see Fig. 10-1), and the molecular and orbital symmetries do not change. Assuming that all of the molecules are in the ground state before excitation, the least energy absorbed in the excitation process (i.e., of the longest wavelength) equals the greatest energy transition in the fluorescence process (i.e., of the shortest wavelength). (See Fig. 10-4 for the absorption and emission spectra of anthracene.) If a significant number of molecules are already in a vibrationally excited state before excitation, some overlap occurs in the absorption and emission spectra because less energy is needed to excite these molecules to the S_1 level.

The types of transitions associated with fluorescence include π^*-to-π and π^*-to-n; σ^*-to-σ transitions are seldom observed because fluorescence spectra from absorptions at wavelengths shorter than 250 nm are not likely to occur—the high energy causes deactivation of excited states by predissociation or dissociation. At 200 nm (equal to 140 kcal/mol) bond ruptures often take place. The molar absorptivity for a π-π^* transition is 100 to 1000 times that for a n-π^* transition. The time involved for the former is of the order of 10^{-7} to 10^{-9} s; that for the latter is 10^{-5} to 10^{-7} s.

Under usual conditions and at a concentration of about 2 μg/mL quinine, the two excitation peaks (250 and 350 nm) and one emission peak (450 nm) will be seen. However, for carefully controlled conditions other peaks appear due to grating monochromator peculiarities and Rayleigh, Tyndall, and Raman scattering. Rayleigh scattering refers to radiation scattered in all directions by elastic collisions (impinging and dispersed radiation are of the same wavelength and are radiated in a random manner). In Raman scattering, the collisions are nonelastic, due to the mixing of the electromagnetic energy with the rotational and vibrational energy of the colliding molecule, and the emerging radiation will be at a different wavelength.

Procedures

A. Preparation of a Calibration Curve: Determination of Quinine in Unknowns and Limit of Detection

Prepare a series of quinine standards from the 100-μg/mL stock standard. Make sequential tenfold dilutions with 0.05 M H_2SO_4. This should give concentrations of 10, 1, 0.1, 0.01, 0.001, 0.0001 μg/mL, etc. Make dilutions until the most dilute solution gives a fluorescence intensity approximately that of the blank (0.05 M H_2SO_4).

Read carefully the instructions for the use of the particular fluorometer to be used. If a spectrofluorometer is available, record the excitation and emission spectra for the 2-μg/mL solution.

If only a filter instrument is available, determine the relative fluorescence intensity for each of the above dilutions, using 0.05 M H_2SO_4 as a blank. Filters for excitation (either at 250 or 350 nm) and for emission at 450 nm should be chosen for these measurements. The emission wavelength is the same (450 nm) for either of the excitation wavelengths (250 and 350 nm). Determine the fluorescence intensity of the unknown quinine solution making proper dilutions if necessary.

Also measure the fluorescence of a sample of quinine tonic water after the following dilutions: pipet 5.00 mL of tonic water into a 250-mL volumetric flask, dilute to the mark with 0.05 M H_2SO_4; pipet 5.00 mL of this solution into a 25-mL volumetric flask and dilute to volume with 0.05 M H_2SO_4. For determination of quinine in urine after ingestion of tonic water, refer to Ref. 6.

Treatment of Data

Plot the relative fluorescence intensity vs. the quinine concentration on 6-cycle log-log graph paper. If the blank has been read separately and water has been used to standardize the instrument, the reading of the blank must be subtracted from the fluorescence readings before plotting.

Discuss any deviation from linearity in the plots. Account for deviations if present. Determine the concentration of the unknown quinine sample and of the tonic water from the calibration curve. Calculate the concentration of quinine in the original unknown (if it was diluted) and in the undiluted tonic water.

Assume the limit of detection is the concentration of material that yields a fluorescence that is 10% more than the emission observed for the blank and calculate its value for quinine. For example, if the water blank has a 20% reading, then the concentration of quinine that would give a reading of 22% would be at the limit of detection. (If the instrument was balanced with the 0.05 M H_2SO_4 blank, repeat the measurements with a water blank to determine the instrument's limit of detection for quinine.)

B. pH Dependence of Quinine Sulfate

Pipet 2.0 mL of the 10-μg/mL standard quinine sulfate solution into a 25-mL volumetric flask and dilute to the mark with a pH 1 buffer solution. Measure the exact pH of the resulting solution with a pH meter. Repeat the above procedure with five other buffer solutions between pH 2 and pH 6. The concentration of quinine sulfate will be the same in each solution. Measure the relative fluorescence intensity of these six solutions.

Treatment of Data

Plot the fluorescence intensity vs. the pH for the six solutions. On the basis of these results, what conclusions can be drawn about the pH dependence of quinine sulfate?

C. Halide Quenching of Quinine Sulfate Fluorescence

Pipet 2.0 mL of the 10-μg/mL standard quinine sulfate solution into a 25-mL volumetric flask, add 1.0 mL 0.05 M NaBr, and fill to the mark with 0.05 M H_2SO_4. Prepare four more solutions with 2.0 mL of 10-μg/mL quinine in each and add 2.0-, 4.0-, 8.0-, and 16.0-mL portions of NaBr successively. Measure the relative fluorescence of the five solutions.

Treatment of Data

Plot fluorescence intensity vs. concentration of bromide ion. Explain the results. Could hydrochloric acid be used to dilute the standard quinine solutions in place of 0.05 M H_2SO_4? Why?

D. Characteristics of Quinine Fluorescence Spectra (A "Dry Lab") (optional)

Read references concerned with interaction of electromagnetic radiation, instrumentation for UV-vis spectroscopy, molecular absorption, and fluorescence spectroscopy. (See references listed in Experiments 6-1 and 10-1 as well as Chaps. 6 and 10 in this manual.)

Because quinine in dilute acid solution has two excitation wavelengths (250 and 350 nm) and one emission wavelength (450 nm), a plot of fluorescence intensity vs. wavelength shows three peaks: 250, 350, and 450 nm. Excitation and emission spectra of quinine at a concentration of 2 μg/mL in 0.05 M H$_2$SO$_4$ are shown in Fig. 10-5.

Treatment of Data

In reference to the spectra in Figs. 10-5, 10-6, and 10-7, answer the following questions:

1 What are the peaks at 250 nm in emission spectra 10-6A and 10-7A due to? What are peaks at 350 nm in emission spectra 10-6B and 10-7B due to?

2 What are the very small peaks at about 275 nm in spectra 10-6A and 10-7A due to? What is the small peak at 395 nm in spectrum 10-7B due to? What effect does this latter peak have on the quinine emission peak at 450 nm in spectrum 6B? What would the effect be, analytically, if the quinine concentration were even lower, say 0.01 μg/mL?

3 What are the peaks at 500 nm in spectra 10-6A and 10-7A due to? The 700-nm peaks in 10-6B and 10-7B?

4 Identify the origin of the peaks at 225, 385, and 450 nm in the "excitation spectrum" of the blank, 10-7C. What effect do these peaks have on the quinine excitation spectrum 10-6C?

5 Compare the quinine excitation spectrum 10-5C with the normal UV absorption spectrum 10-5D and comment on the differences in relative peak heights. The quantum efficiency for quinine fluorescence is very nearly the same, about 0.55, whether the excitation wavelength is 250 or 350 nm. Furthermore, if the quantum efficiencies are the same at these two wavelengths, why is the fluorescence intensity greater with excitation at 350 than at 250 nm (10-5B vs. 10-5A)?

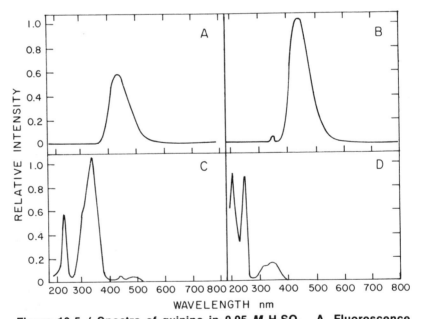

Figure 10-5 / Spectra of quinine in 0.05 M H$_2$SO$_4$. A. Fluorescence spectrum of 2 μg/mL quinine, excitation at 250 nm. B. Fluorescence spectrum with excitation at 350 nm. C. Excitation spectrum of 2 μg/mL quinine, emission monitored at 450 nm. D. Ultraviolet absorption spectrum of 10 μg/mL quinine; the y axis is in units of absorbance A.

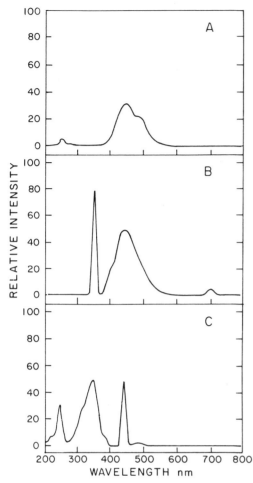

Figure 10-6 / Excitation and emission spectra of 0.03 μg/mL quinine in 0.05 M H$_2$SO$_4$. A. Fluorescence spectrum with excitation at 250 nm. B. Fluorescence spectrum with excitation at 350 nm. C. Excitation spectrum with emission monitored at 450 nm.

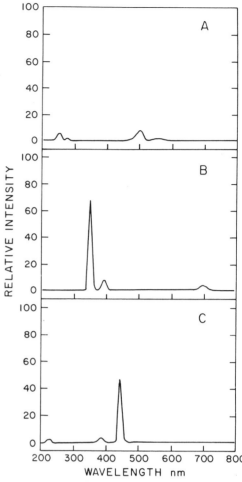

Figure 10-7 / Spectra of the background or blank, 0.05 M H$_2$SO$_4$. A. Fluorescence spectrum with excitation at 250 nm. B. Fluorescence spectrum with excitation at 350 nm. C. Excitation spectrum with emission monitored at 450 nm.

6 What effects would these extra peaks have on the analytical sensitivity or detection limit for quinine analysis if you were using (*a*) a grating-monochromator scanning fluorometer with a 10-nm bandpass and excitation at 350 nm vs. (*b*) a filter fluorometer with a mercury vapor lamp, a primary filter to isolate the 365-nm mercury line, and a short-wavelength cutoff secondary filter which passes all light above 400 nm?

Questions

1 Explain why the radiation from fluorescence is measured 90° from the excitation radiation.

2 Discuss other deactivation processes which compete with fluorescence.

3 Describe how fluorescence excitation spectra are measured.

EXPERIMENT 10-2

Determination of Fluorescein by Fluorometry†

Purpose   Fluorescein will be used to illustrate qualitative and quantitative chemical analysis by use of a filter fluorometer. Some principles and methods of fluorometry will be explored.

References
1   Turner Associates, "Manual of Fluorometric Procedures," Amsco Instrument Co., Carpenteria, Calif., 1968.
2   G. G. Guilbault (ed.), "Fluorescence: Theory, Instrumentation, and Practice," Dekker, New York, 1967.
3   S. Udenfriend, "Fluorescence Assay in Biology and Medicine," Academic Press, New York, 1962.
4   C. E. White and R. F. Argauer, "Fluorescence Analysis. A Practical Approach," Dekker, New York, 1970.
5   A. J. Pesce, C. G. Rosen, and T. L. Pasby (eds.), "Fluorescence Spectroscopy. An Introduction for Biology and Medicine," Dekker, New York, 1971.
6   W. R. Seitz in "Treatise on Analytical Chemistry," P. J. Elving, E. J. Meehan, and I. M. Kolthoff (eds.), 2nd ed., part 1, vol. 7, Interscience, New York, 1981, chap. 4.

Apparatus   Filter fluorometer (such as the Turner Model 110 or 111)
Filter set, primary (2A, 47B, 7-60) and secondary (2A-12, 2A-15, 8, 65A) filters (1% and 10% neutral filters)
Cuvettes (12), 12×25 mm borosilicate
Burets (2), 10 mL (or 10 pipets, 1 through 10 mL)
Conical flasks (9), 25 mL

Chemicals   0.05 M disodium hydrogen phosphate [Na_2HPO_4], dissolve 13.4 g of $Na_2HPO_4 \cdot 7H_2O$ in 1 L of distilled water
Fluorescein, disodium salt (also known as uranine)
Stock solutions:
  1   1 mg/mL disodium fluorescein in 0.05 M disodium hydrogen phosphate
  2   10 μg/mL disodium fluorescein (by 100-fold dilution of 1 mg/mL solution)
  3   0.1 μg/mL disodium fluorescein (by 100-fold dilution of 10 μg/mL solution)

Solutions 1 and 2 are stable in the refrigerator; solution 3 must be used within 1 week.
Student dilutions must be made and used the same day, within 10 h. All solutions must be kept in the dark because they are photosensitive.

Theory   Fluorescein has a high quantum yield (approximately 0.85) such that low concentrations yield intense fluorescence. The latter has a slight pH dependence within

†Adapted from an experiment by G.K. Turner Associates in the "Manual of Fluorometric Procedures," Amsco Instrument Co., Carpenteria, Calif., 1968.

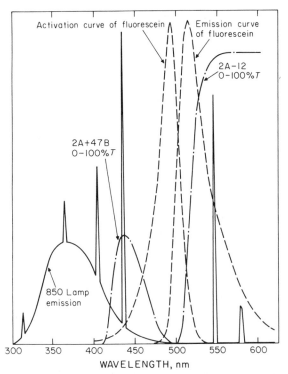

Figure 10-8 / Spectral characteristics of fluo-rescein, the excitation source, and various filters.

the range from pH 5 to pH 11. Use of 0.05 *M* disodium hydrogen phosphate buffer (pH 9) ensures constant emission efficiency.

The Turner Model 110 optical system is illustrated by Fig. 10-2. Two filters must be selected. The primary filter isolates the excitation wavelength, which for fluorescein occurs at 492 nm, with a range between 425 and 525 nm (Fig. 10-8). Note also that the light source of the fluorometer (within the region from 425 to 525 nm) consists mainly of the 436-nm mercury line. To isolate this light, consultation of the operating manual or other source on the characteristics of filters will indicate that a combination of a 2A and a 47B filter will suffice. The emission curve, extending from 475 to 650 nm, has a peak at 516 nm. A sharp-cut filter, that transmits all light which has a longer wavelength, will give good results as a secondary filter. The 2A-12 filter transmits light that is emitted in the longer wavelength half of the emission band. A 2A-15 filter transmits about 10 nm less of the emitted light band.

Procedures

If the 10-μg/mL and 0.1-μg/mL fluorescein solutions are not furnished, prepare these by successive 100-fold dilutions from the 1-mg/mL stock solution, using 0.05 *M* disodium hydrogen phosphate buffer as the diluent. Store in a dark place, or paint the bottles on the outside with black paint.

A. Calibration Curve for the Determination of Fluorescein

Warm up the fluorometer (see the operating manual for the instrument) and read the instructions for operation of the instrument thoroughly before beginning the experiment.

Prepare fluorescein solutions in 25-mL conical flasks as follows:

	CONCENTRATION, μg/mL										
	0.00	0.01	0.02	0.03	0.04	0.05	0.06	0.07	0.08	0.09	0.10
mL fluorescein (0.1 μg/mL)	0	1.0	2.0	3.0	4.0	5.0	6.0	7.0	8.0	9.0	10.00
mL buffer, 0.05 M Na_2HPO_4	10.0	9.0	8.0	7.0	6.0	5.0	4.0	3.0	2.0	1.0	0

This can be done by one of two methods: (1) Fill one buret with buffer and another with 0.1 μg/mL fluorescein and dispense into flasks as indicated, or (2) pipet the stated volumes of each into the flasks. Mix by swirling, then pour each solution into a cuvette, filling $\frac{3}{4}$ full. Keep the solutions out of bright light.

Install the primary and secondary filters and zero the instrument with the dummy cuvette. Adjust the sensitivity with the 0.10-μg/mL standard by use of the range selector and neutral filters (1% and 10%), to the highest possible on-scale reading. (The neutral filters are placed on the secondary side in addition to the secondary filter; their purpose is to reduce light intensity and, in turn, reduce the sensitivity of the instrument.)

Measure and record the % fluorescence of the eleven solutions. Measure the fluorescence of the unknown.

Repeat the measurements of the eleven solutions plus the unknown using a 2A-15 secondary filter (or a No. 58 if the former is not available). Keep the same neutral filters and range setting.

Repeat the measurements after increasing the range selector to the next larger opening (the higher concentrations cannot be measured). Decrease the range selector to the next lower setting from the original, and repeat the measurements.

Treatment of Data

Plot all data on the same sheet of graph paper (fluorescence vs. concentration) and determine the concentration of the unknown from the first curve. Determine the slope of each curve.

Questions

1 How linear were all of the plots? Should they all be within one dial division?
2 Explain how the plots differ in slope.
3 Would you expect a more linear calibration curve at higher concentrations when activated at 436 nm rather than at 492 nm?
4 How much more fluorescence would be obtained by exciting at 492 nm than at 436 nm if an equal amount of energy were available? Would this increase be desirable?

B. Comparison of Secondary Filters

Prepare dilutions of the 10-μg/mL stock solution in a manner similar to the previous dilutions which will now yield concentrations of 1.00, 2.00, 3.00 μg/mL, etc.

Measure the fluorescence of each solution using a 2A-12 secondary filter. (A 1% neutral filter will probably be required.) Repeat with the 8 and the 65A filters together as the secondary filter combination.

Treatment of Data

Plot the results as in Procedure A.

Question

1 Comment on the linearity of the plots—whether or not they are linear and how they compare with each other.

C. Limit of Detection

Dilute 1 mL of the 0.1-μg/mL stock solution of fluorescein with 9 mL of buffer to make a 0.01-μg/mL solution. Increase the sensitivity of the fluorometer until the reading is high on the scale. Measure the fluorescence of the buffer. Is the fluorescence of the blank (buffer) at least 2 dial divisions different? If not, then dilute the solution to 0.001 μg/mL and again measure the fluorescence of the standard and the blank. Continue in this manner of dilution until the blank reading becomes significant. (Readjust the sensitivity with each successive dilution.)

Treatment of Data

Assuming that the limit of detection is the concentration of material which yields fluorescence greater by 10% than the blank, calculate the limit of detection for fluorescein on the instrument used. For example, if the blank reading is 20% when the instrument is standardized with distilled water, then the concentration of fluorescein that would give a reading of 22% would be at the limit of detection.

Calculate the limit of detection of fluorescein.

Question

1 Would activation at 492 nm and measurement of emission with a narrow-pass secondary filter at 514 nm reduce the minimum detectable concentration?

D. Effect of Contamination

Fill a cuvette with tap water and another with distilled (or deionized) water. Use fresh distilled water from an all-glass still that has not been in contact with rubber, cork, etc. Use filter 2A plus filter 47B as the primary filter and filter 2A-12 as the secondary filter, and adjust the sensitivity such that a substantial reading is obtained with tap water. Measure the fluorescence of the distilled water. Repeat the experiment with filter 7-60 as the primary filter (activation is at 365 nm). Place your finger over the distilled-water cuvette and invert several times. Measure the now-contaminated distilled water at the same sensitivity settings as above (use both excitation bands).

Treatment of Data

Explain the effect of contamination from your finger. Does this substantiate statements that purity of reagents is frequently a problem when activating with visible light?

Question

1 Is there a substantial difference in the ratios of the fluorescence of tap water to that of distilled water between the 2A plus 47B filter and the 7-60 filter? If so, what is the explanation?

EXPERIMENT 10-3

Determination of Pharmaceuticals (Acetylsalicylic and Salicylic Acids) by Fluorometry†

Purpose

Acetylsalicylic acid (aspirin), often referred to as ASA, hydrolyzes to salicylic acid (SA), which is present to some extent in most aspirin. This method demonstrates how both of these substances can be accurately determined by fluorometry.

†Adapted, in part, from G. H. Schenk, F. H. Boyer, C. I. Miles, and D. R. Wirz, *Anal. Chem.*, *44*, 1593 (1972).

References

1 G. H. Schenk, F. H. Boyer, C. I. Miles, and D. R. Wirz, *Anal. Chem., 44,* 1593 (1972).

2 C. I. Miles and G. H. Schenk, *Anal. Chem., 42,* 656 (1970).

3 C. A. Parker, "Photoluminescence of Solutions," Elsevier, New York, 1968, pp. 328–344.

Apparatus

Filter fluorometer [such as the Turner Model 110 equipped with a 4-watt low-pressure mercury arc source (emits intensely at 254 nm and weakly at 297 and 313 nm)]

Volumetric flasks, eight 25 mL; four 100 mL

Pipets, one 5 mL, four 10 mL, one 15 mL, one 20 mL

Whatman No. 1 filter paper

Filters

For ASA:

Primary: No. 7-54 far UV and a near UV absorbing plastic filter

Secondary: B & L No. 340 interference filter (peak at 341 nm; bandwidth, 24 nm)

For salicylic acid:

Primary: Narrow-pass Wratten primary filter No. 34A for peak at 325 nm and transmission of 313 mercury line

Secondary: Corning No. 3-71 sharp-cut filter for 465 nm and above transmission

Optional

An absolute spectrofluorometer to obtain corrected spectra for excitation and emission, equipped with 75-watt xenon source and 10.0-mm quartz cuvettes. These spectra are desirable but not required. For details see Ref. 1.

Chemicals

Acetylsalicylic acid (ASA)

Salicylic acid (SA), USP reagent grade

Aspirin tablets

Spectro-quality chloroform

Acetic acid (1% v/v acetic acid in $CHCl_3$)

Benzoic acid, ACS reagent

2.22×10^{-3} M ASA, stock standard (0.40 g/L) in 1% acetic acid–chloroform

5.43×10^{-3} M SA, stock standard (0.75 g/L) in 1% acetic acid–chloroform

2.22×10^{-5} M ASA standard (4.0 μg/mL) in 1% acetic acid–chloroform

5.43×10^{-5} M SA standard (7.5 μg/mL) in 1% acetic acid–chloroform

SALYCYLIC ACID
M.W. 138.12

ACETYLSALICYLIC ACID
M.W. 180.17

Theory

Addition of acetic acid to solutions of acetylsalicylic acid and salicylic acid increases the fluorescence intensity of both substances to analytically useful levels. This is particularly helpful for acetylsalicylic acid (ASA) because its quantum efficiency as compared with quinine bisulfate is about 0.02. An optimal solvent for both ASA and SA is 1% acetic acid–chloroform. A linear calibration curve can be obtained for solutions of ASA up to 5 μg/mL and of SA up to 7.5 μg/mL in 1% acetic acid–chloroform, by use of a simple filter fluorometer. To eliminate variation from tablet to tablet, several aspirin tablets (from 5 to 20) can be ground together and a representiative sample (equivalent to the weight of one tablet) used for analysis.

Procedures

The fluorescence excitation and emission spectra can be obtained on a spectrofluorometer, if available. Spectra for ASA and for SA are presented in Figs. 10-9 and 10-10. A spectrum for a mixture of ASA and SA is shown in Fig. 10-11.

A. Calibration Curves

Standard Curve for ASA

Pipet 5, 10, 15, and 20 mL of the 4.0-μg/mL solution into 25-mL volumetric flasks, successively. Fill each to the mark with a 1% v/v acetic acid–chloroform solution.

Using the 254-nm mercury line for excitation (No. 7-54 filter plus a near UV plastic filter for the primary combination) and a B & L No. 340 interference filter (peak at 341 nm, bandwidth 24 nm) for a secondary filter, obtain readings of % fluorescence for each of the five solutions. Use the 1% acetic acid–chloroform solution as a blank. A 3X setting should be suitable for reading with a Turner Model 110 Filter Fluorometer up to 4.0 μg/mL.

Standard Curve for SA

Pipet 5, 10, 15, and 20 mL of the 7.5-μg/mL SA solution into 25-mL volumetric flasks, successively. Fill each to the mark with 1% acetic acid–chloroform solution.

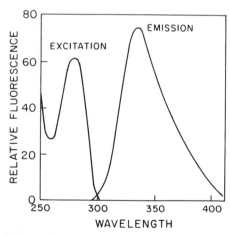

Figure 10-9 / Excitation and emission spectra of 1.0 × 10⁻⁴ *M* acetylsalicylic acid in 1% acetic acid–chloroform determined at 3X sensitivity.

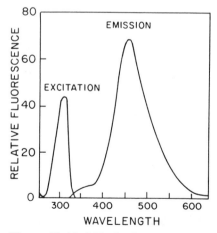

Figure 10-10 / Excitation and emission spectra of 2.23 × 10⁻⁵ *M* salicylic acid in 1% acetic acid–chloroform determined at 3X sensitivity.

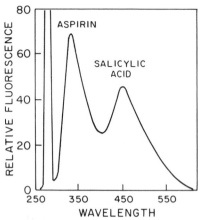

Figure 10-11 / Emission spectrum of acetylsalicylic acid — salicylic acid mixture in 1% acetic acid–chloroform. Excitation at 280 nm. (*Note:* Excitation light scatter centered at 280 nm.)

Using the 313-nm mercury line for excitation (narrow-pass Wratten filter No. 34A plus the No. 7-54 filter for the primary filter) and a Corning No. 3-71 sharp-cut filter for transmission above 465 nm as a secondary filter, obtain readings of % fluorescence for each of the five solutions. A 10X setting should be suitable with the Turner Model 110.

B. Determination of ASA and SA in Aspirin Tablets

Grind three to five aspirin tablets to a powder. Weigh 400 mg of the powder (or the equivalent of one tablet) and transfer to a 100-mL volumetric flask; dissolve in 1% v/v acetic acid–chloroform and dilute to the mark. Filter rapidly through Whatman No. 1 filter paper. Determine the % fluorescence of the salicylic acid by use of the same setting as for the standard curve, as well as the same filters. *Save* the filtered solution for the next step (determination of ASA).

To determine the acetylsalicylic acid content, dilute the above filtered solution 1:1000 with 1% acetic acid–chloroform. This can be done by making three successive dilutions of 10 mL to 100 mL. Read the % fluorescence with the appropriate filters and settings on the instrument. Readings should be made within 1 h from the time of the dissolution of the aspirin. After that interval, the amount of ASA will be low.

Treatment of Data

Prepare standard curves (ASA and SA) on the same sheet of graph paper (% fluorescence vs. concentration). With these, determine the concentration of ASA and SA in the aspirin sample.

Report the amount of ASA and SA in units of milligrams per aspirin tablet. Compare your value of ASA with that indicated on the label of the product.

Questions

1 Are your standard curves linear? If not, where do they begin to deviate?
2 What is the significance of the presence of salicylic acid in your tablets? Is it undesirable?
3 Why do you think the fluorescence changes upon standing?

EXPERIMENT 10-4

Critical Micelle Concentration of Surfactants†

Purpose

The critical micelle concentration (CMC) of sodium dodecyl sulfate (SDS) will be measured by fluorescence emission using Acridine Orange dye as a probe.

References

1　M. Rujimethabhas and P. Wilairat, *J. Chem. Ed., 55,* 342 (1978).
2　B. H. Robinson, N. C. White, C. Mateo, K. H. Timmins, and A. James in "Chemical and Biological Applications of Relaxation Spectroscopy," E. Wyn-Jones (ed.), Reidel, Dordrecht-Holland, 1975, pp. 201–210.

Apparatus

Filter fluorometer or spectrofluorometer
Filters, Zeiss M 436 (wavelength 436 nm) (or filter for 492 nm, the excitation wavelength) and filter to isolate emission wavelength at 525 nm
Volumetric flasks (10), 25 mL
Pipets, 1 mL, 2 mL, 4 mL, 6 mL, 8 mL, 10 mL, 12 mL, 14 mL, 16 mL

Chemicals

5.9×10^{-5} *M* Acridine Orange stock solution, 100 mL
0.0205 *M* sodium dodecyl sulfate stock solution, 100 mL

Theory

Micellar effects on reaction rates and the use of micelles as model systems for the mode of action of carcinogens have been measured directly by conductometric and surface tension methods during the past few years. The present method makes use of a probe, Acridine Orange, so that the critical micelle concentration (CMC) of a surfactant solution can be determined by fluorescence or by UV-visible spectrophotometric measurements. The concentration of the Acridine Orange dye is held constant while the sodium dodecyl sulfate concentration is varied. When the % fluorescence is plotted vs. the concentration of sodium dodecyl sulfate (SDS) a sigmoidal curve is obtained, the midpoint of which is considered to be the CMC. This increase in fluorescence intensity corresponds to the change in environment of the dye from an aqueous solution to the hydrophobic micellar binding site. Below the CMC, an aggregate of dye and surfactant molecules exists, probably in the form of stacks of dye-surfactant salt (Ref. 2). The CMC is the concentration at which micelle formation begins. Micellar colloids are aggregates of many small molecules held together by weak, secondary bonding forces, and are unlike molecular colloids (which are particles of single macromolecules). The colloidal nature of these particles makes possible the adsorption of the fluorescent acridine dye to the nonfluorescent surfactant.

Procedure

Prepare ten solutions by pipetting 2 mL of the Acridine Orange stock solution (5.9×10^{-5} *M*) into each of ten 25-mL volumetric flasks. Pipet 1-, 2-, 4-, 6-, 8-, 10-, 12-, 14-, and 16-mL volumes of the sodium dodecyl sulfate stock solution (0.0205 *M*) successively into nine of the 25-mL flasks. Dilute the ten flasks to the mark with distilled water. Use the one that contains only Acridine Orange as a blank. Measure the % fluorescence of each solution, using an excitation wavelength of 436 nm (or 492 nm, the actual excitation peak) and an emission wavelength of 525 nm. These measurements can also be made on a Spectronic 20 spec-

†Adapted, in part, from M. Rujimethabhas and P. Wilairat, *J. Chem. Ed., 55,* 342 (1978) with permission. Copyright © 1978, Division of Chemical Education, American Chemical Society.

trophotometer or other spectrophotometer at 500 nm by recording absorbance units instead of fluorescence units. The curve will not be as steep, however.

Treatment of Data

Plot the % fluorescence vs. the concentration of sodium dodecyl sulfate. Determine the midpoint of the sigmoidal curve and report it as the CMC. The literature value, as determined by a conductometric method, is 8.1×10^{-3} M at 25°C.

Questions

1 Explain what a micelle is and why it forms.
2 Explain what a colloid is and why it forms.

EXPERIMENT 10-5

Mercury(II) Determination by Oxidation of Thiamine to Thiochrome

Purpose

This experiment illustrates a simple fluorometric determination for mercury(II) which is based on its oxidation of thiamine to yield highly fluorescent thiochrome.

References

1 J. Holzbecher and D. E. Ryan, *Anal. Chim. Acta.*, *64*, 333 (1973).
2 Turner Associates, "Traces," vol. 11, no. 6, G. K. Turner, Amsco Instrument Co., Carpenteria, Calif.

Apparatus

Filter fluorometer (such as the Turner Model 110) or spectrofluorometer
Filters, primary, 7-60; secondary, 2A
Volumetric flasks (8), 10 mL
Pipets, one 0.5 mL, two 1 mL, two 2 mL, one 3 mL, one 4 mL, two 5 mL

Chemicals

Borate buffer (pH 7.7)
3×10^{-5} M thiamine solution (freshly prepared each day)
Unknown mercury solutions (approx. 1 ng Hg per mL) (neutral pH) (Unknowns containing organic material need acid digestion and should be 10 to 500 ng in Hg; this will allow dilution to lower the salt concentration to below 0.02 M.)
Mercury stock solution: mercuric chloride [$HgCl_2$] (200 ng Hg per mL)

Theory

Mercury(II) compounds do not fluoresce; however, many of them can be determined indirectly by measuring the amount of thiochrome present before and after oxidation of thiamine to thiochrome. Only cyanide, iodide, sulfide, and EDTA out of 24 common cations and anions interfere by decreasing the fluorescence. Water and organomercurials can be analyzed readily by this method. Fluorescence intensity is linear over a range from 10 to 200 ng Hg per mL.

note *Thiamine* can be determined by oxidation with *mercuric chloride* in a similar fashion.

Procedure

Prepare standard solutions by pipetting 0.5, 1.0, 2.0, 3.0, 4.0, and 5.0 mL of the mercury stock solution (10 ng/mL) successively into six 10-mL volumetric flasks. Pipet 5 mL of the unknown into a seventh flask. Add 2 mL of borate buffer (pH 7.7) to the seven flasks, as well as to an eighth flask that contains *no* mercury (for the blank). Add 1 mL of thiamine solution (3×10^{-5} M) to each of the eight flasks. Dilute each to the mark with distilled water. Allow the reaction to proceed in the dark for 1 h at room temperature, or 10 min at 90°C. Measure the fluorescence of the cooled solutions with an excitation at 375 nm (using a No. 7-60 filter) and emission at 440 nm (No. 2A filter).

Treatment of Data

Prepare a calibration curve of % fluorescence vs. ng Hg per mL and use this to determine the Hg concentration of the unknown solution.

Questions

1 Write the reaction for the oxidation of thiamine to thiochrome by Hg(II).
2 What part of the thiochrome structure causes it to fluoresce?

11
Nuclear Magnetic Resonance Spectroscopy

Magnetic Properties of Nuclei; Resonance

The transitions involved in NMR spectroscopy are between the spin states of nuclei with nonzero nuclear magnetic moments. Because a nucleus with an odd number of protons has a charge and a spin, it has a magnetic moment. In an assemblage of atoms the orientation of these spins, in space, is totally random. However, if the assemblage is placed in an external magnetic field the nuclear moments are oriented with respect to the field. In the case of a hydrogen nucleus, two orientations of the rotating magnetic vector are possible, one oriented with the external field and the other against it. These two orientations have slightly different energies which can be represented as:

No field   With field

$+\frac{1}{2}$ against field

$-\frac{1}{2}$ with field

ΔE

If the atoms are placed in such a magnetic field, a transition between these two energy states can be brought about by the absorption of a quantum of energy, $h\nu$. The difference in energy between these two levels is exceedingly small (in the radio frequency region of the spectrum). Hydrogen represents an atom with two magnetically induced quantum levels, but atoms with large nuclear spins have numerous energy levels.

An atomic nucleus possesses an intrinsic angular momentum (with an as-

Table 11-1 / Nuclear Spin Quantum Numbers (*I*), Magnetic Moments (μ), and Quadrupole Moments (*eQ*)

Mass Number	Atomic Number	*I*	μ	*eQ*	Examples
Even	Even	0	0	0	C^{12}, O^{16}, S^{32}, S^{34}
Odd	Even (Odd)	$\frac{1}{2}$	$\pm\frac{1}{2}$	0	H^1, C^{13}, N^{15}, P^{31}
Even	Odd	1	$\pm 1, 0$	>0	H^2, N^{14}

sociated magnetic moment), which has values that are integral or half-integral multiples of $h/2\pi$. It is expressed as $I(h/2\pi)$, and I is the nuclear spin quantum number. The magnetic moment (μ) results from the spinning or rotation of electric charge associated with the nucleus. The spin quantum number and magnetic moment are related to the mass number and atomic number as shown in Table 11-1.

Nuclei with $I = 0$, such as C^{12} and O^{16}, do not have NMR transitions. In general, only molecules that include atomic nuclei with spins of $\frac{1}{2}$ yield high resolution NMR spectra. Those with spin quantum numbers of 1 or greater give broadened spectra because of their electric quadrupole moments.

The most important nuclei with $I = \frac{1}{2}$ and high isotopic abundance are H^1, F^{19}, P^{31}; these atoms are readily studied by NMR. C^{13} has a spin of $\frac{1}{2}$ but its sensitivity is limited and its natural isotopic abundance is only 1%.

The energy spacing between the quantum states is dependent on the strength of the magnetic field at the nucleus, and hence,

$$\nu = \frac{\gamma H_0}{2\pi}$$

where
ν = the frequency of radiation
H_0 = the magnetic field
γ = constant, the magnetogyric ratio

For the proton, in a field with $H_0 = 14,092$ gauss, resonance occurs at 60.0 MHz.

NMR spectra are exceedingly sharp because the nucleus of an atom is isolated; collisions affect only the external electrons. High resolution work is limited to liquid samples (including solutions). Gases generally have too few nuclei to observe the signal, and the nuclei in solids are subjected to strong fields within the crystals which broaden the signals severely.

Chemical Shift of Protons

The relation $\nu = \gamma H_0/2\pi$ indicates that all protons go into resonance at the same magnetic field (at fixed frequency). If this actually turned out to be the case, the technique would have little use for structural characterization. In high resolution instruments, separate peaks are obtained for each proton. There is no conflict because the H_0 in the relation represents the magnetic field *at the nucleus*, not the externally applied field. The electrons in an atom or molecule act as a shield and slightly alter the external magnetic field. The shielding process varies for different types of protons and depends on the environment of the nucleus in the molecule. Therefore, a proton in an aldehyde group CHO goes into resonance at a different applied magnetic field than a proton in a CH_3 group.

The separation of resonance frequencies of nuclei from that of an arbitrary standard (generally, TMS, tetramethylsilane) is called the *chemical shift*. It is defined as

$$\delta = \frac{(\nu_{\text{sample}} - \nu_{\text{ref}}) \times 10^6}{\text{oscillator frequency}}$$

where δ is a dimensionless constant with the units of parts per million (ppm). Another convention assigns a value of 10 to TMS, and denotes the shift by τ, where $\tau = 10 - \delta$.

If the electron density of the protons is reduced by an adjacent electronegative group (or magnetic group), then the strength of the induced magnetic field that is in opposition to the applied field is smaller; the protons will be in resonance with the RF oscillator at lower applied field (H). Resonance at lower H is called a down-field chemical shift, whereas an upfield shift is the reverse. Figure 11-1 summa-

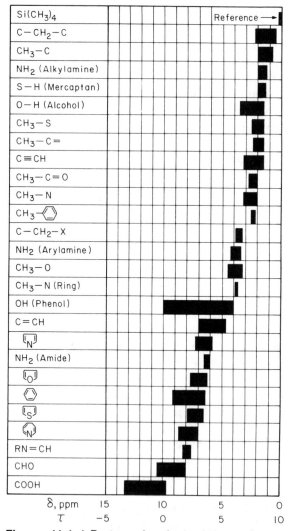

Figure 11-1 / Proton chemical shifts. [*From Chemical and Engineering News, 43, 98 (1965). Copyright © 1965 by American Chemical Society, Washington, D.C. Reprinted with permission.*]

Figure 11-2 / Spin-spin splitting, AB pattern.

rizes representative proton chemical shift values for common organic functional groups.

Spin-Spin Splitting; Coupling Constants

Spin-spin splitting or spin coupling is a phenomenon in which one proton can make small alterations in the shielding of a neighbor proton through electronic bonding. Although proton A and proton B on adjacent carbons (Fig. 11-2) are too far apart to influence the applied field of the other protons through space, proton A can partly polarize the surrounding electronic shells and this polarization can be transmitted through the bonding system to proton B. Through this mechanism, proton B will experience a different applied field, depending upon whether proton A is in the upper or the lower spin-energy state. This will be observed in the spectrum as two peaks (doublet) separated by a value called the coupling constant (J). The intensity of the two peaks, as shown in Fig. 11-2, will be equal because there is an equal probability that the proton A nucleus will be in either spin-energy level.

If there are two magnetically equivalent A protons, then there will be four ways that the two A protons can be aligned, with two ways being degenerate (equivalent). This leaves three possible modes for coupling with proton B. A triplet with an area ratio of 1:2:1 for proton B will be observed (Fig. 11-3). These two examples illustrate that a proton (or protons) coupled by adjacent protons will be split into ($n + 1$) peaks where n is the number of protons on the adjacent atom (or atoms). For example, two A protons coupled with a B proton will split proton B into ($n + 1$) or a triplet. This is the first-order coupling rule which applies as long as the frequency separation ($\Delta\nu$) of the two types of protons (A and B) is much larger than the coupling constant ($\Delta\nu/J \gg 1$). Furthermore, the integrated area under an absorption peak is proportional to the number of nuclei producing the peak, providing the nuclei are not saturated. Integration of the peak areas of proton A and B will give a ratio 2:1, respectively. The integrated areas are in the same ratio as the different kinds of protons.

Another illustration of spin-spin splitting is the spectrum for acetaldehyde. Under low resolution the spectrum has two peaks, with relative areas of 3 and 1, which corresponds to each type of proton. Under higher resolution each of the peaks shows further structure. The CH_3 peak splits into a doublet and the CHO proton exhibits a quadruplet structure. This fine structure is caused by spin coupling. The methyl group is split into a doublet because of interaction of these

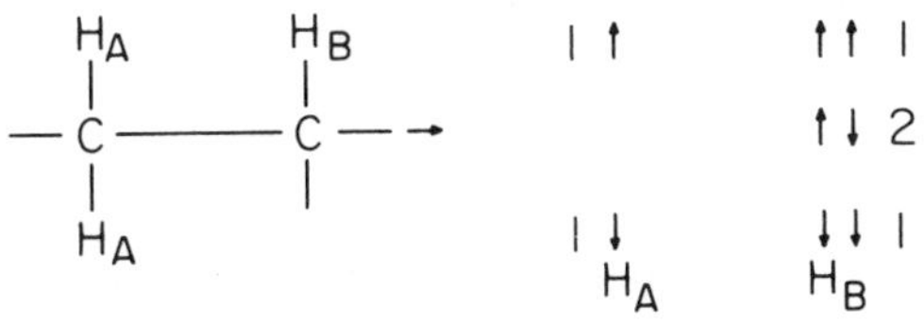

Figure 11-3 / Spin-spin splitting, A_2B pattern.

protons with the two possible spin states of the proton on the aldehyde group which can be either + or −. On the other hand, the aldehyde group is split into a quartet. The quartet arises from the interaction of the three protons on the CH_3 group. These three equivalent protons can give rise to eight possible permutations of spins.

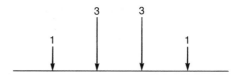

1 and 8 are unique, and give rise to the two end peaks of the quartet. 2, 4, and 7 have the same effective splitting, as do 3, 5, and 6. The resulting quartet appears as

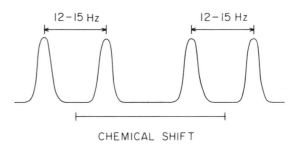

The magnitude of the splitting depends upon a variety of factors. (Some typical coupling constants are shown in Table 11-2.) If two different protons are on the same carbon atom, the magnitude of the splitting (coupling constant) is 12 to 15 Hz and two doublets occur:

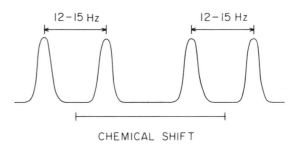

The two hydrogen atoms have different chemical environments and their resonance frequencies are separated by a chemical shift. Each of these two peaks is split by spin coupling to the other proton.

The factors that affect the magnitude of this interaction are:

1 The distance between protons
2 The type of bonding of the atoms between the protons
3 Geometrical considerations

Splitting is not observed between identical protons, and the magnitude (in hertz) is independent of the magnetic field.

NMR Instrumentation

An NMR spectrometer is composed of six basic units (see Fig. 11-4):

1 A *magnet;* permanent magnet or electromagnet capable of generating a strong, stable, homogeneous magnetic field to separate the nuclear energy states.
2 A *transmitter coil;* to furnish RF-irradiating energy; it is placed at right angles to the sweep coils.

Table 11-2 / Coupling Constants

	Range of $J_{\alpha\beta}$ in Hz
Hα, C, Hβ (geminal)	12–15
C—C with Hα and Hβ	2–9
H—C—C—C—H	0
C=C with two H (cis type)	0–3.5
H, C=C, H	11–18
H, H, C=C	6–14
H—C=C—C—H	0.5–2.0
C=C—C=C with H, H	10–13

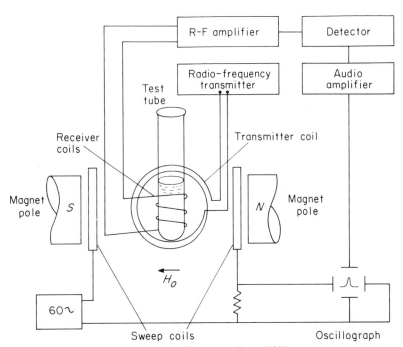

Figure 11-4 / Schematic diagram of an NMR spectrometer.

3 A *sweep generator;* to sweep the magnetic field through the region of resonance to produce the spectrum.

4 A *receiver coil;* around the sample holder to couple the sample with the RF receiver.

5 A *detector;* to process the NMR signals.

6 A *recorder;* to display the spectrum, either in the absorption mode or integral mode.

The sample holder, which is shown in Fig. 11-5, is in the field of the electromagnet with the RF field at right angles to the magnetic field. The magnetic field of the sweep generator coils is swept slowly (increased) to the region of resonance. At resonance a change in nuclear magnetic dipole occurs; a voltage is then induced in the receiver coil, which in turn is amplified, detected, and recorded. The resulting spectrum is a plot of the energy of resonance-transition vs. intensity.

The sample is contained in a thin-walled, precision-bore glass tube with an outside diameter of 5 mm. An air-bearing turbine rotates the sample at a rate of several hundred revolutions per minute to keep the magnetic field homogeneous throughout. Sidebands are produced in the spectrum due to the spinning of the sample because the resonance peaks are modulated at the spinning frequency.

Applications

Some of the common applications and studies that make use of proton NMR include:

1 Quantitative analyses; for example, a mixture of benzaldehyde, ethanol, and toluene; phosphate mixtures; H_2O in D_2O

2 Cis-trans isomerism

3 Effect of hydrogen bonding on chemical shifts

4 Effect of halogen substitution in hydrocarbons

5 Determination of correct structural formulas

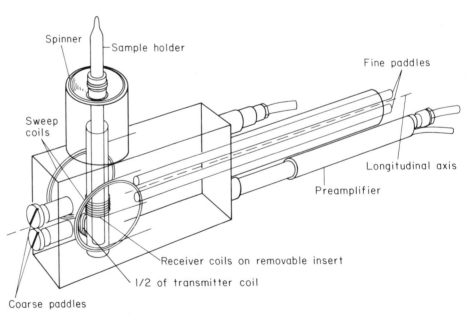

Figure 11-5 / The sample holder for the NMR spectrometer shown in detail.

6 Crystallinity measurements in polymers
7 Solution equilibria and exchange kinetics
8 Protonation equilibria and metal complexation
9 Degree of deuteration
10 Effect of molecular asymmetry
11 Keto-enol tautomerism
12 Structural studies of alkaloids, steroids, fatty acids, etc.
13 Studies of the kinetics for restricted rotation and isomerization

---

EXPERIMENT 11-1
Proton NMR Spectra, Chemical Shifts, and Coupling Constants

Purpose

These experiments introduce some applications of NMR and illustrate chemical shifts and coupling constants. A qualitative analysis of gasoline will demonstrate five different molecular environments for hydrogen. A quantitative determination of ethanol in a mixture of benzene and ethanol provides experience with the integration of spectral traces.

References

1 H. H. Willard, L. L. Merritt, Jr., J. A. Dean, and F. A . Settle, Jr., "Instrumental Methods of Analysis," 6th ed., Van Nostrand, New York, 1981, chap. 11.
2 D. A. Skoog and D. M. West, "Principles of Instrumental Analysis," 2nd ed., Saunders, Philadelphia, 1980, chap. 14.
3 G. W. Ewing, "Instrumental Methods of Chemical Analysis," 4th ed., McGraw-Hill, New York, 1975, chap. 12.
4 H. H. Bauer, G. D. Christian, and J. E. O'Reilly, "Instrumental Analysis," Allyn and Bacon, Boston, 1978, chap. 12.
5 R. M. Silverstein, G. C. Bassler, and T. C. Morrill, "Spectrometric Identification of Organic Compounds," 4th ed., Wiley, New York, 1981, chap. 4.
6 R. H. Cox and D. E. Leyden in "Treatise on Analytical Chemistry," P. J. Elving, E. J. Meehan, and I. M. Kolthoff (eds.), 2nd ed., part 1, vol. 10, Interscience, New York, 1983, chap. 1.

Apparatus

NMR spectrometer, 60 MHz
NMR tubes (10), 5 mm o.d.
Pipets, 0.5 mL, 0.1 mL graduated

Chemicals

Deuterated chloroform [$CDCl_3$]
Absolute ethanol
Benzene
Tetramethylsilane [TMS]
Ethyl chloride, ethyl bromide, ethyl iodide (7% solutions in $CDCl_3$)
Unknown: Mixture of benzene-ethanol
Two samples of premium gasoline

Part I. Effect of Electronegativity on the Chemical Shift

Theory

Discussion of the chemical shift of protons and of spin-spin splitting is presented in the introductory section of Chap. 11 and in the references preceding.

For a series of alkyl halides, an increase in electronegativity results in an increase in the downfield chemical shift in parts per million. From a plot of τ_{CH_2} and τ_{CH_3} vs. the Pauling electronegativity for the ethyl halides of chlorine, bromine, and iodine, the τ values for ethyl fluoride can be estimated.

Procedure

Record the proton NMR spectra of the 7% solutions of ethyl chloride, ethyl bromide, and ethyl iodide in $CDCl_3$, respectively, with a 60-MHz NMR spectrometer. Each solution should contain a few drops of TMS in its NMR tube to serve as a reference.

Treatment of Data

Tabulate δ, τ, and J values for the three compounds, and then plot their τ_{CH_2} and τ_{CH_3} values vs. the Pauling electronegativities for the respective halogen atoms. The electronegativities are:

$$F = 4 \qquad Cl = 3.0 \qquad Br = 2.8 \qquad I = 2.5$$

Questions

1 What conclusions can be made with respect to the effect of electronegative substituents on the chemical shifts of adjacent protons?
2 How sensitive to the number of bonds between the H and the electronegative atom is the preceding effect?
3 Does this substituent effect have any influence on the coupling constants?
4 Sketch an estimated spectrum for CH_3CH_2F and indicate approximate τ and J values.

Part II. Quantitative Determination of Mixtures of Benzene and Ethanol

Theory

The chemical shift for benzene is downfield from any of the chemical shifts of ethanol (benzene, 7.4 ppm vs. TMS; OH of ethanol, 5.2 ppm). Therefore, mixtures of ethanol and benzene can be assayed by quantitative NMR.

Procedure

Record NMR spectra for the following solutions in individual NMR tubes:

2 drops of TMS and 0.5 mL of 10% ethanol in $CDCl_3$
2 drops of TMS and 0.5 mL of 10% benzene in $CDCl_3$
2 drops of TMS and 0.5 mL of an unknown mixture of ethanol and benzene

Carefully integrate each spectrum.

Treatment of Data

Compare the peak areas of the spectra for your unknown with those for individual ethanol and benzene spectra. Determine the percent of each component in the unknown.

Part III. Proton NMR Spectra of Premium Gasolines: A Qualitative Analysis

Theory

Hydrogen in gasoline is present in at least five different molecular environments which can be resolved by NMR. These include: (1) protons attached to aromatic rings, (2) protons attached to unsaturated carbons, (3) CH_2 or CH_3 groups bonded directly to an aromatic ring, (4) CH_2 or CH_3 groups adjacent to an unsaturated carbon, and (5) normal, branched, and cyclic alkanes.

Procedure

Add 2 drops of TMS and 0.5 mL of $CDCl_3$ to two NMR tubes. Place 0.03 mL of one brand of gasoline in one of the tubes, and 0.03 mL of another brand in the second tube. Run an NMR spectrum of each sample. If some of the peaks are of insufficient amplitude, expand the vertical scale 6 times and rerun the spectra.

Treatment of Data

Determine the chemical shifts for each peak in your spectra and identify the type of proton associated with each peak.

EXPERIMENT 11-2

pK_a Values for Organic Bases†

Purpose

The pK_a values of organic bases in aqueous solution can be determined by proton nuclear magnetic resonance. Heterocyclic bases exist in two forms in a protic solvent, and the concentration of each depends upon the value of the equilibrium constant K_a and the pH of the solution. Because the proton chemical shifts of the cationic and uncharged species are different, the change in their relative peak areas as a function of pH can be used to evaluate the pK_a for a given base.

References

1 C. S. Handloser, M. R. Chakrabarty, and M. W. Mosher, *J. Chem. Ed., 50,* 510 (1973).
2 D. D. Perrin, "Dissociation Constants of Organic Bases in Aqueous Solutions," Butterworths, London, 1965.
3 M. W. Mosher, C. B. Sharma, and M. R. Chakrabarty, *J. Mag. Res., 7,* 247 (1972).

Apparatus

Proton NMR spectrometer, 60-MHz
NMR tubes (15 to 20), 5 mm o.d.
Pipets, 10 mL, 1 mL
pH meter with expanded scale and a combination pH-electrode or micro electrodes
Beaker, 20 mL or 30 mL

Chemicals

1 *M* hydrochloric acid [HCl]
Tetramethylammonium chloride, saturated
1 *M* sodium hydroxide [NaOH] in a dropping bottle
Nitrogenous bases or their hydrochloride salts; substituted pyridines such as 4-bromopyridine, 3-chloropyridine, 4-pyridinecarboxaldehyde, isonicotinic acid, lutidines, or picolines

Theory

Heterocyclic bases can exist in two forms in protic solvents

$$\underset{\substack{\text{(BH}^+)}}{\overset{\substack{}}{\boxed{N^+H}}} \underset{}{\overset{H_2O}{\rightleftharpoons}} \underset{\substack{\text{(B)}}}{\overset{}{\boxed{N}}} + H_3O^+ \qquad K_a = \frac{[H_3O^+][B]}{[BH^+]}$$

---

†Adapted with permission from Ref. 1. Copyright 1973, Division of Chemical Education, American Chemical Society.

The pK_a for such a system can be expressed by the relation

$$pK_a = pH + \log \frac{[BH^+]}{[B]}$$

Nearly 100% of the base will be in the uncharged form B at high pH values. The resonance lines of the protonated form BH^+ appear downfield compared with those of the uncharged form; the lines due to the protons that are alpha to the nitrogen shift less (by approximately one-half) than the lines due to the protons at the beta position (see Ref. 3 for an explanation of this difference).

A plot of observed chemical shifts relative to tetramethylammonium chloride (used as an inert standard) vs. pH of the solution results in a titrationlike curve. The chemical shifts in strong acid are taken from the initial flat portion of the curve; the chemical shifts in basic solution are taken from the final flat portion of the curve. The fractional amounts of BH^+ and B can be calculated from the change between the two forms by the relations

$$\text{Fractional amount of } BH^+ = \frac{(\text{observed chemical shift}) - (\text{chemical shift in base})}{(\text{chemical shift in acid}) - (\text{chemical shift in base})}$$

and

$$\text{Fractional amount of } B = 1 - \text{fractional amount of } BH^+$$

The pK_a can be estimated from the relation

$$pK_a = pH + \log \frac{\text{fraction } BH^+}{\text{fraction } B}$$

Procedure

Add approximately 1.5 to 2 g of the nitrogenous base (or its hydrochloride salt) to a beaker and *slowly* add 1 M hydrochloric acid (to avoid fuming and spattering) to dissolve the material; transfer quantitatively to a 25-mL volumetric flask and dilute to the mark with HCl. Pipet 10 mL of this solution into a 20- or 30-mL beaker that contains 1 mL of saturated tetramethylammonium chloride. Determine the pH of the solution with an expanded-scale pH meter to the nearest 0.01 pH unit. Remove a sample (0.5 mL) and place it in an NMR tube. Adjust the remaining solution in the beaker to a higher pH (approximately 0.5 pH unit) by addition of 1 M sodium hydroxide; redetermine the pH accurately. Remove a second sample of 0.5 mL and transfer it to a second NMR tube. Repeat the process of adjusting the pH higher and removing an additional sample in a manner suitable for the particular organic base. For example, a base with a pK_a value between 3 and 5 should have a sample removed every 0.5 pH unit between pH 1 and pH 3, every 0.25 pH unit between pH 3 and pH 5, every 0.5 pH unit between pH 5 and pH 7, and every pH unit between pH 7 and pH 10. Record the NMR spectrum of each sample at a temperature of 37°C with a 60-MHz NMR spectrometer.

Treatment of Data

Calculate the chemical shift of the protons relative to the inert tetramethylammonium chloride standard from each spectrum.

Plot the observed chemical shifts (in parts per million) as a function of pH. From the plot, take a reading of pH and of chemical shift in acid (from the initial flat portion of the curve), one from the rapidly changing steep portion of the curve before the center or one from the steep portion after the center, and one in base (the final flat portion of the curve). Calculate the fraction of BH^+ and of B and the pK_a for BH^+. Report the pK_a value to the nearest 0.01 unit.

Question   1   Would NMR be a suitable technique for determining other types of equilibrium constants such as formation constants of metal complexes and solubility product constants?

EXPERIMENT 11-3

Proton NMR Spectrum of Acetylacetone: A Study of Keto-Enol Tautomerism†

Purpose   The usefulness of the NMR method for the study and measurement of tautomeric equilibria is illustrated.

References   1   R. M. Silverstein, G. C. Bassler, and T. C. Morrill, "Spectrometric Identification of Organic Compounds," 4th ed., Wiley, New York, 1981, chap. 4.
2   H. H. Bauer, G. D. Christian, and J. E. O'Reilly, "Instrumental Analysis," Allyn and Bacon, Boston, 1978, chap. 12.

Apparatus   NMR spectrometer
NMR sample tubes (8), 5 mm o.d.
Test tubes and corks (20), 3 in
Medicine droppers (12)
Test tube rack
Beaker, 25 mL
Erlenmeyer flasks (2), 25 mL (with cork stoppers)

Chemicals   Acetylacetone (2,4-pentanedione) [AcAc]
Trifluoro AcAc
Hexafluoro AcAc
TMS
Carbon tetrachloride [CCl_4]
Cyclohexane
Benzene
Methanol
Chloroform
Acetonitrile and substituted compounds (listed in Procedure C)

Theory   In this experiment the percentage of enolization will be determined for acetylacetone (AcAc).

$$CH_3-\overset{\overset{\textstyle O}{\|}}{C}-CH_2{}^a-\overset{\overset{\textstyle O}{\|}}{C}-CH_3 \rightleftharpoons CH_3-\overset{\overset{\textstyle O}{\|}}{C}-CH^b=\overset{\overset{\textstyle OH^c}{|}}{C}-CH_3$$
$$\text{keto} \qquad\qquad\qquad \text{enol}$$

Because the keto hydrogens (*a*) and the enol hydrogens (*b* and *c*) are magnetically nonequivalent protons, they will have different chemical shifts. Therefore, the keto and enol tautomeric forms can be detected in the sample by proton NMR, as illustrated by Fig. 11-6.

In this study the percentage of enolization of acetylacetone is sought. Just as NMR can be used to determine the ratio of the number of different kinds of pro-

†Experiment developed at University of North Carolina, Chapel Hill, North Carolina.

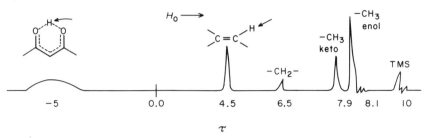

Figure 11-6 / ¹H NMR of acetylacetone.

tons in a single compound, it also can be used to determine the ratio of magnetically nonequivalent protons in two different forms (keto and enol) of acetylacetone. This provides the means to determine the ratio of the two tautomers in a given solution. The peak chosen for the enol form can be either the vinyl or enol methyl peak; for the keto form, either the methylene or keto methyl can be used. Because the integrated areas of these peaks per proton are to be compared, the integrated area of the methylene peak must be divided by two before the percentage of enolization is estimated. As an example, suppose that the integrated area from the automatic integration by the NMR spectrometer is 4 units for the vinyl peak and 2 units for the methylene peak. The percentage of enolization is determined by the relation

$$\% \text{ enol} = \frac{\text{enol area (vinyl)} \times 100}{\text{enol area} + [\text{keto} (-CH_2-) \text{ area}]/2} = \frac{4(100)}{4 + \frac{2}{2}} = \tfrac{4}{5}(100) = 80\%$$

Solvent Effects

The more stable tautomeric form of 1,3-diketones is the enol form; the carbon-carbon double bond of the enol form is in conugation with the second carbonyl group.

Experiments have established that the intramolecular hydrogen bond of acetylacetone stabilizes the enol tautomer by 5 to 10 kcal and the conjugated system further stabilizes this tautomer by another 2 to 3 kcal.

Acetylacetone is present in the enol form to the extent of about 15% in aqueous solutions and about 92% in hexane. The greater enolization in hexane results from stabilization of the enol form by internal hydrogen bonding. In aqueous solutions the carbonyl groups are hydrated, or hydrogen bonded to water molecules, and there is less to gain by enolization. Therefore, polar solvents which stabilize the keto carbonyls will decrease the percent of enolization depending on the degree of their stabilization. An increase in the concentration of a polar solvent in a solution will increase the degree of interaction with acetylacetone.

Substitution

For 1,3-diketones, the nonbonded van der Waals interactions between R and R′ become important. If substitutions of bulky groups are made on the methylene carbon (so-called α-substitutions), then the *steric* hindrance between the bulky group (R″) and R (or R′) results in a reduction of the percent of enolization. Furthermore, the *inductive* properties of the substituent will either increase the electron density in the vicinity of the α-protons and give more keto form, or decrease the electron density and favor the enol form. If the substitution is made on the β-carbon (methyl group) rather than the α-carbon, the inductive effect influences the percent of enolization.

Saturation Effects

Saturation effects can lead to errors in quantitative determinations by NMR. For the simple calculation of percentage enolization, saturation causes a proportional decrease in the integrated areas of the two peaks used in the calculation. Hence, the ratio of the two peak areas remains constant and saturation effects will not cause significant errors. However, in other calculations, these errors may be significant.

Procedures

note Become familiar with the operating controls of the NMR spectrometer. You must be checked in by one of the authorized personnel before you will be allowed to use the instrument.

A. Solvent Effects

You must become familiar with the absorption patterns of the solvents that are used in this experiment. These spectra may be recorded or found in the Varian catalog or in other references.

You are now ready to perform the experiment using the following steps:

1 Prepare 0.2-mole fraction (N_A) solutions of acetylacetone (0.200 g) (AcAc) in these solvents: cyclohexane (0.673 g), CCl$_4$ (1.2311 g), benzene (0.625 g), methanol (0.256 g), chloroform (0.955 g), and acetonitrile (0.328 g). The solutions can be prepared by weighing out the proper quantities in a cork-stoppered 3-in test tube, which is held by a 25-mL beaker (preferably polyethylene) on the balance. Also, prepare a saturated AcAc sample in distilled water and a sample of pure AcAc.
2 Allow the solutions to stand 24 h to come to equilibrium.
3 Fill the NMR sample tubes $\frac{3}{8}$-full with the samples and add to each tube a drop or two of TMS [tetramethylsilane, (CH$_3$)$_4$Si], which serves as an internal standard.
4 Record the NMR spectra of the eight samples. Appropriate instrumental settings for a Varian A-60 are 1000-Hz sweep width, 500-s sweep time, 000-Hz sweep offset, 0.10-milligauss RF field, 0.4 spectrum amplitude, 4-Hz filter bandwidth, and *integrate* at 80 integral amplitude.

Treatment of Data

Calculate the percent of enolization in each solvent by the procedures that are discussed in the Theory section. Compare in tabular form the percent of enolization relative to solvent polarity (i.e., dielectric constant) for each solvent.

B. Concentration Effects

Prepare 0.4, 0.6, and 0.8 N_A concentrations of AcAc in cyclohexane and in acetonitrile. Continue as in steps 2, 3, and 4 in Procedure A.

Treatment of Data

Calculate the percent of enolization for each concentration of AcAc in the two solvents. For each of the two solvents, plot a graph of the percentage enolization vs. mole fraction (N_A) of AcAc.

C. Substitution Effects   Prepare 0.2 N_A concentrations of 3-phenylacetylacetone, dibenzoylmethane, benzoylacetone, 3-methylacetylacetone, 1,1,1-trifluoroacetylacetone, and hexafluoroacetylacetone in chloroform.

Repeat steps 2, 3, and 4 in Procedure A.

Treatment of Data   Calculate the percentage enolization for the various substituted AcAc molecules and note the effects of the different substituents on the degree of enolization.

D. Saturation Effects   Record the NMR spectrum of pure AcAc (with TMS) with the instrument settings that were used in step 4 of Procedure A, except change the RF field from 0.03 to 2.0 milligauss.

Allow the integrator to warm up about the same length of time before integrating each spectrum.

Treatment of Data   Plot the integrated peak areas of one peak at various RF power settings and note the effects of saturation.

Question   1   Explain how steric factors and inductive effects influence the degree of enolization.

EXPERIMENT 11-4

Determination of the Protonation Sequence of Methyliminodiacetic Acid and of the Stoichiometry of the Molybdenum-Methyliminodiacetic Complex by Proton NMR

Purpose   Methyliminodiacetic acid (MIDA) is a metal complexing agent similar to EDTA, with three acidic protons. The relative acidity of these protons is to be determined. In addition, the stoichiometry of the Mo^{VI} complex of MIDA will be determined by observing NMR line shifts as a function of pH or Mo^{VI} concentration.

References   1   J. W. Akitt, "N.M.R. and Chemistry; An Introduction to Nuclear Magnetic Resonance Spectroscopy," Chapman and Hall, London, 1973.
2   S. I. Chan, R. J. Kula, and D. T. Sawyer, *J. Am. Chem. Soc., 86,* 377 (1964).
3   L. V. Haynes and D. T. Sawyer, *Inorganic Chem., 6,* 2146 (1967).
4   R. J. Kula, D. T. Sawyer, S. I. Chan, and C. M. Finley, *J. Am. Chem. Soc., 85,* 2930 (1963)
5   R. J. Kula and D. T. Sawyer, *Inorganic Chem., 3,* 458 (1964).

Apparatus   NMR spectrometer (such as the Varian Model EM 360)
NMR tubes (8), No. 504, 5 mm
Pipets (2), 1 mL
Disposable pipets and bulbs
Beakers, ten 10 mL, one 25 mL

Chemicals   0.5 M MIDA in D_2O, 15 mL
0.5 M sodium molybdate [Na_2MoO_4] in D_2O, 10 mL
6.0 M hydrochloric acid [HCl] in D_2O
6.0 M sodium hydroxide [NaOH] in D_2O
t-Butyl alcohol, a few drops per NMR tube
pH paper (pH 1 to 12)

Part I. Determination of the Acidity of the Protons of MIDA

Theory

At pH 0, MIDA is fully protonated.

$$CH_3 - \overset{+}{\underset{|}{N}} - H \quad \overset{CH_2COOH}{\underset{CH_2COOH}{<}}$$

Because both the carboxylic acid and amine protons are acidic, incremental addition of base will cause sequential chemical shifts of the neighboring C—H protons. This will provide a measure of which is most acidic and which is least acidic. Remember, the chemical shift is directly related to the electron density around the nucleus.

Procedure

Take about 15 mL of 0.5 M MIDA stock solution and pipet 1 mL into a 10-mL beaker. By use of 6 M HCl or 6 M NaOH dissolved in D_2O, adjust the pH of the MIDA solution to zero. Add 1 mL of 0.5 M MIDA to each of two other 10-mL beakers. Adjust the pH of the second to pH 6 and of the third to pH 12. Add a few drops of t-butyl alcohol to each beaker. Fill three 5-mm NMR tubes to a depth of about 50 mm with the various MIDA solutions and record their spectra.

Treatment of Data

Assign all peaks observed in the spectra of MIDA. Remember that butyl alcohol, not TMS, was used as the reference. What is the most acidic proton in MIDA? What is the least acidic proton? Explain.

Question

1 Why are fewer peaks seen in the spectra of MIDA than *could* be expected?

Part II. Determination of the Stoichiometry of the Mo-MIDA Complex

Theory

Below pH 8, Mo^{VI} forms a complex with MIDA. By recording spectra of solutions with fixed amounts of Mo^{VI} and varying amounts of MIDA, the stoichiometry for the complex can be determined.

Procedure

Obtain 10 mL of 0.5 M Na_2MoO_4 stock solution, and five more 10-mL beakers. Use the remaining 0.5 M MIDA stock solution to pipet the following amounts of MIDA and Mo^{VI}.

Beaker	Stoichiometry Mo:MIDA	Na_2MoO_4 (mL)	MIDA (mL)
1	1:$\frac{1}{2}$	2	1
2	1:1	1	1
3	1:1$\frac{1}{2}$	2	3
4	1:2	1	2
5	1:3	1	3

Add a few drops of t-butyl alcohol to each, fill five NMR tubes with the solutions, and record their spectra.

Treatment of Data   What is the stoichiometry of the Mo-MIDA complex? Explain.

Question   1   Some of the spectra obtained in Part II contain no excess MIDA. A comparison of the spectra of bound MIDA and free MIDA (as seen in Part I) shows remarkable differences, including a change in the number of peaks. Explain why the bound and free MIDA exhibit different spectra, remembering that two protons in the same magnetic environment will give only one peak. (*Hint*: Make a model of the free MIDA and of the Mo-MIDA complex.) Does molybdenum bind to the nitrogen or carboxylate groups of MIDA, or does it bind to all three groups?

EXPERIMENT 11-5

Analysis of APC Tablets by Proton NMR†

Purpose   The three components of "APC" tablets (aspirin, caffeine, and phenacetin) will be determined by proton nuclear magnetic resonance spectroscopy.

References   1   T. V. Parke, A. M. Ribley, E. E. Kennedy, and W. W. Hilty, *Anal. Chem., 23*, 953 (1951).
2   M. Jones and R. L. Thatcher, *Anal. Chem., 23*, 957 (1951).
3   D. B. Hollis, *Anal. Chem., 35*, 1682 (1963).

Apparatus   Proton NMR spectrometer, 60 MHz (Varian EM 360, or a similar instrument)
NMR tubes (2), 5 mm
Microspatula (can be made by pounding an iron or copper wire with a hammer; this will fit into an NMR tube)

Chemicals $CDCl_3$
Tetramethylsilane [TMS] (1%) in deuterochloroform [$CDCl_3$]
Aspirin (acetylsalicylic acid), 20 to 40 mg per tube
Phenacetin, 20 to 40 mg per tube
Caffeine, 20 to 40 mg per tube
  To each of the above tubes add 1 mL of 1% TMS ($CDCl_3$); color-code and seal for use by entire class of students.
APC tablet, finely ground (Use no more than 100 mg for analysis due to the solubility of aspirin in $CDCl_3$.)

Theory   Pain-relieving tablets that contain aspirin, phenacetin, and caffeine are common, and are represented on the market by various brand names, such as "APC" and "PAC." As a result, considerable effort has gone into methods for the quantitative determination of each of the three components in the mixture. Published methods include ultraviolet absorption (Ref. 1), infrared absorption (Ref. 2), and

†This experiment is adapted from a similar experiment used at Occidental College by Professor Frank Lambert.

proton nuclear magnetic resonance (Ref. 3). In this experiment only the NMR method will be considered, although it is instructive to do the same determination by all three methods.

Relative amounts of caffeine, phenacetin, and aspirin can be determined by either of two methods: (1) by standard comparison, using the caffeine standard to calibrate the integrator; or (2) by the absolute method, using only the spectrum of the unknown and determining the relative percentage of each component by weight, normalizing so that the % caffeine plus the % aspirin plus the % phenacetin equals 100%.

Procedures

Read specific, simplified operating instructions for the nuclear magnetic resonance spectrometer. A 60-MHz, magnetic-sweep instrument with either external or internal lock is adequate. A compound whose spectrum is familiar (such as ethyl benzene) should be run first to insure that the instrument is operating properly.

A. Determination of the Spectrum of Each Separate Component

Three color-coded tubes that contain 20 to 40 mg of each of the three components (aspirin, caffeine, and phenacetin) as well as 1 mL 1% TMS ($CDCl_3$) have been prepared for this part of the experiment. The spectrum and the integral of the spectrum of the contents of each tube should be taken. Examination of these spectra will allow suitably nonoverlapping peaks to be chosen for the quantitative analysis (after the identification of the peaks that belong to each component) of each component in an APC tablet.

B. Quantitative Analysis of the Unknown

Clean two NMR tubes by rinsing several times with chloroform. Into one tube weigh accurately (to the nearest 0.1 mg) approximately 100 mg of your ground APC unknown. (Do not exceed 100 mg since aspirin is only slightly soluble in $CDCl_3$.) Into the other tube weigh accurately (to the nearest 0.1 mg) approximately 50 mg of pure caffeine. To each tube add 1 mL of 1% TMS ($CDCl_3$); immediately close both tubes.

Using the caffeine standard just prepared, adjust the sensitivity and resolution of the instrument. Take a spectrum and integral of the known caffeine solution.

As quickly as possible and without changing any instrument parameters, replace the caffeine tube with your unknown tube, scan and integrate.

Treatment of Data

From the spectra in Procedure A, determine which peaks belong to each component and determine the number of protons per peak. Select peaks for each component to be used for quantitative analysis.

Determine the relative amounts of caffeine, phenacetin, and aspirin (in percent by weight) by two methods:

1 *Standard comparison method.* Use the caffeine standard to calibrate the integrator, and determine the percent of each component.
2 *Absolute method.* Using only the spectrum of the unknown, determine the relative percent of each component by weight, normalized so that % caffeine plus % aspirin plus % phenacetin equals 100%.

Discuss and compare the results of the two methods.

Question

1 Evaluate NMR as a technique for the determination of aspirin at the trace level. How does it compare with other analytical techniques such as fluorescence, UV-visible absorption spectroscopy, and liquid chromatography?

EXPERIMENT 11-6

Determination of Amino Acid Structure and Sequence in Simple Dipeptides by Proton NMR†

Purpose

Nuclear magnetic resonance spectra of a dipeptide in neutral and in basic solution are recorded. From the spectra of the neutral solution, the identities of the two amino acids in the dipeptide are established, and from the spectral line shifts between the neutral and basic solutions the sequence of the amino acids is determined.

Reference

1   M. Sheinblatt, *J. Am. Chem. Soc.*, *83*, 2845 (1966).

Apparatus

High resolution spectrometer, 60 MHz
NMR tube
Disposable (serum or Pasteur) pipets and bulbs (or an aspirator)

Chemicals

Deuterium oxide [D_2O] supply
9 *M* sodium hydroxide [NaOH] solution
6 *M* hydrochloric acid [HCl] solution
Unknown: Dipeptide, 40 to 50 mg of unknown plus 10 mg DSS (sodium-2,2-dimethyl-2-silapentane sulfonate) or acetonitrile (TMS is not soluble in D_2O.)

Theory

The position of the NMR spectral line of a C—H proton adjacent to an ionizable center depends on the state of ionization of that center. Removal of a proton from an acid group causes the spectral lines of neighboring C—H protons to shift toward higher field strengths, while protonation of a basic group shifts these spectral lines toward lower field strengths. These shifts are caused by inductive changes in electron density, and their magnitudes generally fall off rapidly as the distance between the C—H protons being affected and the ionizable center increases.

Peptides are molecules that contain short sequences of amino acids. The dipeptide contains two amino acids that are linked with an amide or peptide linkage. A full discussion of the type of linkage involved, and the physical effects of the order of the dipeptide (e.g., XY or YX) is found in basic organic and biochemistry texts.

Dipeptides in neutral aqueous solution generally are "zwitterions" with a positively charged group ($-NH_3^{\oplus}$) that has an acidity equivalent to pK_a 8 to 10 and a negatively charged group ($-COO^{\ominus}$) that is equivalent to the conjugate base of an acid with an acidity represented by pK_a 2 to 3. Addition of base to such a neutral solution dissociates the $-NH_3^{\oplus}$ group. Consequently, the spectrum of C—H protons adjacent to the $-NH_3^{\oplus}$ group shifts toward higher field strengths. Conversely, addition of acid to the neutral solution causes protonation of the $-COO^{\ominus}$ group, shifting the spectral pattern of the adjacent protons toward lower field strengths.

Consider a dipeptide XY, where X and Y are two different amino acid residues. If the sequence is XY (X = $NH_3^{\oplus}$—RCO, Y = NHR′COO$^{\ominus}$), addition of base will cause the spectral lines of the group R in the X fragment [denoted R(X)]

---

† From a similar experiment by Dr. Larry Amos and Professor James Sudmeier, University of California, Riverside, Calif.

to shift toward higher field strengths. Also, addition of acid to the neutral solution will cause the spectral lines of the R′(Y) group to shift toward lower field strengths. If the sequence in the dipeptide were reversed to YX (NH$_3$$\oplus$R′CO—NHRCOO$\ominus$), then the spectral pattern of R′(Y) will shift toward higher field strengths with addition of base, while the pattern of R(X) will shift toward lower field strengths with addition of acid. If the correct assignments of peaks are made, then the sequence of amino acids in a dipeptide can be determined by observing peak positions in acidic and/or basic solution, relative to positions in neutral solutions. *Note:* A truly rigorous determination of amino acid sequence would demand examination of both the basic *and* acidic solutions. However, in the interest of economizing on instrument time and expensive compounds, only basic systems will be examined relative to the neutral solutions. This will yield entirely satisfactory results with careful work and strict adherence to procedure.

Procedures

A. Operation of the NMR Spectrometer

Place an acetaldehyde (CH$_3$CHO) reference tube in the tube holder, using the depth gauge; place the tube *carefully* into the probe.

Turn on the spinner to about 30 Hz, following the instructions for the particular instrument that is being used. After setting the necessary controls, in accordance with the manufacturer's instructions, scan the acetaldehyde spectrum; if working in partners, make two copies of the spectrum. Set the instrument for integration and integrate the methyl doublet and aldehyde quartet and record the spectrum.

caution Breaking a probe insert is one of the few *irreversible* things that can be done to an NMR spectrometer. Use great caution not to exert any kind of sideways pressure on the NMR tube. A broken insert can cost $300 and weeks or months of shutdown time on the instrument.

B. Tuning Test

Demonstrate how the spectrometer can be tuned by preparing a spectrum of the aldehyde quartet on a total sweep range of 50 Hz. This broadens out the quartet, permitting a better examination of the peak shape and ringing. For details, refer to the instrument instructions.

C. Procedure for Unknowns

Obtain an unknown from your instructor. Be sure to record its number. Place about 50 mg of unknown in an NMR tube, with some DSS (sodium-2,2-dimethyl-2-silapentane sulfonate) for an internal reference. Add enough deuterium oxide (D$_2$O) from a disposable pipet to make about 50 mm of solution in the bottom of the tube. Put a cap on the tube, and shake to speed dissolution. If the solid does not dissolve completely, add dropwise 6 *M* HCl until the solid is *almost* completely dissolved. *Note:* Adding HCl until the solution is perfectly clear may cause it to end up acidic, instead of nearly neutral, which can lead to an erroneous interpretation of the results. Furthermore, a slightly cloudy sample will not harm the NMR spectrum to a significant degree.

Record the NMR spectrum after the HOD and/or DSS peak has been used to tune the instrument. Set the amplitude so that the largest peak(s) caused by the unknown barely stays on scale; if necessary, let HOD and DSS go off scale. Scan the entire 500-Hz sweep range. Integrate all peaks arising from the unknown —there is no need to integrate HOD or DSS.

Remove the sample tube from the spectrometer and add one or two drops (no more) of the strong NaOH solution. Shake the tube well and take another spectrum. In this spectrum you may see spinning sidebands from the enhanced HOD peak; these are sharp bands, equally spaced on either side of the HOD resonance.

If spin bands are suspected, identify them positively by changing the spin rate noticeably; spin bands move with changed spin rate, while proton resonances do not.

There is no need to integrate this second spectrum unless peaks which overlapped in the neutral solution have now separated; in this case, integrate for positive identification.

special tips and hints

1 At the high spectrum amplitudes used for the unknowns, there will be considerable noise in the base line. On large peaks, this is no problem, but multiply-split small peaks may be obscured. Noise can be diminished by increasing the strength of the RF field, but going too far can lead to saturation and loss of signal. Noise can also be overcome by decreasing the filter bandwidth from 4 to 2 Hz; the effect is to slow the pen speed and thereby cause the pen to be less responsive to spurious noise. High damping is good as long as a sufficiently slow sweep rate is used.

2 The internal reference DSS is similar to TMS, and its principal resonance coincides with that of TMS. However, DSS has a three-carbon side chain, with noticeable resonances around 0.6 and 2.9 δ, and a barely discernible broad resonance around 1.7 δ. Be on the lookout for these to avoid confusing them with the unknown.

D. Shutdown Procedure

The instructor will inform you of the shutdown procedure. *Never shut off any power switches*.

Treatment of Data

The unknowns are dipeptides composed of two of the following amino acids.

Alanine

$$NH_2-\underset{\underset{H}{|}}{\overset{\overset{CH_3}{|}}{C}}-C\underset{OH}{\overset{O}{\diagup}}$$

Glycine

$$NH_2-CH_2-C\underset{OH}{\overset{O}{\diagup}}$$

Leucine

$$NH_2-\underset{\underset{H}{|}}{\overset{\overset{\displaystyle CH_3-\underset{|}{\overset{H}{C}}-CH_3}{\underset{|}{\overset{|}{CH_2}}}}{C}}-C\underset{OH}{\overset{O}{\diagup}}$$

Phenylalanine

$$NH_2-\underset{\underset{H}{|}}{\overset{\overset{\displaystyle\bigcirc}{\underset{|}{CH_2}}}{C}}-C\underset{OH}{\overset{O}{\diagup}}$$

Valine

$$CH_3-\overset{\overset{\displaystyle H}{|}}{C}-CH_3$$

$$NH_2-\overset{\overset{\displaystyle |}{C}}{\underset{\underset{\displaystyle H}{|}}{C}}-\overset{\overset{\displaystyle O}{\parallel}}{C}\diagdown_{OH}$$

Tyrosine

OH

CH_2

$$NH_2-\overset{|}{\underset{\underset{\displaystyle H}{|}}{C}}-\overset{\overset{\displaystyle O}{\parallel}}{C}\diagdown_{OH}$$

The amino acids in the dipeptides are different; i.e., there are none in which the amino acids are the same, such as alanylalanine.

From the spectrum of the unknown in neutral solution, identify the two amino acids of the dipeptide. Label the peaks according to the groups responsible for them. While counting protons, recall that the amine protons are not necessarily seen in the region scanned, and that the amino acids are connected by the peptide linkage,

$$-\overset{\overset{\displaystyle O}{\parallel}}{C}-\overset{|}{\underset{\underset{\displaystyle H}{|}}{N}}-$$

For instance, although alanine and glycine contain, in the free state, a total of 12 protons between them, a dipeptide composed of them contains only 10 protons as a zwitterion, of which 4 are on nitrogens. This leaves only 6 CH-type protons, which simplifies the identification of the unknown from the NMR spectrum.

Once the unknown is identified, decide how the two amino acid residues are connected (X—Y or Y—X?) by considering the shifts of resonances of groups on the carbon atom *alpha* to the $-NH_3^{\oplus}$ group in the zwitterion, upon addition of base. Typical shifts for such resonances between the neutral and basic solution are 30 to 35 Hz.

In writing your report (1) indicate the logic you need to identify the two amino acids; (2) be explicit about which resonances shifted, and their meaning as to the sequence of amino acids in your unknown; (3) show which resonances would have shifted on addition of base if the sequence had been reversed; and (4) include the spectra and a table listing positions of all peaks (except HOD and sidebands) relative to DSS for both solutions.

Questions

1 Why do resonances near the $-NH_3^{\oplus}$ group shift to higher field strengths on addition of base?

2 How would you extend this experiment to determine the sequence of amino acids in a tripeptide composed of three different amino acids, X, Y, and Z?

EXPERIMENT 11-7

Magnetic Resonance Spectroscopy: Titration of Alanine Followed with NMR Detection†

Purpose

This experiment demonstrates an application of proton NMR spectroscopy to the study of a biological titration.

Reference

1   F. J. Waller, I. S. Hartman, and S. T. Kwong, *J. Chem. Ed.*, *54*, 447 (1977).

Apparatus

Proton NMR spectrometer, 60 MHz
pH meter with mini-combination pH electrode
Micro buret
Disposable pipets
NMR tube
Capillary filled with TMS

Chemicals

9 M sodium hydroxide [NaOH], 10 mL
1 M α-alanine and/or β-alanine solutions, 25 mL
Buffers, pH 7 and pH 10 for standardization of pH meter

Theory

An increase in pH causes the chemical shifts of the nonexchangeable protons of an amino acid to move upfield, due to the increased negative charge density in the molecule as the equilibrium shifts to the right for the total titration of α-alanine:

$$\overset{+}{N}H_3-\overset{\overset{\displaystyle CH_3}{\displaystyle |}}{C}H-COOH \underset{}{\overset{OH^-}{\rightleftharpoons}} \overset{+}{N}H_3-\overset{\overset{\displaystyle CH_3}{\displaystyle |}}{C}H-COO^- \underset{}{\overset{OH^-}{\rightleftharpoons}} NH_2-\overset{\overset{\displaystyle CH_3}{\displaystyle |}}{C}H-COO^-$$

Zwitterion   Anion

The chemical shifts of the protons α and/or β to the $\overset{+}{N}H_3$ portion and the pH values of the solution are measured simultaneously as the titration proceeds. External TMS is used as the standard. In α-alanine, the methyl protons appear as a doublet and the α-proton is a quartet. In β-alanine, both sets of methylene protons are triplets.

Procedure

Record the NMR spectrum of approximately 0.5 mL of a 1 M amino acid solution (use a capillary insert of TMS in the NMR tube as a reference). Titrate 15 mL of the amino acid solution with 9 M NaOH in a micro buret; add increments of 0.02, 0.04, 0.04, 0.10, 0.10, 0.10, 0.20, 0.20, 0.20, 0.50, and 0.30 mL (a total of 1.80 mL). After each addition of NaOH run an NMR spectrum and determine the pH of 0.5 mL of the solution. Return the aliquot to the solution and add the next increment of NaOH; repeat until the titration is completed, returning the sample used for the NMR spectrum back to the solution before the next addition of base.

Treatment of Data

Measure the chemical shift values relative to the external TMS standard at the midpoint of the doublet for α-alanine. Plot pH vs. the chemical shift for the methyl protons in α-alanine. Determine the pK_a for $\overset{+}{N}H_3$ from the midpoint of the titration curve.

†Adapted with permission from Ref. 1. Copyright 1977, Division of Chemical Education, American Chemical Society.

EXPERIMENT 11-8

Magnetic Resonance Spectroscopy: Interpretations of ^{13}C NMR Spectra—
A Study of Amygdalin ("Laetrile")

Purpose	Spectra of amygdalin will be used to illustrate the interpretation of ^{13}C NMR.

References

1 T. Cairns, J. E. Froberg, S. Gonzales, W. S. Langham, J. J. Stamp, J. K. Howie. and D. T. Sawyer, *Anal. Chem., 50,* 317 (1978).
2 G. C. Levy, R. L. Lichter, and G. L. Nelson, "Carbon-13 Nuclear Magnetic Resonance Spectroscopy," 2nd ed., Wiley-Interscience, New York, 1980.
3 J. B. Stothers, "Carbon-13 NMR Spectroscopy," Academic Press, New York, 1972.
4 A. Lombardo and G. C. Levy in "Treatise on Analytical Chemistry," P. J. Elving, E. J. Meehan, and I. M. Kolthoff (eds.), 2nd ed., part 1, vol. 10, Interscience, New York, 1983, chap. 2.

Apparatus

NMR tubes
Fritted glass funnel
Beakers, 250 mL

Chemicals

Acetonitrile, nanograde
Ethanol (95% and absolute)
Trifluoroacetic acid, sodium salt (Eastman Chemicals, Rochester, NY)
Trifluoroacetic anhydride (Eastman) (Trifluoracetic anhydride is mixed with an equal volume of acetonitrile that contains 10% sodium trifluoroacetate.)
Petroleum ether, pesticide grade
Amygdalin (Aldrich Chemical Co.) [The material is recrystallized by dissolving 2 g in 100 mL 95% ethanol (heated to boiling). The resulting solution is filtered twice through a fritted glass funnel, prior to its evaporation to 20 mL, and allowed to crystallize. The resulting trihydrate crystals are then dissolved in 100 mL of boiling absolute ethanol and concentrated to less than 10 mL. After recrystallization, the crystals are dried in a 60°C vacuum oven 18 h and stored in a desiccator.]
Epimeric mixture [Add 1 drop of NH_3(aq) (5 mL concd NH_3 diluted to 25 mL) to an amygdalin sample solution (0.4 g per 5 mL).]

Theory

The ^{13}C chemical shift scale is similar to that for protons. If TMS is used as the reference, resonances at higher frequency (at constant applied field) are considered to have positive chemical shifts in parts per million—the same as the proton "δ" scale. Aliphatic carbons occur at higher shielding (lower frequencies) and, therefore, have the smaller chemical shifts. With increasing substitution, and particularly heteroatom substitution, the resonances appear at greater chemical shifts. Olefinic and aromatic carbons occur at 90 to 170 ppm; whereas, carbonyl carbons have chemical shifts at 160 to 220 ppm. In parts per million the carbon scale is greater than 20 times as wide as that of the proton. Because the line widths are similar, this represents a true increase in dispersion.

Addition of a polar substituent to a carbon causes ^{13}C shifts. This inherent sensitivity to substituents makes ^{13}C NMR an exceptional structural characterization method for organic molecules.

Amygdalin (I), a naturally occurring cyanogenetic glycoside found in the kernels or seeds of members of the *Rosaceae* (almond, apple, apricot, cherry, peach, pear, plum, quince), was first isolated in 1830. The structural elucidation of amygdalin (I) and various synthetic approaches were first published in 1923–1924. About 10 years later the biochemistry was discussed by Viehoever and Mack. As a gentiobioside of mandelonitrile, it contains several chiral (asymmetric) centers that can give rise to epimeric mixtures. In particular, the aglycone entity or nonsugar-derived chiral center of mandelonitrile is susceptible to epimerization, particularly under basic conditions, because of the weakly acidic character of the benzylic proton. The naturally occurring amygdalin that is extracted from seeds and kernels has the R configuration.

Laetrile vs. Amygdalin

Amygdalin is not structurally synonymous with laetrile. Laetrile (II) results from (*a*) hydrolysis of amygdalin and subsequent oxidation of the L-mandelonitrile-β-glucoside product with platinum black, or (*b*) by condensation of mandelonitrile with glucose followed by oxidation, or (*c*) by condensation of mandelonitrile with glucuronic acid. The proponents of laetrile and the popular press have confused the situation by their use of the name to describe amygdalin. Presently, the drug that is offered as an anticancer agent is amygdalin (I).

Structures of amygdalin and laetrile are shown in (I) and (II), respectively.

(I)
Amygdalin
$C_{20}H_{27}NO_{11}$
D–mandelonitrile–beta–D–glucoside–beta–D–glucoside

(II)
Laetrile
$C_{14}H_{15}NO_7$
L–mandelonitrile–beta–glucuronic acid

Proton NMR provides evidence that epimers are present in commercial amygdalin samples; methine chemical shift values [6.00 ppm for (R)-amygdalin and 6.09 ppm for (S)-amygdalin] for the benzylic proton attached to the chiral (asymmetric) aglycone carbon atom. The second resonance appears when (R)-amygdalin is epimerized by base-catalysis. Natural abundance broad-band proton-decoupled ^{13}C NMR spectra of amygdalin (0.4 g per 5 mL), of epimerized

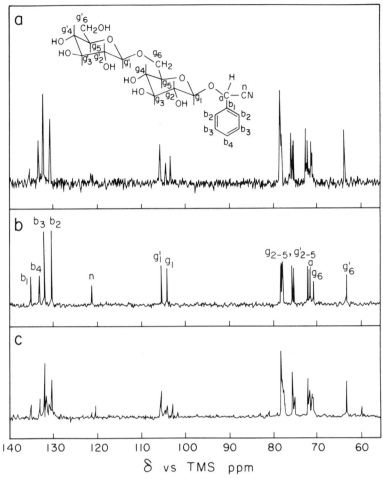

Figure 11-7 / ¹³C NMR of (a) epimerized (from addition of NH₃) (R)-amygdalin, (b) pure (R)-amygdalin, and (c) a commercial pharmaceutical sample from Mexico.

amygdalin (0.4 g per 5 mL), and of a Mexican pharmaceutical sample of an aqueous solution of "amygdalin" (0.8 g per 5 mL) are illustrated in Fig. 11-7. Off-resonance continuous-wave decoupling experiments cause the carbon resonances that are labeled b_1 and n to remain singlets; carbon resonances g_6 and g'_6 to become triplets and all other carbon resonances to split into doublets. When the epimerization experiment is repeated with pure D_2O solvent, the acidic hydrogen atom at the carbon labeled a is exchanged for a deuterium atom. This causes the ¹³C resonance of this carbon to have a substantial decrease in intensity (less nuclear Overhauser effect).

Procedure

The following set of conditions is suggested:

Spectrometer internally locked on the deuterium resonance of the solvent, 1:4 (v/v) $D_2O:H_2O$

The ¹³C fixed-frequency transmitter (22.638 MHz, probe head and preamp used)

Two 8k free induction decays (FIDS) collected at a sweep width of 5000 Hz, yielding an acquisition time of 1.638 s and digital resolution of 0.61 Hz

Signal-to-noise ratio obtained after 2200 4.5-μs pulses (90° pulse, 13.4 μs); total data collection time, about 1 h

Exponential multiplication performed on FIDS prior to transformation by use of a line-broadening constant of 0.5 Hz

Run spectra of (*a*) Aldrich amygdalin standard, base epimerized with ammonia, (*b*) Aldrich amygdalin standard, and (*c*) an injectable sample of amygdalin, if available.

Treatment of Data

Assign each of the resonances on each spectrum, with an explanation as to your assignment. Indicate the evidence for the R and S epimers. Indicate any impurities, if present.

EXPERIMENT 11-9

Magnetic Resonance Spectroscopy: ESR Spectra of Organic Free Radicals

Purpose

ESR is an ideal means to study free radicals that result from irradiation of molecules and that are formed as intermediates in chemical reactions. Free radicals formed from amides and malonic acid that are irradiated with x-rays are investigated.

References

1 E. D. Olsen, "Modern Optical Methods of Chemical Analysis," McGraw-Hill, New York, 1975, chap. 13.
2 H. H. Bauer, G. B. Christian, and J. E. O'Reilly, "Instrumental Analysis," Allyn and Bacon, Boston, 1978, chap. 13.
3 J. E. Wertz and J. R. Bolton, "Electron Spin Resonance: Elementary Theory and Practical Applications," McGraw-Hill, New York, 1972.
4 H. M. Swartz, J. R. Bolton, and D. C. Borg (eds.), "Biological Applications of Electron Spin Resonance," Wiley-Interscience, New York, 1972.
5 M. T. Rogers, S. J. Bolte, and P. S. Rao, *J. Am. Chem. Soc., 87,* 1875 (1965).
6 I. R. Goldberg and A. J. Bard in "Treatise on Analytical Chemistry," P. J. Elving, E. J. Meehan, and I. M. Kolthoff (eds.), 2nd ed., part 1, vol. 10, Interscience, New York, 1983, chap. 3.

Apparatus

Electron spin resonance spectrometer
Quartz sample cell, approx. 3 mm o.d.
X-ray (1-MeV) source for irradiation of samples

Chemicals

Acetamide
Propionamide
N,N-dimethylbutyramide
N,N-diethylacrylamide
N,N-di-*n*-propylacetamide
N,N-diisopropylacetamide
Trimethylacetamide
Malonic acid
Vandyl sulfate (reference)

Theory

Electron spin resonance (ESR), electron paramagnetic resonance (EPR), and electron magnetic resonance (EMR) are terms used synonymously for a method which is similar to nuclear magnetic resonance (NMR). It differs from NMR in that the reorientation of the magnetic moment of an *unpaired electron,* rather than a nucleus, in a magnetic field is observed. Unpaired electrons occur in free radicals, in substances such as O_2, NO, and NO_2, in transition metal species, in semiconductor materials, and in metals.

The basic resonance energy equation for ESR spectroscopy is

$$\Delta E - h\upsilon = gBH$$

where

g = splitting factor (= 2.003 for the free electron)
B = Bohr magnetron = 9.273×10^{-21} erg/G
H = applied magnetic field

The frequency is held constant and the magnetic field varied. At a particular value of the magnetic field, H_R, resonance absorption of energy occurs to give rise to a peak in the spectrum. Frequencies for ESR are usually in the microwave region.

Unpaired electrons frequently show fine structure in their resonances due to coupling with spinning nuclei in their vicinity. Although the principles are much the same as those for spin-coupling of protons, the hyperfine splitting patterns often are unresolved and, therefore, preclude quantitative interpretation.

An unpaired electron in the absence of a magnetic field may exist in the $+\frac{1}{2}$ or $-\frac{1}{2}$ state. When it interacts with an external magnetic field, the energy levels become split, the lowest energy level has the spin magnetic moment aligned in the direction of the magnetic field and corresponds to the quantum number $M_s = -\frac{1}{2}$; the upper level corresponds to the number $+\frac{1}{2}$. Interaction with one proton will further split each of these energy levels, giving rise to two spectral lines. Two interacting protons cause a further splitting, as shown in Fig. 11-8(c), to give rise to three spectral lines of intensity 1:2:1. The following table gives the number of lines and their relative intensities for n equivalent interacting atoms with $I = \frac{1}{2}n$:

Number of Equivalent Atoms with $I = \frac{1}{2}n$	Number of Lines $n + 1$	Relative Intensities of ESR Lines
1	2	1:1
2	3	1:2:1
3	4	1:3:3:1
4	5	1:4:6:4:1
5	6	1:5:10:10:5:1
6	7	1:6:15:20:15:6:1

ESR spectra are usually recorded in the first-derivative form to obtain greater resolution than the normal spectrum provides. Quantitative measurements are from the second integral of the ESR curve. The intensities of the lines may be estimated from the integrations of the curves or by an approximation: intensity = derivative height $\times (\Delta H)^2$, where ΔH is the difference in the applied magnetic field between the peak and the trough of the first-derivative spectrum. Relative intensities of lines in a spectrum can be determined in this manner (see Fig. 11-9). Coupling constants are determined by measuring the difference in applied field between peaks, as indicated in Fig. 11-9.

A block diagram of an ESR spectrometer is shown in Fig. 11-10 and a simple physical schematic is shown in Fig. 11-11.

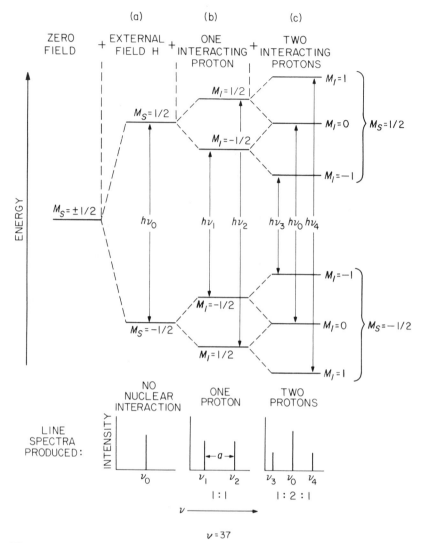

Figure 11-8 / **The splitting of electron spin energy levels by the successive addition of (a) an external magnetic field *H*, (b) one interacting proton, and (c) a second interacting proton. The line spectrum produced is shown immediately below the splitting diagram; all three spectra are at a constant field *H*.** M_s = **electron spin quantum number;** M_I = **nuclear spin quantum number. (From "Modern Optical Methods of Analysis," E. D. Olsen, copyright © 1975 by McGraw-Hill, New York, fig. 13-41, p. 564. Reprinted with permission.)**

Procedures

A. Amides

Tune the ESR spectrometer in accordance with the instructions that are supplied with the instrument.

Irradiate a sample of acetamide in the quartz sample tube for 1 min with x-rays—a 1-MeV source (1×10^7 REPs: 1 REP = 93 erg/g of material). Immediately run an ESR spectrum. Irradiation will produce paramagnetism. Proceed with each amide in the same manner until all the spectra are recorded.

Treatment of Data

The empirical rules for sites of radiation damage to amides are

1 Alkyl substituent on nitrogen is most favored site for loss of a hydrogen atom.

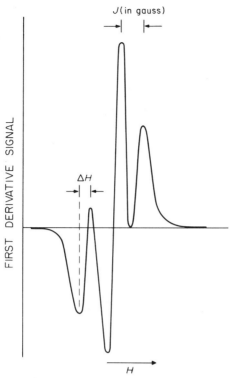

Figure 11-9 / ESR spectrum.

2 In amides and N-substituted amides, N—H bonds never appear to be broken; C—H bond adjacent to nitrogen atom or carbonyl group is usually broken.
3 Hydrogens on carbon atoms adjacent to N or C(O) are not abstracted.

Determine the number of lines in each spectrum, and the relative intensities of these lines. The approximate intensities can be determined from multiplying the derivative height times $(\Delta H)^2$.

Write the structure of each of the free radicals formed from x-irradiation of the amides.

B. Malonic Acid

Irradiate malonic acid for 1 min by use of a 1-MeV source of radiation. Record an ESR spectrum and use vanadyl sulfate as a reference.

Treatment of Data

The empirical rules for sites of radiation damage to acids are

1 Free radical carbon tends to be next to COOH group.
2 Free radical carbon tends to have as few H's as possible directly attached.
3 C—H or C—F bond breaks before C—C bond.

Determine the number of lines in the spectrum and the relative intensities of these lines and the coupling constants.

Write the structure of the malonic acid radical.

Determine the g value for the unpaired electron of the radical species.

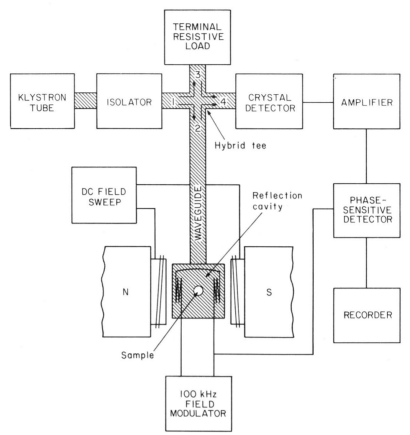

Figure 11-10 / Block diagram of a typical modern ESR spectrometer. (*From "Magnetic Resonance Spectroscopy," H. G. Hecht, copyright © 1967 by John Wiley & Sons, Inc., New York, p. 122. Reprinted by permission.*)

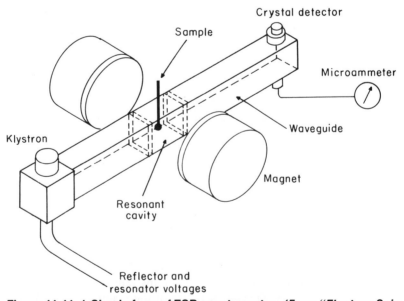

Figure 11-11 / Simple form of ESR spectrometer. (*From "Electron Spin Resonance in Chemistry," P. B. Ayscough, copyright © 1967 by Methuen & Co., Ltd., London, p. 137. Reprinted by permission.***)**

EXPERIMENT 11-10

Magnetic Resonance Spectroscopy: ESR Spectra of Transition Metal Complexes

Purpose

The use of ESR for the characterization of the structure for complexes of paramagnetic metal ions is illustrated.

References

1　A. Serianz, J. R. Shelton, F. L. Urbach, R. C. Dunbar, and R. F. Kopczewski, *J. Chem. Ed., 53,* 394 (1976).
2　R. A. Rowe and M. M. Jones, *Inorg. Syn., 5,* 115 (1957).
3　M. Bersohn and J. C. Baird, "An Introduction to Electron Paramagnetic Resonance," Benjamin, New York, 1966.
4　T. F. Yen (ed.), "Electron Spin Resonance of Metal Complexes," Plenum, New York, 1969.
5　R. Wilson and D. Kivelson, *J. Chem. Phys., 44,* 4440 (1966).
6　R. Wilson and D. Kivelson, *J. Chem. Phys., 44,* 154 (1966).

Apparatus

ESR spectrometer, 9.5 GHz
Quartz sample cell
Heat exchanger immersed in liquid nitrogen

Chemicals

Bis-(acetylacetonato)oxovanadium(IV) [prepared by the one-electron reduction of vanadium(V) ion, through the use of ethanol, see Ref. 2.]
Toluene (deoxygenated by bubbling N_2 through it for 15 min)
DPPH standard

Theory

The spectrum of $V^{IV}O(AcAc)_2$ shows eight lines (vanadium has a spin of $\frac{7}{2}$), when run at room temperature. However, the intensities of the eight lines are not all equal, due to dipolar coupling and spin-orbit coupling. At liquid-nitrogen temperatures, the spectrum splits into two overlapping eight-line spectra. The coupling constants and *g* values can be measured from the spectra, and used to determine the geometry of $V^{IV}O(AcAc)_2$, the structures of its frozen and liquid solutions, as well as the ordering of energy levels in the MO description of the complex.

Procedure

Tune the ESR spectrometer according to the instructions supplied with the instrument. A power attenuation of less than 50 mW and a modulation amplitude less than 2.5 will give the best resolution. Resonance should then occur in the region from 3000 to 4000 G. Prepare a 0.01 *M* solution of $V^{IV}O(AcAc)_2$ in toluene that has been deoxygenated and record its ESR spectrum. Use a heat exchanger (immersed in liquid nitrogen) to pass dry nitrogen directly into the ESR cavity. Record a spectrum of the complex at liquid-nitrogen temperature.

Treatment of Data

Determine the *g* values from the two spectra, as well as the coupling constants. How do the spectra of the liquid and frozen samples compare?

Discuss how these spectra can be used to determine the geometry of $V^{IV}O(AcAc)_2$ and the structures of its frozen and liquid solutions (Refs. 5 and 6).

Question

1　Why is this solution deoxygenated?

III

Separation Methods

12
Gas Chromatography

Prior to the invention of gas chromatography by Martin and Synge, and James and Martin, the separation of small amounts of close-boiling volatile liquids was, to say the least, a difficult operation. For mixtures where chemical methods could not be employed, such liquids could be separated only by fractionation in precision microfractionation columns, and only then if enough material was available, if the boiling points were sufficiently different, and if azeotropic mixtures could be avoided.

Gas chromatography is a surprisingly simple technique with great versatility, and is a proven method for the separation and analysis of volatile mixtures. Its great utility is due to the small amounts of sample that are required, and to the tremendous resolution that can be achieved.

Gas chromatography involves injection of a small amount of sample into a moving stream of gas, which is termed the *mobile phase* or the *carrier gas*. The sample is carried by the gas stream through a column that consists of a tube packed with solid particles. Because the particles are retained in the column, they are referred to as the *stationary phase*. Separation of a sample mixture into its individual components is achieved if the components are retained in the column to different extents. Separation of a sample that consists of components *A* and *B* is illustrated schematically in Fig. 12-1A. Component *A* is retained in the column longer than component *B*, thereby allowing the two components to be separated. The resulting *chromatogram* is shown in Fig. 12-1B. Component *B* is detected as it *elutes* from the column; in turn component *A* is eluted through the detector.

In the *elution* method of chromatography, a carrier gas more lightly adsorbed than any component in the sample is passed through the column into which a sample has been introduced. Each component of the sample will then partition be-

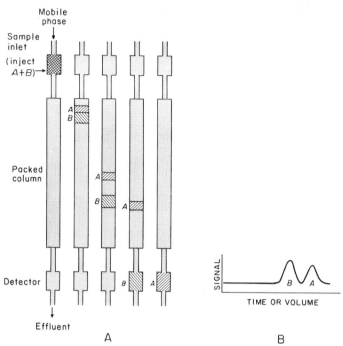

Figure 12-1 / (A) Schematic representation of chromatographic separation of a mixture consisting of two components, *A* and *B*. **(B)** Chromatogram of *A* and *B*.

tween the stationary phase and the gas phase, and that in the gaseous phase will be moved by the flow of carrier gas. As a result, the sample will be carried through the column in a definite time (the *retention* time), which depends upon the affinity of the column packing material for that component, the temperature, and the rate of flow of the carrier gas.

Apparatus

A simplified block diagram of a typical gas chromatograph is shown in Fig. 12-2. The apparatus consists of a pressurized tank of carrier gas, a pressure regulator to control the flow rate of the gas through the chromatograph, a sample inlet, the column, a detector with associated electronics, a recorder, and a flow meter to measure the flow rate of carrier gas. Chromatographs also provide heating for the column, the sample inlet, and the detector. The temperatures of these three components can usually be controlled independently.

Column Materials

There are two methods of elution analysis, *gas-liquid-partition* chromatography and *gas-solid-adsorption* chromatography. In the gas-adsorption method, the column is packed with adsorbents such as activated charcoal, silica gel, or aluminum oxide. Separation of various components depends upon their respective adsorption coefficients. The chief difficulty with gas adsorption is caused by the existence of nonlinear adsorption isotherms for most components. This gives rise to asymmetrical peak shapes (tailing, etc.) which prevent clean separations. For this reason, the gas-liquid-partition method is employed whenever possible.

The gas-liquid-partition chromatography (GLPC, GLC, or GC) column consists of a tube packed with a support material (such as crushed firebrick, Celite, or small glass beads) which is impregnated with a nonvolatile liquid (such as a silicone oil). The sample to be separated is introduced into the column and the tube

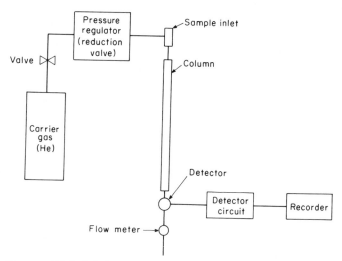

Figure 12-2 / Schematic diagram of a gas chromatograph.

is swept with a suitable carrier gas (such as helium). Each component of the sample divides itself (*partitions*) between a gas phase (in the interstices between the grains of packing) and a liquid phase (the oil coating the grains of packing). The ratio between the concentration of any component in the gas phase and its concentration in the liquid phase (the *partition coefficient*) is a constant (Henry's Law is usually followed at these low concentrations in the liquid phase) whose value depends upon the nature of the species (heat of solution) and the temperature.

In practice, the temperature of the column is adjusted so that the substances that are to be separated have suitable vapor pressures. The components of the sample mixture progress through the column at a rate that is proportional to the relative vapor pressures of the individual components of the mixture. Because the stationary oil is supported on the surface of finely divided particles, the number of times that each component evaporates and redissolves during its passage through the tube is very large, and, accordingly, the separation is extremely efficient (1000 to 3000 theoretical plates are not unusual in a packed column). The most volatile component travels down the tube at the fastest rate and separates from other components of lower vapor pressure; it thus emerges at the end of the column as a separate band.

Common Column Materials†

The ability to separate components in complex mixtures requires the judicious choice and use of a variety of solid-adsorption and liquid-partition columns. The novice chromatographer who consults a catalog from a supply house is presented with an immense array of materials from which to select. Selection of the "best" packing material from such a list can be bewildering. To obviate this situation for liquid-partition phases, two things have been done:

(1) Panels of expert chromatographers have met and selected a small list of "preferred" stationary phases [*J. Chromatog. Sci., 13,* 115 (1975)]. In their view 90% of the separation problems that can be resolved by GC can be accomplished using the phases in the list; these materials should be used if at all pos-

---

†Written by T.W. Gilbert, University of Cincinnati.

sible. Not only does their use aid in the standardization of chromatographic procedures, but these materials possess superior thermal stability. Thus, even though the literature indicates that a given phase will accomplish the desired separation, it is not necessarily the best packing material to buy. Because few columns are purchased and used for only a single purpose, selection of a column that will accomplish the separation and, at the same time, be able to function at temperatures as much as 100° higher is a better approach. There are many phases that were widely used in the past, but are now obsolete.

(2) To assist in the selection of stationary phases, a quantitative system to represent the retention characteristics of stationary liquid phases has been developed. Known as McReynold's Constants, these values replace older purely subjective evaluations. For example, THEED (tetrahydroxyethylenediamine) is known as an "alcohol retarder" because this phase greatly retains this class of compounds relative to other classes of similar molecular weight and boiling point; similarly, TCEP [1,2,3-tris(2-cyanoethoxy)propane] is known as an "aromatic retarder" because benzene (bp, 80°C) and higher aromatics are eluted after all saturated hydrocarbons up to n-decane (bp, 174°C). The McReynold's system quantifies these statements, and makes similarities and differences in phases immediately apparent. Some catalogs and texts contain extensive listings of these constants and explain their use.

The primary list of the six most generally useful preferred liquid stationary phases (in order of increasing overall polarity) is:

1 Dimethylsilicone; T_{max}, 350°C. Examples: OV-101, SP-2100, SE-30.
2 50% phenyl methyl silicone; T_{max}, 375°C. Examples: OV-17, SP-2250.
3 Polyethylene glycol; T_{max}, 225°C. Example: Carbowax 20M.
4 Diethyleneglycol succinate (DEGS); T_{max}, 200°C.
5 3-Cyanopropyl silicone; T_{max}, 275°C. Examples: Silar 10CP, SP-2340.
6 Trifluoropropylmethyl silicone; T_{max}, 275°C. Examples: OV-210, SP-2401, QF-1.

This list is not intended to satisfy every chromatographic need; a secondary list of 18 phases provides finer increments in the polarity range; and a third list of 13 phases has been prepared to accommodate special separation problems. Carborane-silicones, such as Dexsil 300, have extreme temperature stability and can be used routinely at 450°C and for short periods to 500°C. Porous polymer packing materials have no liquid phase as such and have been found useful for applications such as the determination of water in solvents. For example, water is eluted before toluene from Chromosorb 101, and 0.01 to 0.10% water concentrations can be easily determined in this solvent.

Detectors

A means to detect separated components as they elute from the column is an essential part of a gas chromatograph. The two most commonly used detectors are the flame ionization detector and the thermal conductivity detector.

The flame ionization detector is based on the formation of ions when the sample passes through a hydrogen/oxygen flame, i.e., it is burned as it elutes from the column. The ions from the burned sample lower the resistance of the flame. The lowered resistance is accompanied by current in a circuit that applies a large voltage across the flame as shown schematically in Fig. 12-3. Thus, a peak is recorded whenever a component elutes from the chromatograph and undergoes combustion in the flame. The flame ionization detector is currently the most popular detector. It provides excellent sensitivity, a wide range of linear response,

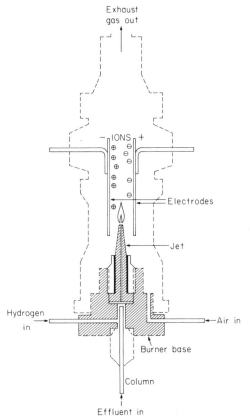

Figure 12-3 / Flame ionization detector.
(Courtesy of Beckman Instruments.)

and good reliability. The detector is widely applicable, because it responds to any compound that "burns" in the flame. Its main disadvantage is destruction of the sample.

The thermal conductivity detector is based on the difference in thermal conductivity of the carrier gas and the sample. Thermal conductivities of several compounds and elements are shown in Table 12-1. This small sampling is repre-

Table 12-1 / Thermal Conductivity of Gases

Gas	$K \times 10^5$, cal/°C · mol
Acetone	2.37
Air	5.83
CO_2	3.52
N_2	5.81
He	34.80
H_2	41.60
CH_4	7.21
Ethane	4.36
Butane	3.22
Methanol	3.45

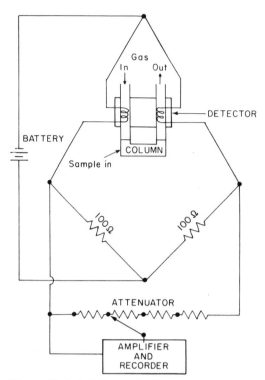

Figure 12-4 / Schematic diagram of thermal conductivity detector. (Adapted from "Principles of Instrumental Analysis," 2nd ed., Douglas A. Skoog and Donald M. West, copyright © 1980 by Saunders College/Holt, Rinehart and Winston. Reprinted by permission of Holt, Rinehart and Winston, CBS College Publishing.)

sentative in that most organic compounds exhibit thermal conductivities quite different than that for He and H_2. Thus, He (or H_2, although the danger of explosion exists) can serve as the carrier gas. The thermal conductivity is measurably altered by the presence of most organic compounds, because they reduce the molar concentration of He in the detector. The thermal conductivity of carrier gas eluting from the column is measured as shown schematically in Fig. 12-4. The inlet carrier gas passes over one heated element (a fine Pt or W wire or a thermistor) and then over a second element after it passes from the column. The pure He carrier gas dissipates ("cools") the elements in proportion to its thermal conductivity. When the sample passes over the heated element, the element responds quickly to the lower thermal conductivity associated with dilution of pure He by the sample and increases in temperature. This temperature change is accompanied by a change in electrical resistance in the element, which is detected by a current in the Wheatsone bridge circuit. The bridge responds to any imbalance in the resistances of the two elements. Because most compounds exhibit much smaller thermal conductivities than He, the detector is essentially a universal detector. Although it was once the most common detector of gas chromatography, it is being steadily supplanted by the more sensitive flame ionization detector.

Many other detectors are also used for GC, albeit they are less common than flame ionization and thermal conductivity. One worthy of mention, however, is the electron-capture detector, which exhibits excellent sensitivity for compounds with electronegative functional groups such as halogens. Consequently, it is used extensively for the detection of pesticides that are halogenated.

Qualitative and Quantitative Measurements

A typical chromatogram is shown in Fig. 12-5. A particular component in a mixture can be identified by its *retention time*, t_R, which is the time between injection and detection. Figure 12-5 shows how t_R is measured from a chromatogram. The retention time of a particular compound is constant for a fixed set of chromatographic conditions (flow rate, temperature, column condition). Qualitative identification is made by comparing t_R of the unknown with retention times of standards that have been injected into the chromatograph. This strategy works so long as components have unique retention times. Confirmation of identity by another technique such as mass spectrometry is judicious, and essential for totally unknown components.

The area under each peak is proportional to the concentration of that component in the original mixture. If the peaks are reasonably sharp and the flow rate is carefully controlled, the peak height is also proportional to concentration. Thus, a calibration curve can be prepared by plotting either peak height or peak area as a function of concentration for a series of standards.

Experiments 12-2 and 12-3 are designed to illustrate quantitative and qualitative measurements by GC.

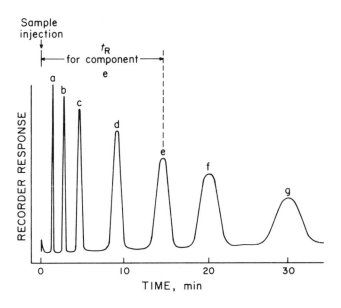

	B.P., °C
a. Acetone	56.5
b. Methyl ethyl ketone	80
c. Diethyl ketone	102
d. Ethyl *t*−butyl ketone	125
e. Isopropyl *t*−butyl ketone	135
f. *n*−propyl *t*−butyl ketone	145
g. Isobutyl *t*−butyl ketone	153

Figure 12-5 / Chromatogram of a mixture separated on a Carbowax 400 column, 100°C, thermal conductivity detector.

EXPERIMENT 12-1

Determination of Optimum Flow Rate in Gas Chromatography

Purpose

The optimum flow rate for the determination of chloroform by gas chromatography is determined by use of the van Deemter equation.

References

1 H. H. Willard, L. L. Merritt, Jr., J. A. Dean, and F. A. Settle, Jr., "Instrumental Methods of Analysis," 6th ed., Van Nostrand, Princeton, N.J., 1981, chaps. 15 and 16.
2 G. W. Ewing, "Instrumental Methods of Chemical Analysis," 4th ed., McGraw-Hill, New York, 1975, chaps. 18 and 19.
3 H. H. Bauer, G. D. Christian, and J. E. O'Reilly, "Instrumental Analysis," Allyn and Bacon, Boston, 1978, chap. 22.
4 D. A. Skoog and D. M. West, "Fundamentals of Analytical Chemistry," 4th ed., Saunders, Philadelphia, 1982, chaps. 27 and 29.
5 D. A. Skoog and D. M. West, "Principles of Instrumental Analysis," 2nd ed., Saunders, Philadelphia, 1980, chaps. 24 and 26.
6 S. Dal Nogare and R. S. Juvet, Jr., "Gas-Liquid Chromatography," Interscience, New York, 1962.
7 H. Purnell, "Gas Chromatography, Wiley, New York, 1962.
8 J. J. van Deemter, F. J. Zuiderweg, and A. Klinkenberg, *Chem. Eng. Science,* 5, 271 (1956).

Apparatus

Gas chromatograph with recorder
Tank of helium gas with pressure regulator
Timer
Soap bubble flow rate meter
Column: Dimethylsilicone, SE-30
Syringe, 10 μL

Chemical

Chloroform (reagent grade)

Theory

Good separation of a given pair of components by gas chromatography depends on the judicious selection of a column substrate (both the solid support and the liquid phase) and on the efficiency of the overall GC system. The relative position of the various components in the sample on the chromatogram is affected by a solute-solvent type of interaction with the column substrate. Solvent efficiency, called relative retention time, is expressed as the ratio of the retention time of a solute compared to the retention time for a material (such as air) which does not interact with the liquid phase. The relative retention time is adjusted for the distribution coefficients of each solute in the solvent, the flow rate of the carrier gas, and the temperature of the column, and these retention times may be used to compare solvent efficiency.

Column efficiency is concerned with the broadening of an initially compact band of solutes as it passes through the column. The broadening is a result of the column design and of column operating conditions. This efficiency is described quantitatively by the height equivalent to a theoretical plate (H), where H is defined as that length of column necessary for the attainment of solute equilibrium between the moving gas phase and the stationary liquid phase. In order to compare column efficiencies, however, it is necessary to specify the solvent, solute, temperature, flow rate, and sample size. The number used to describe the column efficiency is expressed in terms of number of theoretical plates, and is calculated from a recorded single peak. As in distillation separations, a theoretical plate is defined as one perfect equilibrium being established between two phases. It is a theoretical concept used to evaluate column performance, but cannot be used as an absolute measure of a column's separating ability; however, it can be used to compare similar columns or to set standards for packing techniques.

Most factors that affect column efficiency are evaluated by their effect on N, the number of theoretical plates. The height of a theoretical plate is then determined by

$$H = \frac{L}{N}$$

where L is length of column in centimeters and N is the number of theoretical plates. The plate number N of a column may be determined directly from a chromatogram (illustrated in Fig. 12-6) by the relation

$$N = \left(\frac{4t_R}{W_b}\right)^2$$

where t_R is the retention time and W_b is the base peak width as shown in Fig. 12-6. Another common way to determine N makes use of the relation

$$N = 5.54 \left(\frac{t_R}{W_{1/2}}\right)^2$$

where $W_{1/2}$ is the width at half the peak height.

Loss in the number of theoretical plates originates in three places:

1. *Sample injection.* The sample should be injected all at once into the column and should be vaporized immediately. For this reason, some sample chambers are heated. In other words, the sample should enter the packed column as a plug of gas with no tailing. The size of the sample should also be small.

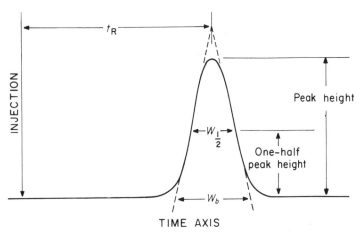

Figure 12-6 / Measurement of t_R, W_b, and $W_{1/2}$.

Overloaded columns not only broaden the peak, but sometimes cause it to be asymmetrical.

2. *Column characteristics.* Aside from the need to select a column that affords selectivity via a wide difference in partition coefficients for the sample components, the efficiency of a given column depends to a large extent upon proper flow rate. For example, consider a sample band in the middle of the column. If the flow of carrier gas were shut off, the band would eventually diffuse throughout the column. Even with carrier gas flowing, this longitudinal diffusion still takes place forward and backward, and results in a broadening of the band. This undesirable diffusion depends upon the time that the band remains in the column, and therefore on the flow rate of carrier gas (also upon column temperature, the nature of the sample, and the packing of the solid support). If the flow rate is increased, the effect of diffusional broadening will be diminished. At higher flow rates, however, a second factor creeps in which increases the broadening; partition equilibrium between the gaseous phase and the liquid phase is no longer maintained. (A thin film of liquid on a smooth surface of the support particles and an increase in temperature aid in attainment of equilibrium.) Therefore, there exists a flow rate for maximum column efficiency as illustrated in Fig. 12-7 (only for a given column at a given temperature).

Eddy diffusion also broadens the band, because the molecules travel through the column along many different paths of somewhat different lengths, with those that take the shortest paths at the front of the band and those that take the longest paths at the back of the band. The effect decreases with smaller and more uniform (in size) support particles, and with more regular packing of the column.

All of these effects are summarized by the van Deemter equation, which in its simplest form is

$$H = A + \frac{B}{\mu} + C\mu$$

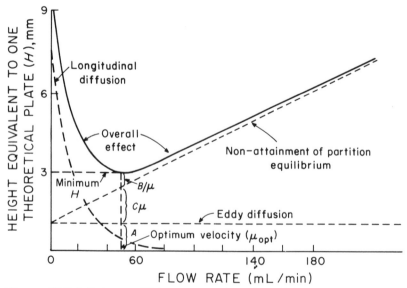

Figure 12-7 / Column efficiency as a function of the flow rate for the carrier gas.

where μ = flow rate
A = eddy diffusion term
B/μ = longitudinal diffusion term
$C\mu$ = mass-transfer rate term

The eddy diffusion term is flow-rate independent and is determined by how well the column has been packed. The problem remaining is to find a flow rate which minimizes the peak broadening due to the mass-transfer rate and longitudinal diffusion. This is done graphically by plotting H vs. flow rate (μ) as shown in Fig. 12-7. The flow rate which has the smallest H is the optimum flow rate for the temperature selected. This graph, known as a van Deemter plot, can also be used to evaluate the A, B, and C terms in the van Deemter equation.

The flow rate at which the overall effect of peak broadening reaches a minimum is μ_{opt}. To determine A, the curve for nonattainment of partition equilibrium is extrapolated back to $\mu = 0$. The value of H at this point is equal to A. The slope of the line used to extrapolate to $\mu = 0$ is the value of C. To find the value of B, the van Deemter equation must be differentiated with respect to μ and set equal to zero.

$$H = A + \frac{B}{\mu} + C\mu$$

$$\frac{dH}{d\mu} = \frac{-B}{\mu^2} + C$$

At μ_{opt}

$$\frac{dH}{d\mu} = 0 \quad \text{and} \quad \frac{-B}{\mu_{opt}^2} + C = 0$$

Therefore,

$$B = C\mu_{opt}^2$$

3. *Volume of detector.* Once the band issues from the column, it should go immediately (e.g., by capillary tubing and close spacing) into a detector of small volume. This is especially important in the case of the sharp peaks which emerge early from the column.

Procedure

Read thoroughly the instructions for operating the gas chromatograph.

Connect the soap bubble flow rate meter (see Fig. 12-8) to the gas exit port (a small metal tube sticking out of the side).

Determine the flow rate by squeezing the soap bulb so that the soap goes above the gas inlet side arm. The gas bubbling through the soap will cause soap bubbles to start rising up the tube. Get 2 or 3 bubbles going up, and use the timer to check the time it takes for a bubble to go from 0 mL to 10 mL. The flow rate is then determined by the relation

$$\text{Flow rate (mL/min)} = \frac{10 \text{ mL}}{\text{time in s}} \times \frac{60 \text{ s}}{1 \text{ min}}$$

Obtain chromatograms of pure chloroform (1.0-μL injections) at six different flow rates. The instructor will suggest a good range of flow rates.

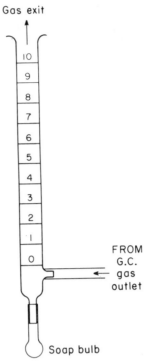

Gas exit

FROM
G.C.
gas
outlet

Soap bulb

Figure 12-8 / Soap bubble flow rate meter.

To insure the equilibration of the gas flow rate in the column, wait several minutes after changing the flow rate before injecting the samples or measuring the flow rate.

Measure the flow rate both before injection and after the sample has gone completely through the instrument.

Treatment of Data

From the flow meter and timer data calculate the flow rate in milliliters per minute for each setting of the flow rate control.

Calculate the number of theoretical plates from the chromatogram. By use of the column length, calculate H at each flow rate.

Construct a van Deemter plot of H vs. μ and determine the optimum flow rate.

Set the instrument flow control at the optimum flow rate for subsequent experiments.

Questions

1 A hydrocarbon gave a GC peak for which $W_{1/2} = 0.60$ s and $t_R = 0.92$ min.
  a Calculate the plate number of the column.
  b Given that the column was 10 m in length, calculate H.
2 Show what effect an increase in the gaseous diffusion coefficient of a solute in the carrier gas would have on the longitudinal diffusion curve in Fig. 12-7.
3 A mixture of four hydrocarbons is to be separated. A van Deemter plot is obtained for one of the hydrocarbons and the flow rate is set at the minimum value in the plot. Does this flow rate also give the minimum value of H for the other three hydrocarbons? Explain.
4 Would μ_{opt} increase or decrease if the column temperature is increased? Explain your answer.

EXPERIMENT 12-2

Quantitative Analysis of Mixtures by Gas Chromatography†

Purpose

A mixture of chloroform and carbon tetrachloride is analyzed by gas chromatography.

References

1 H. H. Willard, L. L. Merritt, Jr., J. A. Dean, and F. A. Settle, Jr., "Instrumental Methods of Analysis," 6th ed., Van Nostrand, Princeton, N.J., 1981, chaps. 15 and 16.

2 G. W. Ewing, "Instrumental Methods of Chemical Analysis," 4th ed., McGraw-Hill, New York, 1975, chaps. 18 and 19.

3 H. H. Bauer, G. D. Christian, and J. E. O'Reilly, "Instrumental Analysis," Allyn and Bacon, Boston, 1978, chap. 22.

4 D. A. Skoog and D. M. West, "Fundamentals of Analytical Chemistry," 4th ed., Saunders, Philadelphia, 1982, chaps. 27 and 29.

5 D. A. Skoog and D. M. West, "Principles of Instrumental Analysis," 2nd ed., Saunders, Philadelphia, 1980, chaps. 24 and 26.

6 S. Dal Nogare and R. S. Juvet, Jr., "Gas-Liquid Chromatography," Interscience, New York, 1962.

7 A. I. M. Keulemans, "Gas Chromatography," 2nd ed., Reinhold, New York, 1959.

8 H. Purnell, "Gas Chromatography," Wiley, New York, 1962.

Apparatus

Gas chromatograph with recorder
Tank of helium gas with pressure regulator
Timer
Soap bubble flow rate meter
Column: Dimethylsilicone (SE-30)
Syringe, 10 μL

Chemicals

Chloroform (reagent grade)
Carbon tetrachloride (reagent grade)

Theory

Gas chromatography is widely used for the quantitative analysis of liquid and gaseous mixtures. The area under a GC peak is proportional to the concentration of that component in the original mixture. Thus, a plot of peak area vs. concentration of known standards provides a calibration curve for quantitative analysis.

A number of methods have been developed to measure chromatographic peak areas. The most commonly used methods are outlined in Table 12-2.

In many instances a linear calibration curve can be obtained by simply plotting peak height vs. concentration. The height is measured from the base line of the chromatograph as shown in Table 12-2.

Quantitative determinations are based on the proportionality between the area under a component peak and the total amount of that component present in the sample. Therefore, a plot of peak area A_1 vs. volume V_1 of a pure component (#1) injected into the column should result in a straight line that passes through the origin with a slope of S_1. Similarly, a second straight line of slope S_2 should be obtained for a second pure component (#2). Calibration curves of this kind can

†From a similar experiment at the University of Cincinnati, courtesy of Dr. T. W. Gilbert.

Table 12-2 / Peak Area Integration Methods

Planimetry   The peak is traced manually with a planimeter, a mechanical device which measures area by tracing the perimeter of the peak. The area is represented digitally on a dial. The precision and accuracy of this method is dependent on the device itself and on the skill of the operator in using it to trace the peak. Precision can be improved by tracing each peak several times, but this makes the procedure very time consuming. Precision error, 4.06%.

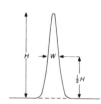

Height × Width at Half Height   Since normal peaks approximate a triangle, the area can be approximated by multiplying the peak height times the width at half height. The normal peak base is not used because large deviations may be caused by tailing or adsorption. The major factor affecting the precision of this method is the accuracy of measuring the peak width, particularly of narrow peaks. The area computed in this way is several percent less than the actual area, but is proportional to sample size under ideal conditions. Reasonable precision can be achieved if the peaks are symmetrical, but poor results are often obtained when the method is used on nonsymmetrical peaks. Precision error, 2.58%.

Cut and Weigh   The peak area is determined by cutting out the chromatographic peak and weighing the paper on an analytical balance. The accuracy of the method depends on the care used in cutting and on the constancy of the thickness and moisture content of the chart paper. A major disadvantage is that the chromatographic data are destroyed. A considerable improvement in accuracy and precision can be obtained by cutting and weighing a Xerox copy of the chromatogram rather than the actual one. This is apparently due to the greater homogeneity and weight of the Xerox paper as compared to the chart paper. This also preserves the chromatogram. Precision error, 1.74%.

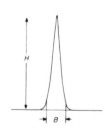

Triangulation   The triangle is constructed by drawing tangents to the slopes of the peak. The base is taken as the intersection of the two tangents with the base line. The height is measured from the base to the point where the tangents intersect. The area is then calculated by the triangle formula: $A = \frac{1}{2}BH$. The accuracy and precision of this method is dependent on many factors, such as the skill employed in constructing the triangle, the sharpness of the pencil used, the shape of the peak, and other variables. Assuming a gaussian peak shape, the triangle formula actually gives only 97% of the true area. Precision error, 4.06%.

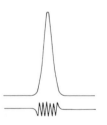

Disc Integrator   A number of electronic, ball and disc, analog, and voltage integrating devices have been designed for direct attachment to recorders so that integration is performed simultaneously with the recording of a chromatogram. Careful adjustment of the recorder's operation is necessary since the accuracy and precision obtainable with the Disc Integrator appear to be limited only by the mechanical performance of the recorder. Precision error, 1.29%.

Electronic Digital Integrator   In this method the chromatographic input signal is fed into a voltage-to-frequency converter, which generates an output pulse rate proportional to the peak height. When the slope detector senses a peak, the pulses from the V-F converter are accumulated and counted, giving a measure of the peak area. The key requirements for such a digital integrator are wide linear range, high count-rate/count capacity, and sensitive, versatile peak detection logic circuits. Precision error, 0.44%.

43	846390
125	489505
153	1219798
202	935264
322	1079453
506	719635
	5290045 T

Source: From Quantitation Techniques Peak Area Integration Methods, Varian Aerograph Chromatography Catalog, 1972. Reprinted by permission of Varian Associates, Inc.

be used for the analysis of mixtures. However, the accuracy obtained is seriously limited by the accuracy with which sample volumes can be measured. Because extremely small sample volumes are usually used, the precision and accuracy with which these volumes can be measured are quite low.

When dealing with binary mixtures, this limitation can be circumvented by constructing a calibration curve in which the area ratio of the peaks for the two components is plotted against a function of the sample composition. The area ratio has the advantage that it should be independent of the sample volume. If R represents the volume fraction of component #2 in the binary mixture, then

$$R = \frac{S_2 A_2}{S_1 A_1 + S_2 A_2}$$

which rearranges to

$$\frac{1}{R} = \frac{S_1 A_1}{S_2 A_2} + 1$$

Therefore, if the area ratio (A_1/A_2) is plotted against $1/R$, a straight line of slope S_1/S_2 should result which passes through the point $A_1/A_2 = 0$, $1/R = 1$. (For example, a prepared known sample that contains 40% of component #2 and 60% of component #1 shows that $1/R = 2.5$. The corresponding measured area ratio A_1/A_2 is plotted at this value of $1/R$. By taking other known volume ratios and measuring their area ratios, a calibration curve can be constructed.) Accordingly, when the area ratio of an unknown mixture is measured, the corresponding value of $1/R$ can be read from the calibration curve and the composition can be calculated.

One further simplification can be made. Because the measurement of peak areas is time consuming and difficult (unless the GC is equipped with an integrator), the ratio of peak heights may be used in place of peak area ratios. The difference is nearly insignificant for components which elute fairly quickly (i.e., narrow elution bands), but appreciable error will be introduced if this substitution is made for components with large retention times where the elution bands are broad.

Procedure

Adjust the flow rate of the chromatograph to the optimum value for $CHCl_3$ as determined in Experiment 12-1.

Prepare (by pipet) a standard series of mixtures of $CHCl_3$ and CCl_4 that contain 30, 40, 50, 60, and 70% $CHCl_3$ by volume.

safety caution

Use a rubber pipet bulb; chlorinated solvents are poisonous to the liver.

Obtain chromatograms of each of the above standards.

Obtain a sample of an unknown $CHCl_3$–CCl_4 mixture from the instructor and obtain its chromatogram.

Treatment of Data

Construct a calibration curve from the peak heights of the $CHCl_3/CCl_4$ standards as outlined in the Theory section.

Use the calibration curve to calculate the % $CHCl_3$ in the unknown.

Questions

1   Explain the order of elution of CH_3Cl and CCl_4 from the chromatograph.
2   Do you expect CH_2Cl_2 to elute faster or slower than $CHCl_3$?

EXPERIMENT 12-3

Resolution and Qualitative Identification of Hydrocarbons by Gas Chromatography†

Purpose
The retention times of a homologous series of hydrocarbons are measured. Peak resolution is calculated and an unknown hydrocarbon is identified. Hydrocarbon components of gasoline are identified.

References
1 H. H. Willard, L. L. Merritt, Jr., J. A. Dean, and F. A. Settle, Jr., "Instrumental Methods of Analysis," 6th ed., Van Nostrand, Princeton, N.J., 1981, chaps. 15 and 16.
2 G. W. Ewing, "Instrumental Methods of Chemical Analysis," 4th ed., McGraw-Hill, New York, 1975, chaps. 18 and 19.
3 H. H. Bauer, G. D. Christian, and J. E. O'Reilly, "Instrumental Analysis," Allyn and Bacon, Boston, 1978, chap. 22.
4 D. A. Skoog and D. M. West, "Fundamentals of Analytical Chemistry," 4th ed., Saunders, Philadelphia, 1982, chaps. 27 and 29.
5 D. A. Skoog and D. M. West, "Principles of Instrumental Analysis," 2nd ed., Saunders, Philadelphia, 1982, chaps. 24 and 26.
6 S. Dal Nogare and R. S. Juvet, Jr., "Gas-Liquid Chromatography," Interscience, New York, 1962.
7 A. I. M. Keulemans, "Gas Chromatography," 2nd ed., Reinhold, New York, 1959.
8 H. Purnell, "Gas Chromatography," Wiley, New York, 1962.
9 R. F. Cassidy, Jr. and C. Schuerch, *J. Chem. Ed., 53,* 51 (1976).

Apparatus
Gas chromatograph with recorder
Tank of helium gas with pressure regulator
Timer
Soap bubble flow rate meter
Column, 4-ft $\times \frac{1}{4}$-in column [with nonpolar stationary phase OV-101 or DC 200, 6% on Chromosorb G (available from Supelco, Inc., Supelco Park, Bellefonte, Pa. 16823) or other suitable column]
Syringe, 10 μL

Chemicals
Pure hydrocarbons, pentane, hexane, heptane, octane, nonane
Sample of unleaded gasoline

Theory
Qualitative identification of components in a mixture by gas chromatography usually is based on retention time t_R which is the time required for the sample to pass through the chromatograph. Measurement of t_R from a chromatogram is illustrated in Fig. 12-9. Because many compounds may exhibit similar retention times, confirmation of the identity of a separated component by a second technique such as mass spectrometry is judicious.

In order for a mixture to be separated successfully, the peaks for the individual components must be resolved from each other. The extent of separation of two

†From a similar experiment at the University of Cincinnati, courtesy of Dr. T. W. Gilbert.

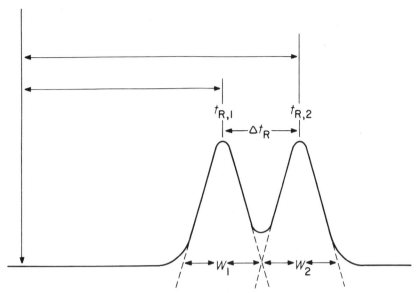

Figure 12-9 / Measurement of t_R and peak resolution.

adjacent, approximately gaussian peaks, can be expressed in terms of the resolution R of a column

$$R = \frac{t_{R,2} - t_{R,1}}{0.5(W_1 + W_2)}$$

with the terms defined in Fig. 12-9.

The peaks are considered to be fully resolved when $R = 1.5$, but for practical purposes $R = 1$ (approximately 98% separation) usually is adequate. The stationary liquid phase fixes the numerator in the above equation for R, while the column efficiency governs the denominator.

The qualitative identification of unknown peaks in a multicomponent chromatogram can be simplified by the use of two relationships: (1) For compounds in the same family (e.g., *n*-alkanes, 2-methylalkanes, or 2,2-dimethylalkanes), a plot of log retention time vs. number of carbon atoms should yield a straight line. (2) For compounds of similar chemical characteristics (e.g., the aliphatic saturated hydrocarbons), a plot of boiling point vs. log retention time is a smooth curve.

Procedure

Set the flow rate of the chromatograph as specified by the instructor or as determined by a van Deemter plot for one of the hydrocarbons.

Obtain individual chromatograms of the available pure hydrocarbons. Be sure to mark each injection time on the chart paper.

Obtain a chromatogram of your unknown sample.

Obtain a chromatogram of a sample of unleaded gasoline.

Treatment of Data

Construct plots of log retention time vs. carbon number and log retention time vs. boiling point from the chromatograms of the pure hydrocarbons.

From the retention times of the components in the unknown sample, identify as many peaks as possible by referring to your log retention time plots.

Calculate the resolution of the adjacent peaks in the chromatogram of your unknown.

From the retention times of the components in the gasoline sample, identify as many peaks as possible by referring to your log retention time plots.

Questions

1 A homologous series of primary alcohols is to be determined by gas chromatography. Given that the retention time for ethanol is 1 min, predict the retention times for 1-propanol, 1-butanol and 1-pentanol.

2 What column do you recommend for the separation of the alcohols in Question 1?

3 List six factors that affect resolution and describe how they can be controlled.

4 What length of column would be required to just resolve two peaks with retention times of 92 and 107 s, respectively, using a column with an H of 1.60 cm/plate.

EXPERIMENT 12-4

Study of a Consecutive Reaction: Kinetics and Gas Chromatography†

Purpose

The kinetics of the methanolysis of diethyl malonate via two successive acid-catalyzed ester exchange reactions is to be studied by gas chromatography.

References

1 S. Dal Nogare and R. S. Juvet, Jr., "Gas-Liquid Chromatography," Interscience, New York, 1962.

2 A. A. Frost and R. G. Pearson, "Kinetics and Mechanism," 2nd ed., Wiley, New York, 1961

3 R. S. Juvet, Jr., and F. M. Wachi, *J. Am. Chem. Soc., 81,* 6110 (1959).

4 R. S. Juvet, Jr., and J. Chiu, *J. Am. Chem. Soc., 83,* 1560 (1961).

Apparatus

Gas chromatograph with recorder
Tank of helium gas
Column, 5 ft $\times \frac{1}{8}$ in (8% di-*n*-nonyl phthalate or 30% w/w dioctyl phthalate on Chromosorb "W" or 7.97% polyethylene glycol 600 on Chromosorb "W" or 7% polyethylene glycol 400 on Chromosorb "P")
Syringe, 5 μL
Conical flask, glass stoppered, 50-mL

Chemicals

Methanol, reagent grade
Diethyl malonate (commercial grade)
p-Toluenesulfonic acid monohydrate

Theory

The methanolysis of diethyl malonate involves two successive acid-catalyzed ester exchange reactions, which are each first order in ester and alcohol. Although a good example of a series consecutive reaction, it is more complicated than the ideal case in that each of the reactions is (*a*) second-order overall, (*b*) in-

†From a similar experiment developed by C. N. Reilley at the University of North Carolina, Chapel Hill.

volves an equilibrium process, and (c) is acid-catalyzed. The two reactions can be expressed by the relations

$$
\begin{array}{cccc}
\underset{A}{\begin{array}{c} COO\underline{CH_2CH_3} \\ | \\ CH_2 \\ | \\ COO\underline{CH_2CH_3} \end{array}} + \underline{CH_3}OH & \xrightarrow{\ H^+\ } & \underset{B}{\begin{array}{c} COO\underline{CH_3} \\ | \\ CH_2 \\ | \\ COO\underline{CH_2CH_3} \end{array}} + CH_3CH_2OH
\end{array}
$$

$$
\begin{array}{cccc}
\underset{B}{\begin{array}{c} COO\underline{CH_3} \\ | \\ CH_2 \\ | \\ COO\underline{CH_2CH_3} \end{array}} + \underline{CH_3}OH & \xrightarrow{\ H^+\ } & \underset{C}{\begin{array}{c} COO\underline{CH_3} \\ | \\ CH_2 \\ | \\ COO\underline{CH_3} \end{array}} + CH_3CH_2OH
\end{array}
$$

The complexities that arise in the rate law can be removed by the use of a large excess of methanol. This causes the methanol concentration to remain constant and the reaction rates to be independent of methanol concentration. Thus, the reaction becomes pseudo-first-order with respect to ester concentration. Pseudo-first-order conditions also eliminate the complexities that arise from equilibrium considerations. The large methanol concentration forces the reaction to proceed largely to product, which is to say that the rates of the reverse reactions will be small compared to the forward reactions. This is especially true in the early states of the reaction when there is only a small concentration of *B, C,* or ethyl alcohol through which the reverse reactions can proceed. The system has been simplified to a pair of series pseudo-first-order forward reactions that are opposed by relatively slow second-order reverse reactions.

A third consideration is that each reaction is acid-catalyzed. This means that while the concentration of acid affects the rate of each reaction, it remains constant. This dependence may be quite complicated, and depends on the mechanisms involved and such parameters as effective acid strength in organic solvents. Thus, the rigorous kinetic expression is complicated. However, the system can be analyzed in terms of apparent rate constants for a constant catalyst concentration.

Procedures

During the study of the methanolysis of diethyl malonate, the column temperature should be controlled within 1°C. The injection port temperature is maintained at 300°C and the inlet pressure for the helium carrier gas is set at 15 psi.

A. Trial Run

Prepare a solution that contains all three esters by mixing 15 mL diethyl malonate, 30 mL methanol, and 0.5 g *p*-toluenesulfonic acid monohydrate. Allow the mixture to stand $\frac{1}{2}$ h, after which time the mixture should contain satisfactory amounts of all three esters. (This preparation can be done by the instructor and used by the entire class.) Satisfactory resolution and analysis time should be achieved at a column temperature of 120°C, and a carrier flow rate of 100 mL/min. The proper temperature and flow rate can be determined by trial runs. The alcohols present are retained only slightly longer than the air peak. Of the esters, dimethyl malonate is eluted first, methylethyl malonate second, and diethyl malonate third. The third ester should elute within 10 min.

Kinetic data are obtained from changes in the individual peak dimensions, which are proportional to changes in concentration of the respective components. All three ester peaks are of sufficient sharpness to allow the use of the peak height method. All peak heights must be standardized to some unit injection volume. Injections are made with a calibrated microsyringe. The syringe needle is placed into the reaction solution and exactly 5 μL of sample are withdrawn. The sample and air are then quickly injected with the needle retained in the injection port for several seconds.

B.   Kinetic Run

Into a 50-mL glass-stoppered conical reaction flask, weigh 12.0 g (approximately 15 mL) of reagent grade methanol, and 5.0 g of commercial grade diethyl malonate. Record several chromatograms to check injection reproducibility and reagent purity (diethyl malonate may contain small amounts of both the methylethyl and dimethyl ester). From their recorded peak heights, determine the amounts of the impurities and subtract as a blank from subsequent data. Stopper the reaction flask tightly to prevent loss of methanol and place in a water bath at 45°C. A 1.71-g sample of p-toluenesulfonic acid monohydrate (2.05 mol % relative to the ester), dried in a vacuum desiccator to remove any excess water, is then added with stirring. A stop watch is simultaneously activated and the reaction is kept sealed except for removal of samples for analysis. The same volume (5 μL) is withdrawn and analyzed approximately every 10 min for 3 h.

Treatment of Data

Peak heights for each of the three esters are measured for each 5-μL sample, and corrected by subtraction of the previously determined blanks for ethylmethyl and dimethyl malonate. The fraction of material in each form (A, B, and C) at a given time is calculated from the relations

$$\frac{A}{\text{Total}} = \frac{h_A}{h_A + h_B + h_C}$$

$$\frac{B}{\text{Total}} = \frac{h_B}{h_A + h_B + h_C}$$

$$\frac{C}{\text{Total}} = \frac{h_C}{h_A + h_B + h_C}$$

where h_A, h_B, h_C are the peak heights for components A, B, and C.

Kinetic plots are prepared from the peak height vs. time for each component. The apparent psuedo-first-order rate constant for the first reaction is then calculated from the A curve by use of the relation

$$k_1' = \frac{2.303}{t} \log \frac{h_0 - h_\infty}{h_t - h_\infty}$$

where k_1' is the apparent pseudo-first-order rate constant, and h_0, h_∞, and h_t are the peak heights of A at time equal zero, infinity, and t, respectively. Values of k_1' should be determined for several sets of peaks during the initial stage of the reaction. From these an average value for k_1' at the 2.05 mol % acid concentration should be determined.

An approximate value of the pseudo-first-order rate constant for the second reaction k_2' can be obtained by use of the relationship that expresses the maximum buildup of the intermediate in terms of the ratio of the rate constants.

$$\frac{B_{\max}}{A_0} = K^{K/(1-K)}$$

where $K = k_2/k_1$. If we make the semi-quantitative assumption that the fraction of any component present is equal to its peak height divided by the sum of all of the peak heights, B_{max}/A_0 appears from the plot to be equal to approximately 0.5. (This assumption is in error, for at any time on the plot the sum of all component heights is not exactly equal to the initial height A_0 of diethyl malonate.) Solution of the above expression by successive approximations indicates that k_1' is about twice the value of k_2'. Include the kinetic plot with your final report, and include the values for k_1' and k_2' as well as the temperature of the reaction system.

Questions

1 How slow must a reaction be in order for it to be amenable to study by gas chromatography?

2 Why is the temperature of the injection port set at 300°C whereas the column temperature is maintained at 120°C?

EXPERIMENT 12-5

Gas Phase Chromatography; Study of the Mechanism for an Elimination Reaction; Dehydration of Butanol†

Purpose

This experiment will serve to illustrate the use of gas phase chromatography for the study of reaction products and reaction mechanisms.

Reference

1 M. S. Newman, "Steric Effects in Organic Chemistry," Wiley, New York, 1956, pp. 322–346.

Apparatus

Round-bottom flask (three-necked), 500 mL
Equilibrating separatory funnel, 125 mL
Water-cooled condenser
Drying tube filled with $CaCl_2$
Heating mantle, 500 mL
Variac or other voltage regulator
Graduated centrifuge tube, 15 mL (in acetone–dry ice bath for sample collection)
Thermometer, 0 to 120°C
Glass and rubber tubing
Gas chromatograph equipped with a silicone-oil column
Helium gas with regulator valves
Recorder for gas chromatograph
Hypodermic syringe, 2 mL

Chemicals

Sulfuric acid (70%)
1-Butanol
2-Butanol

Theory

Gas phase chromatography is a particularly useful tool for the analysis of small samples containing mixtures of several components which are either gases or vol-

†From a similar experiment at the University of California, Riverside, Calif., courtesy of Dr. G. K. Helmkamp.

atile liquids. Because of the high separation efficiency of the method (usually several thousand theoretical plates), very close boiling mixtures can be separated and analyzed. This unique advantage of sensitivity and specificity has provided the chemist with a means of studying reactions which were not previously considered because of the difficult analytical problems.

The dehydration of butanol provides an example of applying gas chromatography to the study of reaction products. The two isomers of normal butanol (1-butanol and 2-butanol) may be dehydrated with sulfuric acid to give a mixture of butenes.

$$CH_3CH_2CH_2CH_2OH + H_2SO_4 \xrightarrow{105°} CH_3CH_2CH=CH_2 + (cis \text{ and } trans)CH_3CH=CHCH_3$$

$$CH_3CH_2-\underset{\underset{OH}{|}}{CH}-CH_3 + H_2SO_4 \xrightarrow{105°} CH_3CH_2CH=CH_2 + (cis \text{ and } trans)CH_3CH=CHCH_3$$

Some isobutylene is also formed but appears in the same peak as the butene-1. The relative amounts of the possible reaction products are determined by a gas chromatogram of the gases evolving from the reaction vessel. From this analysis some conclusions on the effect of steric hindrance and the reaction mechanisms are possible.

By comparing the relative amounts of the products from the dehydration of 1-butanol with those from 2-butanol, conclusions may be made concerning the mechanisms for the elimination reaction.

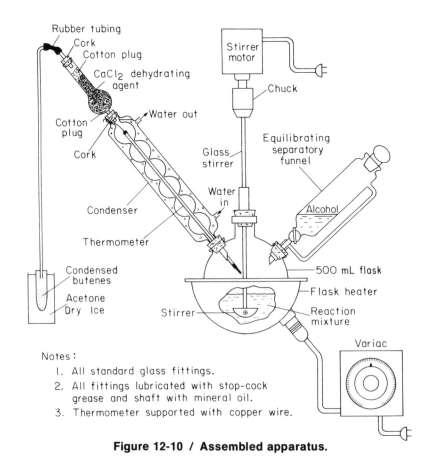

Notes:
1. All standard glass fittings.
2. All fittings lubricated with stop-cock grease and shaft with mineral oil.
3. Thermometer supported with copper wire.

Figure 12-10 / Assembled apparatus.

Procedure

Assemble the glass reaction system as shown in Fig. 12-10, using stopcock grease on the glass joints and mineral oil on the shaft of the stirrer.

Turn on the helium flow, gas chromatograph, and recorder; set the instrument to operate at room temperature. Consult the instructor for specific operating instructions for the gas chromatograph and recorder.

Place 100 mL of 70% sulfuric acid in the round-bottom flask; make sure all connections are tight and that water is flowing in the condenser. Check that the sample trap is in place in the dry ice bath. Place about 25 mL of 1-butanol in the separatory funnel, making sure that the stopcock is closed. Start the stirrer, turn the Variac on, and adjust the Variac's voltage until the sulfuric acid solution refluxes at approximately 105°. After the temperature of the refluxing mixture has stabilized, slowly open the stopcock on the separatory funnel so that the 1-butanol is admitted at a rate of approximately 45 to 60 drops/min. After approximately 2 mL of the products have condensed in the sample trap, insert the syringe needle into the rubber tubing just above the drying tube and collect a 2-mL gas sample (see Fig. 12-10). Inject this sample immediately into the gas chromatograph with the recorder running and mark the point of sample injection on the recorder chart. Three peaks besides the air peak (if there is one) should be observed. If time permits, take another gas sample and record its gas chromatogram for a duplicate check. Cool, empty, and clean the reaction and sample systems; then reassemble the system. Repeat using 2-butanol in place of 1-butanol.

Again clean the system; shut off the recorder, gas chromatograph, and helium flow.

Treatment of Data

The order of appearance of peaks for a gas chromatogram is roughly the same as the order of increasing boiling points of the components. The boiling points for the butene isomers are

Butene-1	$-5°C$
Isobutylene	$-6°C$
Butene-2(trans)	$1°C$
Butene-2(cis)	$2.5°C$

(*Note*: The cis and trans values are frequently reversed in the literature.) For each of the gas chromatograms, measure the area of the individual peaks. Tabulate the areas for 1-butanol and for 2-butanol, and also tabulate the ratios of the trans butene-2 area to the butene-1 area and the cis butene-2 area to the butene-1 area for the two alcohol isomers.

From the tabulated data indicate which butene isomer is the most favored from the dehydration of 1-butanol, and also which is the most favored isomer from the dehydration of 2-butanol. Referring to the ratio of areas in your table for the cis and trans isomers resulting from the two alcohols, postulate and defend a mechanism for the dehydration of butanol with sulfuric acid. The discussion should account for all the reaction products for the two alcohols and should provide some explanation for the relative amounts.

Questions

1　How does your proposed mechanism differ for the dehydration of 1-butanol and 2-butanol?
2　How do you account, mechanistically, for the production of some isobutylene?
3　If the reaction temperature was raised 30°C, predict the effect on the relative amounts of the butene isomers. Explain and justify your prediction.
4　Suggest a column material which might resolve the isobutylene from butene-1.

13

High Performance Liquid Chromatography (HPLC)

High performance liquid chromatography (HPLC) makes use of a pump to deliver a mobile phase solvent at a uniform rate at pressures that are typically from 500 to 5000 psi. The most obvious advantage of HPLC over gravity liquid chromatography is that samples can be separated much more quickly. In addition, samples that are not volatile or that would thermally decompose in gas chromatography can be rapidly and routinely separated. Consequently, this powerful analytical method is complementary to gas chromatography.

HPLC is accomplished by injection of a small amount of liquid sample into a moving stream of liquid (termed the *mobile phase*) that passes through a column packed with particles of a *stationary phase*. As in gas chromatography, separation of a mixture into its components depends on different degrees of retention of each component in the column. (See Fig. 12-1 for a pictorial presentation of elution chromatography.) The extent to which a component is retained in the column is determined by its partitioning between the liquid mobile phase and the stationary phases. A variety of HPLC separation techniques that utilize different stationary and mobile phases have been developed.

Adsorption Chromatography

Adsorption chromatography utilizes a solid stationary phase of a polar nature such as particles of hydrated silica or alumina. The mobile phase and the solute (components in the sample) are in competition for active adsorption sites on the stationary phase particles. Thus, more strongly adsorbed components are retained longer than weakly adsorbed components. Because more polar compounds adsorb on a polar surface to a greater degree than do less polar compounds, retention in the column is related to sample polarity. A generalized polarity scale for various classes of compounds is shown in Table 13-1.

**Table 13-1 / Compound-Class
Polarity Scale†**

Fluorocarbons
Saturated hydrocarbons
Olefins
Aromatics
Halogenated compounds
Ethers
Nitro compounds
Esters ≈ ketones ≈ aldehydes
Alcohols ≈ amines
Amides
Carboxylic acids

†Listed in order of increased
retention.

Retention also can be controlled by the polarity of the mobile phase, which competes with sample components for adsorption sites. Thus, a more polar mobile phase will more effectively displace adsorbed solute molecules and cause the retention time to decrease. Table 13-2 ranks solvents in the order of their strength of adsorption on the adsorbent alumina. Such a scale is called an eluotropic series. A solvent of higher polarity will displace one lower in the polarity scale.

**Partition
Chromatography**

This form of HPLC partitions the solute between the liquid mobile phase and a second, immiscible liquid that is coated on or bonded to solid particles as the stationary phase. Compounds that partition more strongly into the stationary liquid phase are retained longer in the column. This type of chromatography is termed *normal phase* if the stationary phase is more polar than the mobile phase and *reverse phase* if the mobile phase is more polar than the stationary phase. The stationary phase can be a liquid coated on solid support particles. *Bonded-phase* columns have the stationary phase chemically bonded to the solid support and are the

**Table 13-2 / Eluotropic Series for
Alumina** (Silica has a similar
rank ordering.)

Solvent	Relative Polarity
n-Pentane	0.00
Isooctane	0.01
Cyclohexane	0.04
Carbon tetrachloride	0.18
Xylene	0.26
Toluene	0.29
Benzene	0.32
Ethyl ether	0.38
Chloroform	0.40
Methylene chloride	0.42
Tetrahydrofuran	0.45
Acetone	0.56
Ethyl acetate	0.58
Aniline	0.62
Acetonitrile	0.65
i-Propanol	0.82
Ethanol	0.88
Methanol	0.95
Acetic acid	large

most popular column for partition chromatography. For example, *n*-octadecane can be bonded directly to silica by attachment to surface hydroxyl groups to form what is termed a C_{18} column. The extent of partitioning of a solute into the stationary phase can be controlled by varying the solvent polarity.

Ion-Exchange Chromatography

This type of HPLC is based on the partition of ions between a polar liquid phase and a stationary phase with ion-exchange sites. The ion-exchange sites are typically immobilized in small beads of resin that are formed by a cross-linked polymer. Bonded-phase columns in which the ion exchanger is bonded to small particles of silica also are available. Cations are separated on cation exchange resins which contain negatively charged functional groups such as $-SO_3^-$ and $-COO^-$. Anions are separated on anion exchange resins which contain positively charged functional groups such as $-CH_2N^+(CH_3)_3$, a quaternary ammonium ion. Separation is based on ions partitioning into the ion-exchange phase to varying degrees. The selectivity of a resin for an ion is determined primarily by the charge on the ion and its hydrated radius. Resin affinity increases with increasing charge density.

Apparatus

A schematic drawing of a typical high performance liquid chromatograph is shown in Fig. 13-1. The apparatus consists of a container of mobile phase, a pump capable of pressures up to 4000 psi or greater, a valve for injecting sample (usually 10- to 500-μL volumes), the column (sometimes thermostatted), a detector, electronics associated with the detector, and a recorder.

UV-visible absorbance is the most commonly used mode of detection. Such detectors enable the effluent from the column to flow through an 8- to 10-μL spectrophotometric cell for detection of compounds at a particular wavelength. Electrochemical and fluorescence detectors often are used to achieve lower detection limits. The other commonly used detector is based on measurement of the differential refractive index.

Qualitative and Quantitative Measurements

A typical liquid chromatogram is shown in Fig. 13-2. Each component in a mixture can be qualitatively identified by its retention time t_R which is the time between injection and detection. As with gas chromatography the retention time of a particular compound is constant for a fixed set of chromatographic conditions (flow rate, temperature, column condition). Qualitative identification is made by

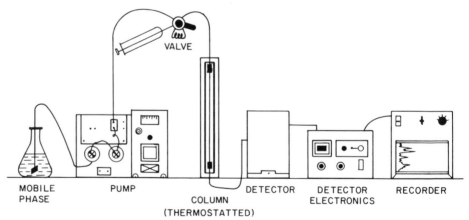

Figure 13-1 / Schematic diagram of a high performance liquid chromatograph. (Reprinted with permission of Bioanalytical Systems, Inc., West Lafayette, Ind.)

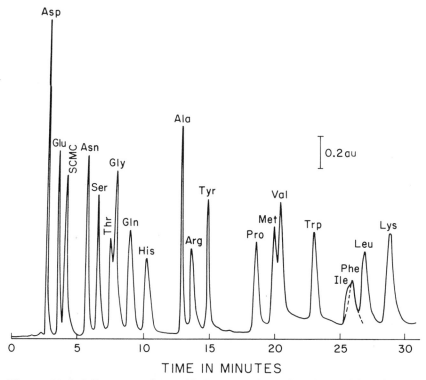

Figure 13-2 / Reverse phase (C$_{18}$) separation of amino acids. (*From I. Kato, W. J. Kohr, and N. Laskowski, Federation of American Societies for Experimental Biology Abstract #2592, 1977. Reprinted by permission.*)

comparing t_R of the unknown with retention times of standards that have been injected into the chromatograph. This strategy works so long as components have unique retention times.

The area under each peak is proportional to the concentration of that component in the original mixture. If the peaks are reasonably sharp and the flow rate is carefully controlled, the peak heights are approximately proportional to concentration. Thus, a calibration curve can be prepared by plotting either peak height or peak area as a function of concentration for a series of standards. Methods for the measurement of peak areas are described in Table 12-2.

EXPERIMENT 13-1

Determination of Pharmaceuticals by High Performance Liquid Chromatography: Determination of Caffeine in Beverages†

Purpose
Reverse phase HPLC is used to determine the concentration of caffeine in coffee, tea, Coca-Cola, and Pepsi-Cola.

References
1 Advances in Chromatography and Lab Automation, "Chromatography Review," *Spectra-Physics*, June 1978.

†Adapted from an experiment at the University of California, Riverside.

2 L. C. Column Report, "Nonaqueous Reversed-Phase Chromatography," DuPont Instruments, 1978.

3 Altex High-Performance Chromatography Catalogue, Altex Instruments, 1978.

Apparatus

High performance liquid chromatograph with UV detector (254 nm) (such as a Bioanalytical Systems Student LC)

Reverse phase C_{18} column (such as LiChrosorb RP18)

Volumetric flasks, five 100 mL, four 20 mL

Syringe, 50 μL

Pipets, 2 mL, 4 mL, 8 mL (one of each)

Chemicals

Caffeine, reagent grade for standards

20% Methanol:80% water solvent, 1 L

Coffee

Tea

Coca-Cola

Pepsi-Cola

Theory

Reverse phase HPLC is used to determine the concentration of caffeine in coffee, tea, Coca-Cola, and Pepsi-Cola. The traditional method for the determination of caffeine is via extraction with spectrophotometric quantitation. Use of the liquid chromatography system permits a fast and easy separation of caffeine from other substances such as tannic acid, caffeic acid, and sucrose found in these beverages (Ref. 3). Five standard solutions of caffeine are prepared and injected into the HPLC. In addition, the beverages coffee, tea, Coca-Cola, and Pepsi-Cola are prepared as indicated in the following section and injected into the HPLC. From the resulting chromatograms, measurements of retention time t_R and peak areas are made. If the flow rate and pump pressure are held constant throughout the entire experiment, t_R may be used as a qualitative measure and the peak area as a quantitative measure. A calibration curve for peak area vs. concentration of the caffeine standards can then be employed to determine the concentration of caffeine in the four beverages.

Procedures

A. Caffeine Standards

1 Into five clean and dry 100-mL volumetric flasks weigh out accurately the following quantities of caffeine: 2.5 mg, 5.0 mg, 7.5 mg, 10.0 mg, and 12.5 mg.

2 Dilute to the mark with the previously prepared 20% methanol:80% H_2O solvent, adjusted to approximately pH 3.50. This is the same solvent to be used as the mobile phase.

3 Shake the five caffeine solutions adequately to insure dissolution and then degas each for 5 min before injection into the chromatograph.

4 Turn the pump and detector on. Set the pump flow rate at 2.3 mL/min and the detector sensitivity at 0.08 AUFS (absorbance units full scale). Turn the recorder on and set at the slow speed rate. Prior to injection of the standards into the column allow the mobile phase (20% methanol:80% H_2O) to pass through the column for 5 to 10 min. Simultaneously record the detector response to insure that there are no substances left on the column from previous experiments.

5 With the syringe provided, remove a 25-μL (or more) aliquot from the least concentrated standard and inject it into the HPLC while in the LOAD position. The injection port delivers only a 20-μL aliquot to the column; however, a 5-μL or more excess is needed for loading.

6 Next, turn the valve on the injection port from the LOAD to the INJECT position and simultaneously push the MARKER button on the detector. This will cause a deflection on the recorder, and provides a record of when the sample is injected.

7 Allow the peak due to the caffeine to be recorded and then follow the same procedure (steps 3 to 6) again for the least concentrated sample. Continue until you have recorded three chromatograms for the least concentrated sample.

8 With the next concentrated sample, follow steps 5 to 7 until you have obtained three chromatograms.

9 Continue steps 3 to 8 for the last three standards using the least concentrated of the three first, and increase the concentration stepwise. For the caffeine standards a total of 15 chromatograms should be recorded.

Treatment of Data

The t_R can be used as a diagnostic tool to determine qualitatively the presence of a substance in a chromatographic mixture. For the 15 chromatograms recorded for the caffeine standards, measure the distance from the marker to the peak maximum. This distance is directly related to the retention time because the recorder is driven at a constant rate; therefore, recording the t_R in units of length is adequate.

The peak area is directly related to the concentration. Integrate peak areas by use of one of the methods that are outlined in Table 12-2.

In tabular form record for the caffeine standards the concentration, retention times, and peak areas. Also, record the mean and standard deviation of the times and areas.

Plot the main peak area for each concentration vs. the concentration in milligrams per milliliter for the caffeine standards.

B. Caffeine in Coffee and Tea

The instructor will supply coffee and tea prepared for this experiment. Into one clean, dry 20-mL volumetric flask pipet 2 mL of coffee, and into another clean, dry 20-mL flask pipet 4 mL of tea.

Dilute each volumetric flask to the mark with 20% methanol:80% water adjusted to pH 3.50.

Follow steps 5 to 9 in Procedure A for each sample (record three chromatograms for each.).

Treatment of Data

From the average retention time (length) for the caffeine peak determine which peak on the coffee and tea chromatograms is due to caffeine.

Integrate the caffeine peaks for the coffee and tea samples.

Take the average peak area for the coffee and tea samples and interpolate from the calibration curve the concentration of caffeine in milligrams per milliliter for each beverage. Include corrections for dilution in the calculations.

C. Caffeine in Cola Beverages

Pour 10 to 15 mL of Pepsi-Cola into a small clean, dry beaker. Pour this into another clean, dry beaker. Pour the Pepsi back to the original beaker. Continue pouring back and forth until the bubbling ceases. The soda is now adequately decarbonated. With two additional beakers, follow the same procedure with Coca-Cola.

Into a clean, dry 20-mL volumetric flask pipet 8 mL of Pepsi-Cola (decarbonated), and into a second 20-mL flask, pipet 8 mL of Coca-Cola.

Dilute each volumetric flask to the mark with 20% methanol:80% water solvent adjusted to pH 3.50.

Follow steps 5 to 9 in Procedure A for each sample, recording three chromatograms for each.

After the last chromatogram, flush the column with 50 mL of 20% methanol:80% water (not at pH 3.50).

Treatment of Data

Make the same determinations as in Procedure B for the Pepsi-Cola and Coca-Cola samples.

In tabular form record the beverage and the concentration of caffeine present.

Questions

1 Explain the rationale for using a reverse phase C_{18} column for the determination of caffeine.

2 Would the construction of a calibration curve based on peak height (rather than area) give accurate results for the determination of caffeine in this experiment?

3 Could an ion-exchange column be used for the determination of caffeine? Explain.

4 Diagram the UV detector and explain how it works.

EXPERIMENT 13-2

Separation of Urinary Compounds by High Performance Liquid Chromatography†

Purpose

Uracil, uric acid, xanthine, allopurinol, and nicotinamide in urine are separated by HPLC and identified with UV spectroscopy.

References

1 D. W. Bastian, R. L. Miller, A. G. Halline, F. C. Senftleber, and H. Veening, *J. Chem. Ed., 54,* 766 (1977).

2 F. C. Senftleber, A. G. Halline, H. Veening, and D. A. Dayton, *Clin. Chem., 22,* 1522 (1976).

Apparatus

Liquid chromatograph, 100-μL sample injection system, UV monitor (254 nm)

Column (30 cm × 4 mm) packed with 10 μm silica particles bonded with octadecylsilane (μ-Bondapak C_{18}, Waters Associates)

Volumetric flasks, five 100 mL, one 10 mL

Pipets, one 5 mL, four 1 mL, four 2 mL, four 3 mL, four 4 mL

Pipet (grad.), 2 mL

Nalgene membrane filter, 0.2 μm

Centrifuge

Chemicals

Stock solutions of
Uracil, 2.43 mg/100 mL
Uric acid, 11.74 mg/100 mL
Xanthine, 4.67 mg/100 mL
Allopurinol, 5.43 mg/100 mL

†Reprinted in part with permission from D. W. Bastian, R. L. Miller, A. G. Halline, F. C. Senftleber, and H. Veening, *J. Chem. Ed., 54,* 766 (1977). Copyright © 1977, Division of Chemical Education, American Chemical Society.

Nicotinamide, 18.09 mg/100 mL
> (made up in doubly distilled water and with several drops of 40% NaOH to enhance solubility; refrigerate until used. Uric acid must be made up the day used.)

Urine samples

Unknown samples that contain the above components (plus uracil)

1:1 methanol:water

$ZnSO_4$ solution, 100 g/L

0.1 M NaOH

Methanol, pure

0.05 M acetic acid, buffered to pH 4.5

Theory

The reverse phase type of packing contains a nonpolar, chemically bonded stationary phase such as the saturated octadecyl hydrocarbon group (C_{18}). Nonpolar components, which have a high affinity for the C_{18} environment, elute with relatively long retention times whereas more polar materials (including ionic compounds) elute early. This experiment illustrates how five clinically important urinary compounds (uracil, uric acid, xanthine, allopurinol, and nicotinamide) can be separated and determined by reverse phase HPLC. The separation is achieved by use of isocratic elution and ultraviolet detection. The order of elution is shown by the chromatogram in Fig. 13-3.

Procedure

Uracil is used as an internal standard because of its stability and its short retention time.

Pipet 5.00-mL aliquots of the uracil stock solution into each of five 100-mL volumetric flasks. Pipet 1.00 mL of each of the other stock solutions (uric acid, xanthine, allopurinol, and nicotinamide) into one of the 100-mL flasks that contains the uracil. Pipet 2.00 mL of each of the solutions into a second 100-mL flask that contains uracil. Pipet 3.00 mL of each into the third flask, 4.00 mL of each into the fourth flask, and 5.00 mL of each into the fifth flask. Dilute each flask to the mark with doubly distilled water. Pipet 5.00 mL of uracil into the unknown flask and dilute to the mark if this has not been done previously by the instructor.

Run chromatograms of each of the five mixtures and of the unknown in accordance with the instructions for the instrument; 0.05 M acetic acid buffered to pH 4.5 is used as the eluent. (It should be degassed or stirred continuously to remove gases.) Set the flow rate at 2.0 mL/min (the column inlet pressure should be about 2000 psi). If necessary, the collected samples can be identified by UV spectrophotometry with the mobile phase (0.05 M acetic acid) as a spectrophotometric reference. The absorption maxima are uracil, 258 nm; uric acid, 292 nm; xanthine, 267 nm; allopurinol, 251 nm; and nicotinamide, 262 nm.

Prepare a urine sample as follows: Pipet 5.0 mL of urine into a centrifuge tube that contains 15 mL of distilled water. Precipitate the urinary proteins by the addition of 1.5 mL $ZnSO_4$ (100 g/L) and 0.8 mL of 0.1 M NaOH. Centrifuge for 10 min at 5000 rpm. Filter the supernatent liquid through a 0.2-μm Nalgene membrane filter (Sybron Corp., Rochester, N.Y.) to remove any remaining particulate matter. Dilute 1 mL of the filtrate to 10 mL and inject the sample directly into the column without further pretreatment. Stop the run after 20 min past the injection, purge the column with 1:1 methanol-water for 15 min and follow with pure methanol. Individual compounds can be identified on the resulting chromatogram by UV spectrophotometry and compared with retention data published in Ref. 2. Because several of the peaks contain more than one component, quantitative determination would be difficult.

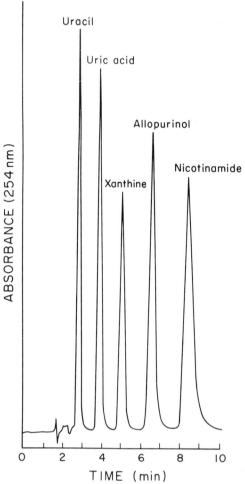

Figure 13-3 / Liquid chromatogram for the separation of 0.12 μg uracil, 0.47 μg uric acid, 0.19 μg xanthine, 0.22 μg allopurinol, and 0.74 μg nicotinamide. [*From D. W. Bastian, R. L. Miller, A. G. Halline, F. C. Senftleber, and H. Veening, J. Chem. Ed., 54, 766 (1977). Copyright © 1977, Division of Chemical Education, American Chemical Society. Reprinted by permission.*]

Treatment of Data

Determine the efficiency of the column by calculating the number of theoretical plates N, using the retention time t_R and the base peak width W_b:

$$N = 16\left(\frac{t_R}{W_b}\right)^2$$

Typical results for the reverse phase column vary from 1500 to 2000 plates for a flow rate of 2.0 mL/min.

Prepare working curves for allopurinol, xanthine, nicotinamide, and uric acid (plot the ratio of the peak area of the component relative to the peak area of the internal standard vs. concentration). All four curves should be plotted on the same graph.

Determine the components in your unknown and the concentration of each from the standard curves.

Identify the individual compounds in the urine sample from the UV spectrophotometry and a comparison with published retention data.

Questions

1 Are any of the urinary compounds studied in this experiment amenable to separation on an ion-exchange column? Explain.

2 For which urinary compound is the HPLC method used in this experiment most sensitive?

3 How could this experiment be performed to get better resolution between the peaks for uric acid and uracil?

EXPERIMENT 13-3

Determination of Antibiotics by High Performance Liquid Chromatography†

Purpose

The antibiotics tetracycline and penicillin V are determined by high performance liquid chromatography with UV detection.

References

1 K. Tsuji, J. H. Robertson, and W. F. Beyer, *Anal. Chem., 46,* 539 (1974).

2 K. M. Lotscher, B. Brander, and H. Kern, *Varian Assoc. Applications Notes,* vol. 9, no. 1, Palo Alto, Calif. (1975).

3 T. Ottaka and M. Yaguchi, *Liquid Chromatography at Work, No. 5,* Varian Assoc., Palo Alto, Calif. (1973).

4 K. Loeffler, *Liquid Chromatography at Work, No. 17,* Varian Assoc., Palo Alto, Calif. (1973).

5 Liquid Chromatography Catalog, Rainin Instrument Co., Inc., Mack Road, Woburn, Mass. 01801 (1979–80).

Apparatus

Liquid chromatograph (equipped with a 280-nm UV monitor and 268-nm monitor)

Stainless steel column, 1.5 m × 2.2 mm i.d. (packed with Pellionex SCX, 30 to 40 μm for tetracyclines)

Column, 25 cm × 2.1 mm i.d. (packed with MicroPak CH, 10 μm for penicillin V)

Syringe, 10 μL

Volumetric flasks (5), 10 mL

Chemicals

U.S.P. Reference Standards: Tetracycline HCl; oxytetracycline HCl; doxycycline HCl

Unknowns of varying concentrations of a tetracycline

0.1 M NH$_4$NaHPO$_4$, pH 8.2 (mobile phase for HPLC)

Samples of tetracycline HCl powders (commercial)

Penicillin V acid (1% solution), in 50% aqueous methanol in water with 0.01 M phosphoric acid [H$_3$PO$_4$]

†Reprinted in part with permission from K. Tsuji, J. H. Robertson, and W. F. Beyer, *Anal. Chem., 46,* 539, copyright 1974 American Chemical Society; and K. M. Lotscher, B. Brander, and H. Kern, Varian Assoc. Applications Notes, vol. 9, no. 1, Palo Alto, Calif. (1975), reprinted by permission of Varian Associates, Inc.

Methanol (20%) in water with 0.01 M H_3PO_4 (mobile phase for penicillin V)

Samples of penicillin V powders

Unknown samples of varying concentrations of penicillin V

Part I. Tetracyclines

Theory

The tetracyclines are broad spectrum antibiotics produced by certain species of streptomyces. Gas chromatographic determination of tetracyclines requires time-consuming and inconvenient derivatization. The liquid chromatographic determination of tetracyclines is made difficult by the many polar groups on the tetracycline molecule. However, they are readily soluble in dilute acids and in fact are usually available as the water-soluble hydrochlorides. This solubility allows cation exchange chromatography to be used for the determination of these compounds. The strong cation exchanger used in this method has a fixed negative charge (sulfonic acid). Structures of some tetracyclines as shown illustrate their diversity of functional group.

OXYTETRACYCLINE (Tetramycin)   DOXYCYCLINE (Vibramycin)   TETRACYCLINE (Achromycin, Tetracyn)   CHLORTETRACYCLINE HYDROCHLORIDE (Aureomycin)

Figure 13-4 illustrates a chromatogram for a mixture of oxytetracycline, tetracycline, and doxycycline that was separated by use of a cation exchange column. The oxytetracycline is almost completely resolved from the tetracycline, and the tetracycline is completely resolved from the doxycycline.

Procedures

A. Qualtitative Determination of One or More of the Tetracyclines

Dry the tetracycline (TC) hydrochloride U.S.P. Reference Standard at 60°C at <5 torr for 3 h. Cool and weigh 4 mg accurately (use a microbalance) into a 10-mL volumetric flask. Just prior to the analysis, dissolve the TC in the mobile phase solution (0.1 M NH_4NaHPO_4, pH 8.2) and dilute to the mark. Repeat this procedure for each of the other tetracyclines.

Record chromatograms of each of the solutions at 45°C and with a flow rate of 0.67 mL/min; use a 280-nm UV monitor and attach an integrator if available. Inject 1.0- or 1.5-μL samples into the column. A 1-mV recorder with a chart speed of 1 cm/min should suffice. Record chromatograms of an unknown TC or mixture of TC's.

B. Quantitative Determination of Tetracycline

Prepare standards of U.S.P. TC Reference Standard that range from 0.090 mg/mL to 0.90 mg/mL. Record chromatograms for each standard solution. Next, obtain an unknown TC from the instructor and record a chromatogram.

C. Analysis of Commercial Tetracycline Products

Obtain a tetracycline hydrochloride sample from one of the pharmaceutical companies. Record a chromatogram.

Treatment of Data

Determine the retention times of the various tetracycline compounds. Determine the retention time of the unknown peak(s) and identify the compound(s).

Measure the area under each peak for both the standards and the unknown for

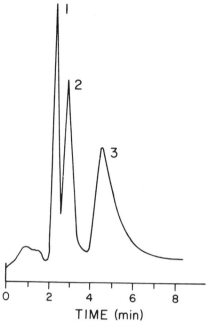

1:Oxytetracycline HCl
2:Tetracycline HCl
3:Doxycycline HCl

Instrument : Varian LC
Column : Pellionex SCX, 30-40 μm
  1.5 m/2.2 mm i.d.
Temperature : 45°C
Mobile Phase : 0.1 M NH$_4$NaHPO$_4$
  pH = 6.2
Flow Rate : 40 mL/hr (50 atm)
Detector : 210 nm UV
Sensitivity : 0.32 aufs
Sample Size : 1 μL
Recorder : 1 mV
Chart Speed : 1 cm/min

Figure 13-4 / HPLC cation exchange separation of tetracyclines. [*From K. M. Lotscher, B. Brander, and H. Kern, Varian Assoc. Applications Notes, vol. 9, no. 1, Palo Alto, Calif. (1975). Reprinted by permission of Varian Associates, Inc.*]

the quantitative determination. Prepare a calibration curve and determine the amount of the TC in the unknown.

From the area under the peak of the commercial tetracycline powder, determine the amount of TC in the sample from your calibration curve.

Submit the chromatograms along with the analytical data in your report.

Questions

1 Explain why the tetracyclines must be derivatized prior to determination by gas chromatography.

2 Draw the structure of the cation exchangers used for the separation in this experiment.

Part II. Penicillin

Theory

Penicillin (biosynthesized by *Penicillium notatum* and related species) that is produced in modern corn-steep liquor media which contains β-phenylethylamine is mainly benzylpenicillin or penicillin G. This form has become the most widely used and the standard by which others are compared. Phenoxymethylpenicillin or penicillin V is produced by *Penicillium chrysogenum* when the media contains phenoxyacetic acid. The structures for penicillin G and penicillin V both contain a lactam ring.

PROCAINE PENICILLIN G

PHENOXYMETHYL PENICILLIN
(Penicillin V, Pen Vee, V Cillin)

Penicillin V (phenoxymethyl penicillin) may be determined by reverse phase HPLC. The solutions to be injected may be prepared directly in acidified aqueous methanol. A totally porous small particle (10μm) reverse phase column (MicroPak CH) is used so that relatively large samples can be injected without overloading the column. Large amounts are necessary due to the low molar absorptivity at the absorption maximum (268 nm).

Procedure

Prepare a series of solutions of penicillin V in 50% aqueous methanol:50% water (0.01 M phosphoric acid) that have concentrations from 0.5% to 1.5%; use U.S.P. Reference Standard material, if available.

Obtain chromatograms of each solution; use a 10-μL sample. The column temperature should be 60°C. The mobile phase used is 20% methanol in water with 0.01 M phosphoric acid at a flow rate of 1 mL/min. Use a 268-nm UV detector, and a 10-mV recorder at a chart speed of 0.5 cm/min.

Obtain a chromatogram of a commercial penicillin powder or unknown provided by the instructor.

Treatment of Data

Determine the area of the chromatographic peak for each of the standard solutions and the unknown. Prepare a calibration graph and calculate the amount of penicillin in the unknown or commercial powder.

Submit chromatograms along with the report.

Questions

1 Do you expect penicillin V to be amenable to determination by gas chromatography? Explain.
2 How can sensitivity be improved for the HPLC determination of compounds with very low molar absorptivities in the UV-visible spectrum?

EXPERIMENT 13-4

Separation of Aromatic Hydrocarbons by High Performance Liquid Chromatography

Purpose

A series of aliphatic benzenes will be separated by reverse phase liquid chromatography.

References

1 *Chromatographic News #32*, Alltech Associates, 2051 Waukegan Road, Deerfield, Ill. 60015 (1979).
2 H. H. Bauer, G. D. Christian, and J. E. O'Reilly, "Instrumental Analysis," Allyn and Bacon, Boston, 1978.

Apparatus

Liquid chromatograph with UV monitor
Column, 250 mm × 4.6 mm i.d. (packed with RSil-C18-HL, 10 μm)
Syringe, 10 μL

Chemicals

Set 1:
Pure samples of benzene, toluene, ethyl benzene, *n*-propylbenzene, and *n*-butylbenzene (20% water in methanol, mobile phase)

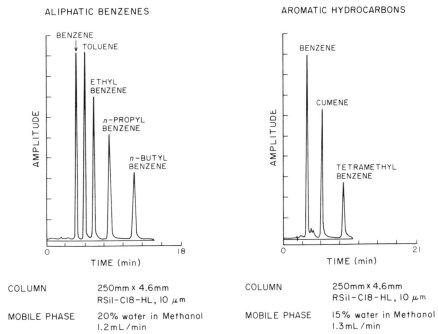

Figure 13-5 / HPLC separations of aromatic hydrocarbons. [*From Chromatographic News #32, Alltech Associates, Deerfield, III. (1979). Reprinted by permission.*]

Set 2:
    Benzene, cumene, and tetramethylbenzene (15% water in methanol, mobile
        phase)
Unknown mixture of Set 1 or Set 2

Theory

The theory of HPLC has been discussed in the text of the chapter. The aromatic hydrocarbons listed in Sets 1 and 2 involve separations of practical interest. As illustrated by Fig. 13-5, excellent resolution can be achieved in their separation by HPLC.

Procedure

Prepare a standard solution of each compound to be determined in the set that has been assigned. They are to be made up to a concentration of 10 mg/mL of mobile phase solvent.

Follow the instructions given for the instrument. The chromatograms can be obtained at room temperature (23° to 25°C); use a flow rate of 1.2 to 1.3 mL/min and a sensitivity of 0.44 AUFS for the UV detector. Inject 5-μL samples of the standard solutions (10 mg/mL) successively and, finally, inject a sample of the unknown.

Treatment of Data

Determine the retention time of each peak from the known standards. Identify the unknown peak(s).

Make an estimate of the amount of each component in the unknown from the areas under the peaks for the standards.

Questions

1 Explain the order of elution observed in the two chromatograms in Fig. 13-5.
2 Do you expect benzoic acid to be strongly or weakly retained by the column used in this experiment?

EXPERIMENT 13-5

Separation of Carbohydrates by High Performance Liquid Chromatography†

Purpose
HPLC offers an excellent alternative to gas chromatography for the quantitative analysis of carbohydrates, with the added advantages of (*a*) high resolution, (*b*) short analysis time, and (*c*) elimination of additional sample preparation steps. The experiment will illustrate how common carbohydrates such as maple syrup and artificial sweetners can be characterized.

References
1 Whatman, Inc., "Liquid Chromatography Applications Report 76-2E, Carbohydrate Analysis," 9 Bridewell Place, Clifton, N.J. 07014.
2 F. M. Rabel, A. G. Caputo, and E. T. Butts, *J. Chromatography, 126,* 731 (1976).

Apparatus
High performance liquid chromatograph
Column, 4.6 mm × 250 mm. (PXS-1025 PAC, from Whatman, Inc.)
Syringe, 10 μL
Detector on HPLC, refractive index (a short-wavelength UV detector is more sensitive, especially for saccharin)

Chemicals
Maple syrup samples [acetonitrile/0.0025 *M* sodium acetate [$NaC_2H_3O_2$], pH 5.0 (85:15), mobile phase]
Artificial sweetener that contains saccharin and lactose (such as "Sweet 'N Low"®) [acetonitrile/0.0025 *M* potassium dihydrogen phosphate [KH_2PO_4], pH 5 with KOH (75:25), mobile phase]
Soybean extract [acetonitrile/0.0025 *M* KH_2PO_4, pH 5 w th KOH (70:30), mobile phase]
Fructose, glucose, sucrose, saccharin, lactose, stachyose, raffinose and malic acid standards

Theory
The theory underlying HPLC is discussed in the text of the chapter.
Natural products can contain dangerous or illegal adulterants. Maple syrup is one such product; its primary ingredient is sucrose with smaller concentrations of fructose and glucose (Fig. 13-6). Artificial sweeteners, such as Sweet 'N Low®, contain saccharin and lactose, as illustrated by Fig. 13-7.
It has been shown during the determination of carbohydrates in soybean extract that stachyose sometimes harbors a flagellated organism that can cause gastric distress in humans. Production control procedures must ensure a minimum concentration of stachyose (Fig. 13-8).

Procedures
The following operating conditions are suggested for this experiment:

Column temperature, ambient
Flow rate: For maple syrup, 1.8 mL/min; for saccharin, 1.53 mL/min; for soybean extract, 1.33 mL/min
Detection: Refractive index or UV in series with R.I. for saccharin

A. Maple Syrup
Prior to injection of the standards into the column, allow the mobile phase [acetonitrile/0.0025 *M* sodium acetate, pH 5.0 (85:15)] to pass through the column for

†Adapted in part with permission from Whatman, Inc., "Liquid Chromatography Applications Report 76-2E, Carbohydrate Analyses."

OPERATING CONDITIONS—

Column:
    PXS-1025 PAC: 4.6 mm x 250 mm

Column Temperature:
    Ambient

Mobile Phase:
    Acetonitrile/0.0025 $\underline{M}$ NaC$_2$H$_3$O$_2$, pH 5.0 (85:15)

Flow Rate:
    1.8 mL/min

Pressure:
    425 psi

Detection:
    RI

Peaks—

a. Solvent front
b. Fructose
c. Glucose
d. Unknown
e. Sucrose

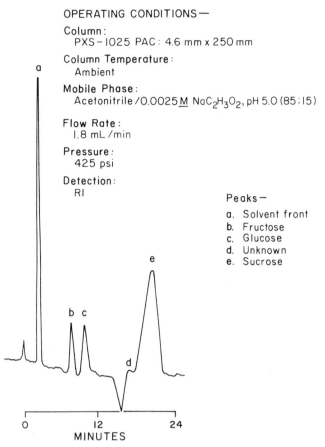

Figure 13-6 / Carbohydrates in maple sugar syrup on Partisil-10 PAC. (From Whatman, Inc., "Liquid Chromatography Applications Report 76-2E, Carbohydrate Analyses." Reprinted by permission.)

5 min. Set the flow rate and pressure as indicated above for maple syrup. Inject each of the standards [fructose, glucose, sucrose, and malic acid (usually present in syrup)] in turn, and, finally, the sample of maple syrup.

B. Artificial Sweetener

Prior to injection of the standards, allow the mobile phase [acetonitrile/0.0025 M KH$_2$PO$_4$, pH 5.0 with KOH (75:25)] to pass through the column for 5 min. Record the detector response to insure that there are no substances left on the column from the previous experiment. Set the pump flow rate at 1.53 mL/min and inject the standards (saccharin and lactose), followed by the unknown sweetener.

C. Carbohydrates in Soybean Extract

Prior to injection of the standards, allow the mobile phase [acetonitrile/0.0025 M KH$_2$PO$_4$, pH 5.0 with KOH (70:30)] to pass through the column for 5 min or until no more substances from previous experiments elute. Inject the standards (sucrose, raffinose, and stachyose) in turn, and then the unknown soybean extract.

Treatment of Data

Measure the distance from the injection mark to the peak maximum for each of the standards and samples. The distance is directly related to the retention time because the recorder is driven at a constant rate; therefore, recording the t_R in units of length is adequate. Tabulate the retention times and identify each carbohydrate.

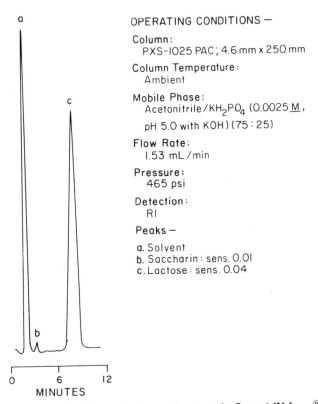

OPERATING CONDITIONS —

Column:
PXS-1025 PAC; 4.6 mm x 250 mm

Column Temperature:
Ambient

Mobile Phase:
Acetonitrile/KH$_2$PO$_4$ (0.0025 $\underline{M}$,
pH 5.0 with KOH) (75:25)

Flow Rate:
1.53 mL/min

Pressure:
465 psi

Detection:
RI

Peaks —

a. Solvent
b. Saccharin: sens. 0.01
c. Lactose: sens. 0.04

**Figure 13-7 / Saccharin and lactose in Sweet 'N Low®
artificial sweetener on Partisil-10 PAC. (From Whatman, Inc., "Liquid Chromatography Applications Report
76-2E, Carbohydrate Analyses." Reprinted by permission.)**

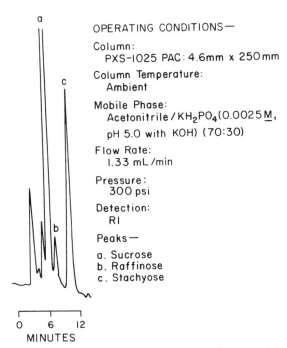

OPERATING CONDITIONS—

Column:
PXS-1025 PAC: 4.6mm x 250mm

Column Temperature:
Ambient

Mobile Phase:
Acetonitrile/KH$_2$PO$_4$(0.0025 $\underline{M}$,
pH 5.0 with KOH) (70:30)

Flow Rate:
1.33 mL/min

Pressure:
300 psi

Detection:
RI

Peaks—

a. Sucrose
b. Raffinose
c. Stachyose

**Figure 13-8 / Carbohydrates in soybean extract
on Partisil-10 PAC. (From Whatman, Inc., "Liquid
Chromatography Applications Report 76-2E, Carbohydrate Analyses." Reprinted by permission.)**

The peak area is directly related to the concentration. Determine the peak
areas by integration.

In tabular form, record the concentrations, retention times, and peak areas for
each standard. Also indicate the retention time, peak area, and the calculated concentration of each component in the sample. (A rough quantitative measure of the
concentration of each component in each sample can be made by comparing the
peak areas of the standard with the peak area of the unknown.)

Questions

1 Explain why a refractive index detector is used for part of this experiment,
rather than a UV detector.
2 Would the measurement of peak height, rather than area, give good quantitative results in this experiment?

14
Exclusion Chromatography

Exclusion chromatography (also called gel permeation chromatography) is a liquid chromatographic method that separates molecules according to differences in molecular dimensions. Complex mixtures of proteins and enzymes which can have widely different molecular weights often can be separated by exclusion chromatography. The method is a gentle separation procedure which does not usually denature labile molecules. If separation on the basis of molecular size is not sufficient, gel chromatography can be used sequentially with other techniques such as ion exchange or adsorption chromatography.

The principle of an exclusion column is best understood by making the assumption that spherical particles are used as the stationary (packed column) phase. These highly hydrated particles contain pores or holes of relatively uniform size. A solution of a protein is allowed to flow through the packed column of the material, which can be prepared in different particle sizes. Because the pore dimensions are of the order of magnitude of those for large molecules, the column can separate the spherical protein molecules by size; different-sized protein molecules will vary in their ability to enter the gel pores (Fig. 14-1). The smaller molecules will enter many pores, and, hence, be held up in the column, while the larger ones, too large to enter many pores, will not be held up and will elute quickly from the column (Fig. 14-2A to D). In any column chromatographic process all molecules in the sample mixture will spend an equal amount of time in the mobile chromatographic phase before they are eluted from the column, because they all move at the same rate when they are in this phase. The difference in elution time arises because the smaller molecules tend to spend time in contact with the stationary phase (as well as with the mobile phase), while the larger molecules spend no time in the stationary phase and never stray from the mobile phase of the col-

SEPHADEX PARTICLE

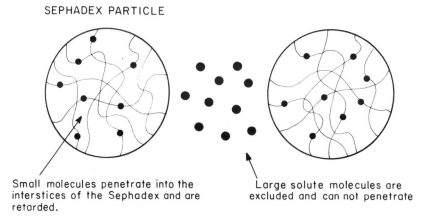

Small molecules penetrate into the interstices of the Sephadex and are retarded.

Large solute molecules are excluded and can not penetrate

Figure 14-1 / Determination of molecular weight on a Sephadex column.

umn. Thus, elution time for larger molecules is largely dependent on the flow rate in the column, while time spent in the stationary phase adds to the elution time for smaller molecules.

The differences in elution volume for molecules of different molecular size in a gel permeation column allow chromatographic separation of a mixture of large molecules, such as proteins and enzymes. To the extent that in a series of large molecules the molecular size of these molecules varies directly with the molecular weight, the elution volume (V_e) can be related to the molecular weight of a sample compound. The column is calibrated with a series of compounds of known molecular weight (the weight range is chosen to "bracket" that suspected for the unknown sample). The elution volume of each of these is plotted against its molecular weight to produce a calibration graph (such as in Fig. 14-3). This allows the molecular weight of the unknown sample to be determined from its elution volume from the calibrated gel permeation column.

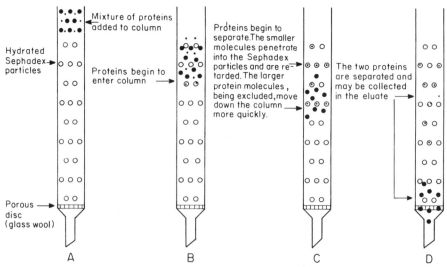

Figure 14-2 / Separation of proteins of different sizes on a Sephadex column.

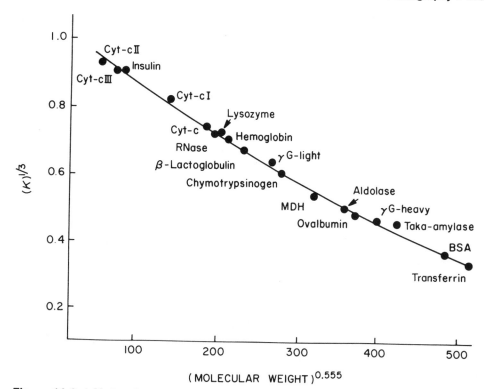

Figure 14-3 / Molecular weights of a series of proteins as a function of K. MDH, malate dehydrogenase; BSA, bovine serum albumin. Sample: 10 mg of each protein dissolved in 0.5 mL of the eluent, maintained at pH 8.6 for 4 h. Medium: Bio-Gel A-5m. Bed: 1.5× 85 cm. Elunt: 6 *M* guanidine hydrochloride, 0.1 *M* 2-mercaptoethanol. Flow rate: 3.0 mL/h · cm² of cross-sectional bed area. The data are displayed according to the method of Porath [J. Porath, *J. Pure Appl. Chem.*, **6**, 233 (1963).]. [*From "Gel Chromatography" (A Laboratory Manual), Bio-Rad Laboratories, Richmond, Calif. (1971). Reprinted courtesy of Bio-Rad Laboratories.*]

The penetration of a large molecule into the gel phase is best envisioned as a partition process. Hence, gel permeation elution volumes must be related to a partition coefficient K,

Time to elute = time in mobile phase + time in gel phase

Multiplying both sides by the mobile phase flow rate,

Elution volume (V_e) = volume to elute a nonpartitioning material (V_0)
$$+ K \times \text{volume of gel phase } (V_s)$$

$$V_e = V_0 + KV_s$$

$$K = \frac{V_e - V_0}{V_s}$$

V_s = total volume of gel $(V_{tot}) - V_0$

The value of K is not directly related to the molecular volume or weight of the macromolecule. Since for a given column the values of V_s and V_0 will be the same for a series of compounds, a plot of V_e against molecular weight yields an effective calibration curve for molecular weight measurements. Actually, the log of the molecular weight is plotted because this quantity is expected to vary linearly with K (or V_e).

EXPERIMENT 14-1
Exclusion Chromatography: Characterization of Hemoglobin

Purpose

The purpose of the experiment is to determine the apparent molecular weight of hemoglobin by gel permeation chromatography.

References

1 "Gel Chromatography" (Laboratory Manual), Bio-Rad Laboratories, Richmond, Calif. (1971).
2 N. E. Heftmann, "Chromatography," 3rd ed., Van Nostrand Reinhold, New York, 1975, chap. 14.
3 B. L. Karger, L. R. Snyder, and C. Horvath, "An Introduction to Separation Science," Wiley, New York, 1973, chap. 15.
4 H. H. Willard, L. L. Merritt, Jr., J. A. Dean, and F. A. Settle, Jr. "Instrumental Methods of Analysis," 6th ed., Van Nostrand, Princeton, N.J., 1981, chap. 18.
5 G. W. Ewing, "Instrumental Methods of Chemical Analysis," 4th ed., McGraw-Hill, New York, 1975, chap. 20.
6 H. H. Bauer, G. D. Christian, and J. E. O'Reilly, "Instrumental Analysis," Allyn and Bacon, Boston, 1978, chap. 21.

Apparatus

Buret, 50 mL (with a small piece of glass wool pushed to the bottom)
Beaker, 250 mL
Graduated cylinder, 25 mL
Test tubes (35)
Glass wool

Chemicals

0.5 M sodium chloride [NaCl]–0.1 M sodium acetate [NaC$_2$H$_3$O$_2$] buffer, pH 6
Sephadex G-75 slurry and Sephadex G-25 slurry (both at concentrations of 15 g per 100 mL buffer)
Dextran blue 2000 solution, 3 g per 100 mL buffer
Myoglobin solution, 1 g per 100 mL buffer
Hemoglobin solution, 1 g per 100 mL buffer
Dextran:myoglobin mixture (1:1), 1 mL of mixture per student

Theory

A gel column of Sephadex, a trade name for a commercially available modified dextran, is used in this experiment. The Sephadex dextran is "cross-linked"; i.e., the polysaccharide chains are joined by smaller intermediate chemical links to form a three-dimensional matrix (something like a ball of steel wool in which the steel fibers are joined at regular intervals to their nearest neighbors). It is the special regularity of the cross-linking, and the consequently uniform hole sizes, in this dextran material that makes it suitable for the gel permeation experiments. After the preparation of the column it is calibrated with two knowns, Dextran Blue 2000 (a soluble polysaccharide too large to fit into the gel holes and thus, nonpartitioning) and myoglobin. From the straight line resulting from the plot of the elution volumes for these two compounds vs. the logs of the molecular weights, the apparent molecular weight of hemoglobin can be determined.

Procedure

Take a small piece of glass wool and push it to the bottom of a dry 50-mL buret with a long length of glass rod; this glass-wool plug serves as a bed for the Sephadex packing and prevents its escape from the buret. Close the stopcock and pour buffer solution (0.5 M NaCl–0.1 M NaC$_2$H$_3$O$_2$) up to the 25-mL mark. Obtain some Sephadex G-75 in a medium-sized beaker. Stir the slurry well and quickly pour the mixture into the buret; fill the column to the top. Open the stopcock all the way and continue to add the well-stirred slurry; keep the buret filled to the top as much as possible. A homogeneous column results only when the buret is kept full with a well-stirred slurry. If all of the Sephadex is allowed to settle and an additional amount is added, a line of demarcation will form in the column. The denser, whiter layer of gel that starts at the bottom and moves up the buret is the packed Sephadex. When this layer reaches the 28-mL mark, fill the buret to the top and then let all the gel settle out. The column should now be packed high enough for this experiment. When the clear buffer reaches a point about 3 mL above the top of the packed gel, close the stopcock and add a small amount of Sephadex G-25 slurry so that a protective layer 1-mL thick will be deposited on the top of your G-75 column. The layer will act as a bed to prevent the G-75 column from being disturbed when the sample is added.

note Never let the column dry out; a layer of buffer should always be on top of the column. Allow the buffer to run down close to the gel but never actually past the top so that air could get into the column.

Now that the column is packed, close the stopcock and remove as much of the buffer above the G-25 as possible without disturbing the gel bed. The last bit of liquid can be removed by opening the stopcock long enough to allow the liquid level to drop to the exact level of the G-25 gel without letting air hit the column.

Apply (slowly) exactly 1 mL of hemoglobin solution to the top of the G-25 gel with a 1-mL volumetric pipet. The pipet should be held a little above the column bed but not against the side of the buret, and the liquid should be allowed to flow out very slowly so that it does not cause an uneven sample band.

The stopcock is now opened and the sample allowed to flow into the column. As soon as the liquid is no longer above the G-25, close the stopcock again; place a graduated cylinder (25 mL) under the buret, then add enough buffer (without Sephadex) to fill the buret. This must be done slowly and carefully for the first few milliliters so as not to disturb the sample. Open the stopcock and allow the liquid to flow out into your graduated cylinder. Refill the buret with buffer as needed to maintain the liquid level above the top.

The hemoglobin sample will move slowly down the column and the band will begin to spread out. When the front of the band gets to the 45-mL mark on the buret, remove the graduated cylinder, note the volume eluted at that point, and begin to collect 1-mL samples in the previously prepared test tubes. Number the test tubes as you collect eluent in them. Continue collecting these fractions until there is no more visible color in the fractions. (This should be no more than 20 tubes; if it exceeds 20 tubes, stop collecting.)

Repeat the above procedure for a 1-mL sample of a solution that contains equal volumes of myoglobin and Dextran 2000. Observe the separation and record the volume required for elution of each peak maximum. This should be done by use of the second set of 15 marked test tubes.

Line up the numbered test tubes of each sample in order against a white background and determine the location of the most intense fraction. Add the number (and, therefore, the volume in milliliters) of this most intense fraction to the volume that was collected in the graduated cylinder. For Dextran Blue 2000 this will be called the void volume (V_o). The void volume is the volume of buffer required to elute a compound that does not penetrate the pores of the packing—a nonpartitioning sample. Dextran Blue 2000 has a molecular weight of 2×10^6 daltons and G-75 excludes any material with a molecular weight that is greater than 50,000 daltons. Thus, the apparent molecular weight of Dextran Blue 2000 on this column is greater than 50,000 daltons.

Note the total volume of the gel (G-75) in the column by use of the buret graduations. This volume is $V_o + V_s$. Because the value for V_o has been measured, V_s can be found from this total volume.

Cleanup

After the last sample has been eluted, remove the G-25 gel layer with a disposable pipet and discard it. Using small portions of buffer added to the top of the buret, cover tightly with a piece of Parafilm® and shake horizontally. Allow the solution to run into the beaker after the Parafilm is removed. Repeat. When all of the gel has been removed, flush out the glass wool plug by opening the stopcock, attaching a rubber hose to the tip of the buret, connecting the hose to a water faucet, and *gently* turning on the water. An alternative to this procedure is to hold the tip of the buret in a container of water and apply suction to the top of the buret, drawing out the glass wool. Pour the G-75 gel into the sideshelf beaker provided. Do not return G-75 to the G-25 beaker or vice versa!

Treatment of Data

Record data for the elution volumes for the three samples.

Sketch your gel column and give dimensions.

Plot a graph of elution volume vs. log of the molecular weight. For this, Dextran Blue 2000 will be considered to have a molecular weight of 50,000 (the exclusion limit of the column is 50,000), and the myoglobin will be considered to have a weight of 17,000. According to theory, these points can be connected by a straight line. From this line, determine the apparent weight of hemoglobin. The actual weight is about 65,000. Do you observe any dissociation into subunits? How much, if the dissociation behaves as a dimer, into monomers?

By use of the volumes measured for V_o (V_e for Dextran Blue 2000) and V_s (the total column volume less V_o), calculate the value for K for hemoglobin and myoglobin.

Questions

1　What would be the value of V_e with your column for a compound with a molecular weight of 25,000 daltons?
2　Why do different columns have different V_o values?
3　Discuss any parts of this experiment that were particularly difficult.

15
Ion-Exchange Chromatography

Ion-exchange resins have found wide application in many scientific fields since their recent introduction. In analytical chemistry they provide a method of separation and of equivalent exchange by which it is possible to solve many difficult problems.

An ion-exchange resin is an organic high polymer which contains ionic functional groups. A typical resin for exchanging cations is illustrated by sulfonated polystyrene cross-linked with divinylbenzene (Fig. 15-1). The resin is polymerized in the form of spherical beads of various sizes. The beads are placed in a long glass tube giving an ion-exchange column in which the chemical separations or equivalent exchange can be carried out. The high-molecular-weight polymeric material might be considered as a highly insoluble, immobile, polybasic anion. The percentage of cross-linking in the polymer is controlled (by the amount of divinylbenzene added) so that the resin molecule has a permeable network configuration through which water as well as ions are able to move.

If a solution containing NaCl is poured through an ion-exchange column containing a sulfonated polystyrene resin, the following reaction takes place:

$$Na^+ + Cl^- + H\ Res \rightarrow Na\ Res + H^+ + Cl^-$$

where H Res represents the immobile polymer on the "hydrogen cycle." The sodium ions have been replaced by an equivalent amount of hydrogen ion. The important result is that although NaCl was poured into the column the effluent is HCl. The amount of NaCl that can be converted into HCl depends on the capacity of the resin and the amount of resin in the column. Once the resin has absorbed

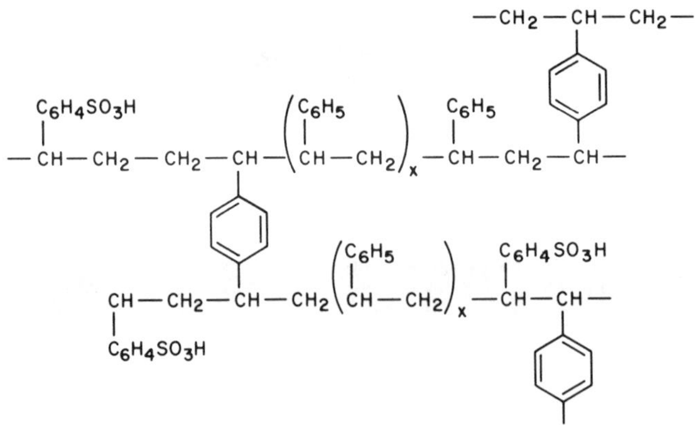

Figure 15-1 / Sulfonated polystyrene cross-linked with divinylbenzene.

its maximum capacity of sodium ions, the equilibrium can be reversed and the resin regenerated. This is done by passing a concentrated solution of acid through the exchange column, resulting in the reaction

$$H^+ + Cl^- + Na\ Res \rightarrow H\ Res + Na^+ + Cl^-$$

Anion-exchange resins, in which basic amine groups instead of sulfonate groups are incorporated into the polymer, can also be prepared. If such a resin is on the hydroxyl cycle, various anions can be exchanged for the hydroxyl ions; e.g.,

$$Na^+ + Cl^- + Res\ OH \rightarrow Na^+ + OH^- + Res\ Cl$$

In this case the sodium chloride is converted to sodium hydroxide.

Ion-exchange resins have many uses. They are widely used in water softening, in which the water is passed through a "mixed bed" consisting of both cation and anion exchangers. The cation resin is put on a sodium cycle, and the anion exchanger is on the chloride cycle. In this fashion all cations are converted to sodium ion and all anions to chloride, thus completely removing the hardness of the water. When the resins have exchanged their maximum capacity of ions they can be reverted to the original condition (regeneration) by passing a strong solution of NaCl through the "mixed bed."

Ion-exchange resins have been used in many aspects of analytical chemistry.

1   Small amounts of metals (e.g., Cu^{2+} in milk) can be concentrated from dilute solutions by pouring a large volume of the solution through a column and later removing the ions with a small volume of eluting agent.

2   The resins have been used to separate desirable or undesirable cations (e.g., NH_4^+ from urine and separation of streptomycin from fermentation broth) and anions (phosphate from ferric ions) before applying a particular method of analysis.

3   Many separations are possible by use of a technique called "ion-exchange chromatography." This principle is discussed in Experiment 15-2.

4   *Conversion of salts to acids or bases.* A simple application of ion exchange is the determination of the amount of dissolved salt in a solution by passing it

through a cation exchanger on hydrogen ion cycle or an anion exchanger on hydroxyl ion cycle, and titrating the acid or base which is liberated. This method will determine only total salt concentrations and is not selective, but some useful applications have been made. Deliquescent salts can be standardized in this way. The method is also useful for the determination of total salt content in industrial and laboratory water.

EXPERIMENT 15-1
Ion-Exchange Resins and Chelometric Titrations

Purpose

We shall now illustrate how a combination of analytical techniques can serve to simplify an otherwise difficult analysis. The unknown is a mixture of H^+, Na^+, Mg^{2+}, and Zn^{2+}, and an analysis for each constituent is desired.

Aliquot 1. H^+ is titrated with base in the presence of the other ions.

Aliquot 2.   The sample is passed through a cation-exchange resin on acid cycle, and all salts are converted into an equivalent quantity of their acid, H^+. This total corresponds to the sum of H^+, Na^+, Mg^{2+}, and Zn^{2+}.

Aliquot 3.   The solution is adjusted to pH 10 and the sum of Mg^{2+} and Zn^{2+} is determined by titration with EDTA.

Aliquot 4.   1 g KCN is added to mask the Zn^{2+} and the solution is adjusted to pH 10. The Mg^{2+} is then titrated selectively in the presence of Zn^{2+} with EDTA.

From the above data the quantity of each ion present may be calculated (Na^+ by difference).

References

1   N. E. Heftmann, "Chromatography," 3rd ed., Van Nostrand Reinhold, New York, 1975, chaps. 12 and 13.

2   B. L. Karger, L. R. Snyder, and C. Horvath, "An Introduction to Separation Science," Wiley, New York, 1973, chap. 12.

3   D. A. Skoog and D. M. West, "Fundamentals of Analytical Chemistry," 4th ed., Saunders, Philadelphia, 1982, chaps. 12 and 29.

4   D. G. Peters, J. M. Hayes, and G. M. Hieftje, "Chemical Separations and Measurements," Saunders, Philadelphia, 1974, chaps. 6 and 17.

5   W. J. Blaedel and H. T. Knight, *Anal. Chem., 26,* 741 (1954).

Apparatus

Ion-exchange column (bottom half of an old 50-mL buret works satisfactorily)
Pipet, 10 mL
Volumetric flask, 100 mL
Titration flasks (wide-mouth Erlenmeyer) (2), 250 mL
Buret, 50 mL
Beaker, 100 mL
Graduated cylinder, 100 mL
Backwashing flask

Chemicals

Unknown sample (contains approximately 4 mmol each of H^+, Na^+, Mg^{2+}, and Zn^{2+} in a 100-mL volumetric flask). Dilute to mark and shake.

$6\ M$ HCl

$0.1\ M$ ammonia buffer, in NH_3 and $0.05\ M$ in NH_4NO_3

$0.100\ M$ NaOH

Indicator, Eriochrome Black T, CI203 (1 part solid dye ground with 100 parts NaCl)

Dowex 50-X8, 50-100 mesh ion-exchange resin

$0.0200\ M$ disodium dihydrogen ethylenediaminetetraacetate

Methylene blue–methyl red mixed indicator (pH change at 5.4) (lavender = acid form, green = base form)

KCN, solid

Theory

In this experiment, a mixture of salts and acid are placed in the top of a column, and the acid eluted is titrated with base to give the sum of acid and salts present. This analysis allows the determination of one salt (NaCl) provided the analysis of the other constituents (H^+, Zn^{2+}, Mg^{2+}) is known. It is important to recognize that 2 mol of H^+ are freed for each mole of divalent metal ions, such as Zn^{2+} and Mg^{2+}.

Chelometric Titrations: Titrations with Ethylenediaminetetraacetic Acid (EDTA)

The introduction of ethylenediaminetetraacetic acid,

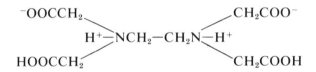

as a titrant in analytical chemistry has simplified the determination of dilute concentrations of metal ions. This reagent and the other chelons are unique in that they form soluble but very slightly dissociated complex ions with many polyvalent cations and consequently can be used effectively as titrating agents. Procedures, for example, have been developed for the volumetric determination of the ions for alkaline earths, rare earths, Mn, Fe, Cu, Hg, Cr, Co, Ni, Zn, Ce, Pb, Al, Ga, In, Tl, Pd, Bi, Zr, Sc, and others; but the first and probably still the most practiced is the determination of hardness (i.e., Ca^{2+} and Mg^{2+}) in water.

Titration of Zinc and Magnesium

Ethylenediaminetetraacetic acid, abbreviated as EDTA or H_4Y, forms a series of salts on neutralization with sodium hydroxide. One of these salts, $Na_2H_2Y \cdot 2H_2O$, is commercially available in a high degree of purity and is used in the preparation of the titration solution.[†] When this solution reacts with zinc and magnesium ions, the reactions can be represented by the following equations:

$$H_2Y^{2-} \rightleftharpoons HY^{3-} + H^+ \qquad pK_3 = 6.16$$

$$HY^{3-} \rightleftharpoons Y^{4-} + H^+ \qquad pK_4 = 10.26$$

$$Y^{4-} + Zn^{2+} \rightleftharpoons ZnY^{2-} \qquad \log K = 16.4$$

$$Y^{4-} + Mg^{2+} \rightleftharpoons MgY^{2-} \qquad \log K = 8.7$$

[†]By proper purification and drying, this compound can be prepared sufficiently pure to serve as a primary standard. (See Ref. 5.)

Use of Metal Ion Indicator

A very satisfactory indicator for the titration of the sum of Zn^{2+} and Mg^{2+} at pH 10 is Eriochrome Black T (also called CI203, its color index number).

It is a tribasic acid and forms colored soluble complexes with both zinc and magnesium ions. During a titration, the EDTA reacts first with the free zinc ions, then with the free magnesium ions, and, finally, with the magnesium in the indicator complex.

$$Mg^{2+} + HIn^{2-} \rightleftharpoons MgIn^- + H^+$$
$$+ \qquad \text{(blue)} \qquad \text{(pink)}$$
$$HY^{3-}$$
$$\Updownarrow$$
$$MgY^{2-}$$
$$+$$
$$H^+$$

When the end point is approached, the solution turns from a pink to a purple to a pure blue. The final pure blue is the end point. The position of the equilibrium in the above equation shifts to the left as the magnesium ions react with EDTA, thereby causing the solution to become pure blue when a quantity of EDTA equivalent to the magnesium content has been added. In other words, titrate until all of the pink tinge (Mg^{2+}-indicator complex) is completely discharged and the addition of an extra drop of EDTA does not produce a color change. Since the magnesium-indicator complex is wine red in color and the free indicator is blue between pH 6.5 and 11.5,

$$H_2In^- \underset{}{\overset{\substack{pH \\ 6.5}}{\rightleftharpoons}} HIn^{2-} \underset{}{\overset{\substack{pH \\ 11.5}}{\rightleftharpoons}} In^{3-}$$
$$\text{(red)} \qquad\qquad \text{(blue)} \qquad\quad \text{(red)}$$

the color of the solution changes from wine red to blue at the end point.

A pH of 10 is best for the titration of the sum of zinc and magnesium. In more alkaline solutions, zinc and magnesium hydroxides may be precipitated, whereas in more acid media, the magnesium is not bound strongly enough to the indicator to give the wine red compound. The optimum pH can be readily obtained and maintained by the addition of a sufficient amount of an ammonia–ammonium chloride buffer.

Titration of Magnesium Only

For this titration the solution is buffered at pH 10 and an excess of KCN is added to complex the zinc so tightly that EDTA cannot titrate it. In this way the magnesium is titrated alone. The CN^- thus acts as a "masking agent" for Zn^{2+}. By the use of such masking agents, EDTA can be used as a more selective titrant. In a similar manner CN^- will mask Ni^{2+}, Co^{2+}, and Cd^{2+}, and triethanolamine will mask Fe^{3+}, Al^{3+}, and Mn^{2+}. Zn^{2+} and Cd^{2+} can be selectively "demasked" by the addition of formaldehyde.

Another method by which the EDTA titration can be made more selective is pH selection. Since EDTA is an acid, the H^+ concentration enters into the equilibrium expressions for the stability of the metal chelates. Thus the alkaline earths which form relatively weak complexes with EDTA can only be titrated in alkaline solution where the H^+ concentration is low:

$$HY^{3-} + Mg^{2+} \rightleftharpoons MgY^{2-} + H^+$$

and the above equilibrium is shifted to the right. However, metals like Bi^{3+} which form very strong complexes with EDTA can be titrated in acid solution. The common indicators are similarly affected, and thus it is possible to titrate many metal ions (rare earths, transition and heavy metal ions) in acid solution without interference from the alkaline earths.

Procedures

You may find that time can be saved by titrating your solutions while the ion-exchange column is in operation.

A. Determination of Acid and Salt Equivalents

First, to ensure that the ion-exchange column is on the "hydrogen cycle," pass 25 mL of 6 M HCl through the column. Do not use a vacuum or pressure to hasten this exchange as many exchange processes are slow reactions. Be careful also that the liquid level in the column never falls below the upper level of resin. This would give rise to channels in the resin and the exchange efficiency decreases. If the solution level has fallen below the top of the resin, the column should be backwashed (see Fig. 15-2). Place the tubing of the backwashing flask on the bottom of the column and force water up through it until the resin is in the upper part of the column. Remove the flask and tap the bottom of the column with a rubber

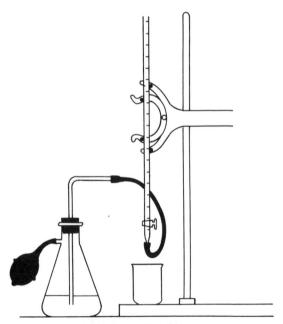

Figure 15-2 / Backwashing flask.

stopper as the resin settles back. Replace the glass-wool plug on the top of the resin.

Rinse the excess HCl completely from the column by adding five or six 25-mL portions of distilled water. Allow each 25-mL portion to pass through the column before adding the next portion. *Discard washings.* From time to time, test a few drops of the solution as it issues from the column with litmus paper. If the solution is acid, continue washing the column with more distilled water.

When you are satisfied that your column is properly rinsed, pipet a 10-mL aliquot of your sample into the cup of your exchange column and collect the effluent in a clean 250-mL titration flask. Pass four 25-mL portions of distilled water through the column, allowing each portion to pass through before adding the next portion. Add a few drops of mixed indicator to the collected effluent, and titrate with 0.1 M NaOH. This allows determination of total acid and salt equivalent.

Repeat this process (including the acidification of the column with HCl) using a second 10-mL aliquot of your sample.

B. Determination of Acid Equivalents

Add a 10-mL aliquot of your sample to a clean 250-mL titration flask, add a few drops of mixed indicator, about 75 mL of distilled water, and titrate with 0.100 M NaOH. Repeat for a duplicate check.

C. Determination of Magnesium plus Zinc

Pipet a 10-mL aliquot of your sample into a 250-mL titration flask, add approximately the correct amount of 0.1 M NaOH to neutralize the acid present (this was determined in the above step), and finally add approximately 50 mL of ammonia buffer and a small amount of solid CI203 indicator. The solution will then have a pink color. Titrate with 0.0200 M EDTA until the solution turns from pink to a purple to a pure blue. The final pure blue is the end point. Repeat for duplicate checks.

D. Determination of Magnesium

Pipet a 10-mL aliquot of your sample into a 250-mL titration flask, neutralize with the required amount of 0.1 M NaOH, add 50 mL of ammonia buffer, approximately 1 g of KCN, a small amount (the smaller the amount, the sharper the end point) of solid CI203 indicator, and 50 mL of distilled water. Titrate with 0.0200 M EDTA until a pure blue color is obtained. Repeat for duplicate checks.

Treatment of Data

Calculate the number of millimoles of HCl, NaCl, ZnCl$_2$, and MgCl$_2$ present in your 100-mL sample.

Questions

1 In the acetate–acetic acid system the concentration of free acetate ion is dependent on (*a*) the total analytical amount of acetic acid and acetate ion added, (*b*) the ionization constant for acetic acid, and (*c*) the pH of the medium. The total analytical concentration of acetate A* is

$$A^* = [Ac^-] + [HAc] \tag{1}$$

and from the ionization constant:

$$[HAc] = \frac{[H^+][Ac^-]}{K_i} \tag{2}$$

Substitute Eq. (2) into Eq. (1) and solve for $[Ac^-]$.

  a Plot the dependence of $[Ac^-]$ on $[H^+]$ for $A^* = 0.1\ M$.

  b Derive an equation for the "effective stability constant" (which is really a constant only at a given pH) for a metal-acetate complex which has an absolute stability constant of K.

$$K = \frac{[MeAc]}{[Me^+][Ac^-]}$$

$$K_{eff} = \frac{[MeAc]}{[Me^+]A^*} = \ ?$$

2 Considering the stepwise acid-ionization constants for EDTA (K_1, K_2, K_3, and K_4) and the stability constant K for the Zn^{2+}–EDTA complex, derive an equation for the effective stability of the zinc complex in terms of the $[H^+]$, K, and K_1, K_2, K_3 and K_4. Repeat these calculations for the Mg^{2+}–EDTA complex, and plot on the same graph. Plot K_{eff} vs. pH from pH 4 to 12. What other equilibria are involved in the basic, buffered solution?

3 The analysis for which constituent is probably most erroneous? Explain.

EXPERIMENT 15-2
Iron-Cobalt Separation by Ion Exchange

Purpose

The purpose of the experiment is to demonstrate that iron(III) and cobalt(II) ions may be separated and recovered by the use of an ion-exchange resin. The analyses are done spectrophotometrically for both cobalt(II) and iron(III), or spectrophotometrically for cobalt(II) and amperometrically† by continuous monitoring for iron(III).

References

1 J. T. Stock, *J. Chem. Ed.*, *51*, 491 (1974).
2 J. T. Stock and M. A. Fill, *Analyst*, *71*, 142 (1946).
3 K. A. Kraus and G. E. Moore, *J. Am. Chem. Soc.*, *72*, 4293 (1950).

Apparatus

Spectrophotometer
Ion-exchange column (a 50-mL buret)
Graduated cylinders, 50 mL, 500 mL
Pipets, 1 mL, 4 mL, 5 mL, 10 mL, and 20 mL
Volumetric flasks, twenty-two 100 mL, two 250 mL, one 1000 mL
Amperometric flow-through cell (optional)

Chemicals

12 *M* HCl distilled [This material must be free of metal ions, which can be assured by either distilling CP HCl and retaining all the distillate, or by passing the HCl (concd) over the exchanger one time to remove the metal ion contamination (primarily Fe^{3+}).]

---

†Amperometric portion of experiment adapted from J. T. Stock, *J. Chem. Ed.*, *51*, 491 (1974). Copyright © 1974, Division of Chemical Education, American Chemical Society. Reprinted with permission.

4 M HCl (prepared by diluting the above)

Hydroxylamine solution [5% (aqueous)]

Orthophenanthrolene [0.3% (aqueous)]

2 M sodium acetate

Ammonium thiocyanate [50% (aqueous)]

Acetone, CP

Dowex 1, X-8, 50-100 mesh ion-exchange resin for column

Congo-red indicator paper

Iron(III) stock solution, 5 g $FeCl_3 \cdot 6H_2O$ diluted with 4 M HCl to 1 L (this need not be distilled acid) (The diluted solution contains 1.0 mg/mL of iron.)

Cobalt(II) stock solution, 12 g $CoCl_2 \cdot 6H_2O$ diluted with 4 M HCl, as above, to 1 L (This contains 3.0 mg/mL of cobalt.)

Iron-cobalt sample solution, 100 mL of each of the two stock solutions are mixed to form this solution

0.1 M HCl

Theory

Kraus and Moore have used anion-exchange resins to effect the separation of a mixture of Fe^{3+}, Mn^{2+}, Co^{2+}, Zn^{2+}, Ni^{2+}, and Cu^{2+}. Like most cations, all these metals form complex ions with chloride ion

$$M^{n+} + mCl^- \rightarrow MCl_m^{n-m}$$

It is noted that when m is greater than n, the complex ion is an anion and as such can be adsorbed on an anion-exchange resin. The relative amounts of the metal M^{n+} existing as a cation and as an anion are determined by the concentration of chloride. At low chloride concentrations the cation M^{n+} is the predominant species, while at high chloride concentrations the metal exists as the anion. Different metals have different stability constants; for example, in 6 M HCl zinc is in the form $ZnCl_4^{2-}$ while nickel exists as Ni^{2+}.

If a mixture of the above-mentioned metals is in a solution of 12 M HCl, all the metals are in the form of anions except Ni^{2+}. So if this mixture is poured into an anion-exchange column, all the metals except nickel would be adsorbed by the resin. The nickel would pass through the resin and be separated. If the column is then treated with 6 M HCl, the Co^{2+} complex (whose stability constant is of such value that this metal exists as a cation) ceases to be adsorbed on an anion exchanger and is eluted from the column. In like manner, each of the metals may be separately eluted by gradually decreasing the concentration of chloride and changing the anions into nonadsorbable cations. Zinc, which forms the strongest chloride complex, reverts to a cation only at very low chloride concentrations and hence is eluted last. The relative chloride concentrations at which each of the metals is eluted are shown in Fig. 15-3.

Figure 15-4 presents adsorption characteristics for the various ions of the elements and is useful in predicting chromatographic separations. D_v is the distribution coefficient for an ion between the column solution and the ion-exchange resin. When log $D_v = 1$, one column volume of eluting solution is required to elute the ion "peak" from the column; larger volumes are required when log D_v is greater than unity.

The elution of a solute species from an ion-exchange column can be monitored, either continuously or by the collection and examination of successive fractions of the eluent. Continuous monitoring allows the elution to be detected while it is occurring and is a rapid method.

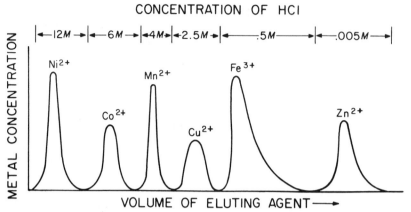

Figure 15-3 / Chromatographic separation of metal ions on an anion-exchange column.

In suitable media, iron(III) gives a limiting current at a rotating platinum electrode (RPE) that is maintained at zero potential with respect to the saturated calomel electrode (SCE). No interference is caused by cobalt(II), which is not electroactive under these conditions. The SCE is not convenient for use in a small flow-through system, but a fixed electrode–stirred solution system can be easily constructed. Ag/AgCl (saturated) in KCl (saturated) has a potential that is approximately 50 mV more negative than that of the SCE and can be used in place of the SCE.

The flow-through system is shown schematically in Fig. 15-5; the four holes in the rubber stopper lie in a circle, rather than in line. A 20-mm o.d. specimen tube cut down to 25 mm high forms the cell and contains the 10-mm long microstirrer bar. This is merely a piece broken from a sewing needle and sealed within melting-point tubing. All of the glass tubes in the stopper are 4.5 mm o.d. Inlet tube A terminates a few millimeters above indicator electrode B. This is a 20-mm length of 0.5-mm diameter platinum wire that is sealed axially through the end of

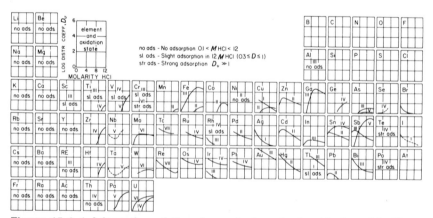

Figure 15-4 / Adsorption of the elements from hydrochloric acid. [*From "Trace Analysis," J. H. Yoe and H. J. Koch (eds.), copyright © 1957 by John Wiley & Sons, Inc., New York.*]

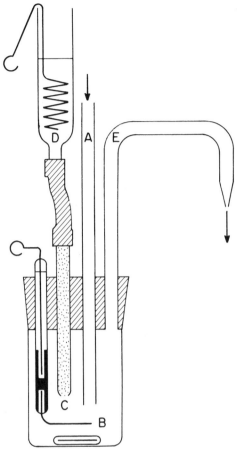

Figure 15-5 / Amperometric flow-through cell. [*From J. T. Stock, J. Chem. Ed., 51, 491 (1974). Reprinted by permission.*]

its support tube. The projecting 10 mm of wire is then bent at right angles, as shown. Connection is made by a copper wire that dips into a short column of mercury in contact with the inner end of the platinum wire. To prevent accidental spillage of mercury, the copper wire is wedged with a wood splinter and sealed in with a bead of epoxy cement.

The extremities of salt bridge tip C are constricted to 2 mm i.d. This helps to retain the 4% agar-agar in saturated KCl that fills the tip. By repeated gentle squeezing, all air bubbles are expelled from the rubber tube that connects the tip to the saturated KCl reservoir D. This contains a 15-cm length of approximately 0.7-mm diameter silver wire that is coiled and bent as shown. Before insertion into D, the coil is anodized in KCl solution to produce a light coating of AgCl. If the usual large-area amperometric SCE is to be used in place of the Ag–AgCl system, the tip of the SCE salt bridge is merely inserted into the KCl solution in D. Entrapment of air in the cell is avoided by making the lower end of exit tube E flush with the underside of the stopper.

Although the flow-through cell can be permanently attached to the ion-exchange column, it is simpler to attach the system after the cobalt fractions have been collected directly from the column inlet.

Procedures

Follow Procedure A or B.

A. Spectrophotometric Detection

Elute the column with two 30-mL portions of water, then acidify by eluting twice with 20-mL portions of distilled 4 *M* HCl. In all elutions allow the solution level to fall nearly to the resin—not, however, into the glass-wool plug!

Pipet 1.00 mL of the sample solution onto the resin. The pipet tip should be introduced into the top of the buret as far as possible to avoid having the sample stick to the sides. It may be necessary to rinse carefully with 1 mL of acid to wash the solution down onto the resin. Allow the sample to sink into the resin.

Add 1 mL of distilled 4 *M* HCl and allow this to sink into the column. Repeat this step. Wait 5 to 10 min.

Add to the column 28 mL of the distilled 4 *M* acid and withdraw it in 4-mL portions into individual 100-mL volumetric flasks. These will now contain the cobalt.

Add 28 mL of distilled water to the column and withdraw in 4-mL portions into individual 100-mL volumetric flasks. These flasks will contain the iron.

Preparation of Standard Iron Solutions

Dilute 5.00 mL of the iron stock solution to the mark in a 250-mL volumetric flask with distilled water.

Into four clean 100-mL volumetric flasks pipet portions of this solution as follows: 5 mL, 10 mL, 15 mL, and 20 mL. (*Do not dilute at this point: see steps 1 to 4 below.*) This gives standard solutions containing, respectively, 0.1, 0.2, 0.3, and 0.4 mg of iron per 100 mL.

Development of Color in Iron Solutions and Measurement

Each of the samples withdrawn from the column and each standard is treated as follows:

1   Add 4.0 mL of 5% hydroxylamine.
2   Place a small square of congo red paper in the solution. It will turn blue. Add sodium acetate solution drop-wise until it just turns red again.
3   Add 4.0 mL *o*-phenanthrolene solution.
4   Dilute to mark with distilled water.

Prepare a solution containing all of the above but no iron to use for 100 %*T* setting.

Measure *A* or %*T* for all the solutions, using a colorimeter or a spectrophotometer at a wavelength of 500 nm.

Make an appropriate plot. Since standards are in terms of milligrams per 100 mL the values for the elution portions taken directly from the plot may be added up to give the amount of iron recovered directly. Determine the total iron recovered.

Preparation of Standard Cobalt Solution

Proceed as directed for the iron standards, taking identical portions of the cobalt stock solution. (*Do not dilute final solutions at this point; see below.*) These solutions will be of the following concentrations: 0.3, 0.6, 0.9, and 1.2 mg cobalt per 100 mL.

Development of Color in Cobalt Solutions and Measurement

Prepare 1 L of a solution 3:1 by volume of acetone and 50% NH_4CNS.

Dilute each standard cobalt solution and each cobalt sample to the mark with this solution. The color is said to be unstable, so this dilution and subsequent measurements should be accomplished systematically.

Measure $\%T$ or A for all solutions, using either a colorimeter or a spectrophotometer at a wavelength of 620 nm.

Determine the total cobalt recovered.

Treatment of Data	Plot the elution curve for iron and cobalt, graphing milliliters eluted on the abscissa and metal concentration (milligrams per 100 mL) on the ordinate. Note the point where the eluting agent was changed. Determine the percentage recovery of iron and of cobalt. Discuss the reasons for not getting 100% recovery of the individual metals and suggest what happened to the unrecovered metal ions.
B. Electrochemical Detection	Follow Procedure A until all the cobalt has been eluted from the column. Fill the amperometric flow-through cell with 0.1 M HCl, then attach it to the ion-exchange column and adjust the magnetic stirrer until the tiny bar spins smoothly. It is desirable to elute at an essentially constant rate of approximately 1 mL/min. (The buret stopcock can be modified to facilitate fine control.) After the flow-through cell is attached, add 28 mL of distilled water to the column, connect the electrodes to a microammeter, and read the current after each milliliter of eluate passes through. If desired, the cell may be calibrated, so that the amount of iron can be compared with the amount of iron placed on the column. First, pass 0.1 M HCl through the cell to establish the small and essentially constant residual current. Then pass through the cell a set of standard iron solutions [$<50\ \mu g$ of iron(III) per milliliter of 0.1 M HCl]. This allows the linearity of the response to be checked.
Treatment of Data	Plot the elution curve for iron(III) and graph the current in microamperes vs. eluate in milliliters. Determine the percent recovery of iron. Plot the elution curve for cobalt, with the cobalt concentration (milligrams per 100 mL) on the ordinate vs. the milliliters eluted on the abscissa. Determine the percent recovery of cobalt.
Questions	1 Discuss the function of HCl in the Kraus-Moore separation procedure. 2 Use the data in Fig. 15-4 to predict the separation observed in Fig. 15-3.

16
Liquid-Solid Chromatography

Liquid-solid chromatographic methods of separation, discovered by Tswett in 1906, have been applied extensively to the separation of organic molecules. The method is extremely valuable, since compounds which are similar enough to defy separation by any other method are often resolved by quite simple chromatographic procedures. For example, the fact that carotene exists in three isomeric forms (α, β, and γ) was never realized until such a mixture was subjected to chromatographic separation. Previously carotene was thought to be a single compound. Similarly, although azobenzene was expected to have cis and trans configurations, their separation was not accomplished until G. S. Hartley (*J. Chem. Soc., 1938,* 633) accidentally resolved them on a chromatographic column during a study of azo dyes.

Liquid-solid chromatographic methods depend on small differences in adsorbability of molecules, even though they are structurally quite similar. If a mixture of two slightly polar organic molecules X and Y are dissolved in a nonpolar solvent such as hexane, they can be adsorbed on a polar solid such as silica gel (Fig. 16-1a). When more hexane is poured through the column, neither compound will migrate, because there is a greater interaction between the solute and polar solid than there is between the solute and nonpolar solvent. However, if benzene (which is more polar than hexane) is passed through the column, X and Y will tend to be displaced and may migrate down the column. Because of differences in adsorbability as well as solvent interaction, the two compounds will move at different rates, as shown in Fig. 16-1b. The latter process is called the "development of the chromatogram." If too polar a developer is used, such as ethyl alcohol,

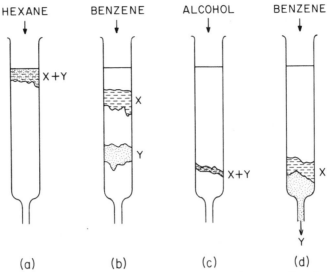

Figure 16-1 / Pictorial representation of separation by gravity flow liquid chromatography.

both X and Y will proceed rapidly down the column, as shown in Fig. 16-1c, and no separation occurs. The ideal developing agent is one which gives a separation and moves the leading substance rapidly enough so that it can be "eluted" off the column, as shown in Fig. 16-1d, and be recovered.

The relative interaction of the solute with the solvent and the adsorbant determines the usefulness of a particular chromatogram, which is controlled to a major extent by relative polarities. For this reason solvents and adsorbants are ranged in graded series, as shown in Tables 16-1 and 16-2.

The data of these tables are a great aid in the correct choice of solvent, developer, and column material for chromatographic analysis. For example, if a mixture of substances is so strongly adsorbed on an aluminum oxide column that development does not occur, a more polar solvent or a less polar column should be used.

Table 16-1 / Graded Series of Eluents

| Petroleum ether
Carbon
  tetrachloride
Cyclohexane
Carbon disulfide
Ether
Benzene
Esters
Chloroform
Alcohols
Water
Pyridine
Organic acids | Increasing
polarity

↓ | Increasing ability
to desorb polar
molecules

↓ | ↑

Increasing adsorbability
of solute |

Table 16-2 / Graded Series of Adsorbants

Sucrose, starch	Increasing ability to adsorb polar molecules	Increasing polarity
Talc		
Sodium carbonate		
Calcium phosphate		
Magnesium oxide		
Silica gel		
Aluminum oxide		
Charcoal		
Fuller's earth	↓	↓

EXPERIMENT 16-1

Chromatographic Separation of cis and trans Azobenzenes Determined Spectrophotometrically†

Purpose

The equilibrium constant for cis and trans azobenzene is measured by chromatographically separating the two isomers and then determining them spectrophotometrically.

References

1 N. E. Heftmann, "Chromatography," 3rd ed., Van Nostrand Reinhold, New York, 1975, chaps. 4 and 5.
2 B. L. Karger, L. R. Snyder, and C. Horvath, "An Introduction to Separation Science," Wiley, New York, 1973, chap. 13.
3 G. S. Hartley, *J. Chem. Soc., 1938*, 633.

Apparatus

Chromatography column, 25–30 cm (a 50-mL buret is satisfactory)
Glass wool
Beaker, 250 mL
Graduated cylinders, 10 mL, 100 mL
Volumetric flasks (2), 100 mL
Spectrophotometer (such as Spectronic 20)
Cuvettes (2)
Conical flasks (2), 125 mL

Chemicals

Aluminum oxide, 100 mesh
Petroleum ether
Azobenzene in petroleum ether, 3 to 4 g/L
Ethyl ether
Methanol
0.2 M NaOH

Theory

Azobenzene exists in two different geometrical configurations, as shown below.

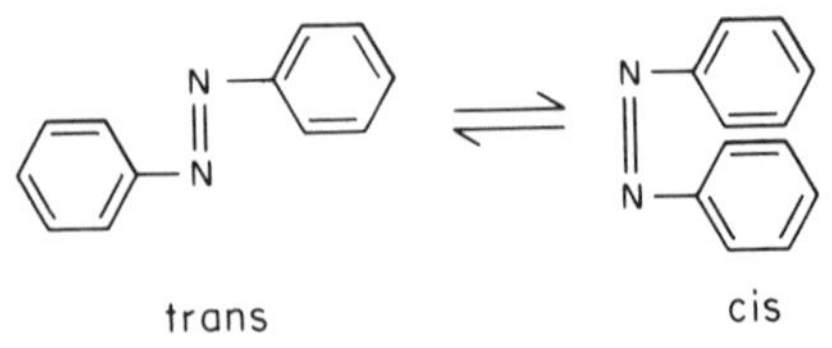

trans   cis

†From a similar experiment at Cornell University, courtesy of Dr. W. D. Cooke.

As in the case of most cis-trans isomers, the cis form has a lower melting point and a higher dipole moment and is much more soluble in polar solvents. One form is labile (usually the cis) and the other is stable. With solid azobenzene the trans isomer is stable, as proved by x-ray diffraction studies. In solution, absorption of light can convert the trans form into the cis isomer. Two reactions then proceed:

Light absorption $\qquad$ trans $\overset{h\nu}{\rightleftharpoons}$ cis

Thermal (very slow) $\quad$ trans $\overset{\Delta}{\leftarrow}$ cis

At equilibrium, 15 to 40% (depending on solvent) of the azobenzene is in the cis form.

Equilibrium is attained slowly, so it is possible to separate both species on a chromatographic column. On a polar column, such as aluminum oxide, and with petroleum ether as the solvent, the cis modification, being more polar, is more strongly absorbed. On a hydrophobic column such as carbon black, however, the trans isomer is more strongly absorbed and the solvent used is methanol.

Procedure

safety caution   Petroleum ether is highly inflammable.

Take about 5 to 8 mL of 100-mesh aluminum oxide and prepare a column of about 5 to 8 cm in height by first placing a glass-wool plug at the bottom of the column, then filling the column to a height of 20 cm with petroleum ether, and finally adding *slowly* from a beaker a slurry of the packing material suspended in petroleum ether. This slow addition should prevent channels from being formed. Cover the top of the column with a small glass-wool plug. Do not allow the solvent level to fall below the top of the aluminum oxide at any time until the experiment is completed.

Add 3 mL of the solution of azobenzene in petroleum ether (60°/90°) to the column. Allow the sample to just enter the column, then rinse down the sides of the column with small portions of petroleum ether (60°/90°). Do not let the level of the solvent fall below the top of the aluminum oxide. Wash the column with fresh solvent. Separation into the two bands should occur quickly. Note the distance in millimeters that the leading edge of the trans isomer has moved after 20 mL of petroleum ether has been eluted through the column.

To move the trans band somewhat faster, the polarity of the solvent can be increased by the addition of 20% ethyl ether to the petroleum ether. Note the movement of the band with 25 mL of the mixed solvent and compare the movement in both solvents. Just before the trans isomer leaves the column, discard the excess. Elute the trans isomer from the column and collect the fraction in a 100-mL volumetric flask; make up to volume with petroleum ether (30°/60°). To elute the cis isomer, a more polar solvent is necessary. Make a 5% solution of methanol in petroleum ether and elute the cis isomer; collect into a 100-mL volumetric flask and dilute to volume with petroleum ether (30°/60°).

Analyze each fraction by use of a spectrophotometer. Petroleum ether (30°/60°) should be used as a blank to set the instrument to 100 %T at 445 nm. Read the %T for each solution and convert to absorbance. Make up a standard solution of 3 mL of azobenzene diluted to 100 mL with petroleum ether (30°/60°). Take a reading of the standard solution. 445 nm is the isosbestic point for the cis-trans spectral mixture.

Treatment of Data

Calculate the equilibrium constant for the conversion of the cis into the trans form of azobenzene. Calculate the percent recovery of the two isomers.

$$K_{eq} = \frac{[\text{cis}]}{[\text{trans}]}$$

$$A_{\text{sample}} = A_{\text{cis}} + A_{\text{trans}}$$

$$\% \text{ recovery} = \frac{A_{\text{trans}} + A_{\text{cis}}}{A_{\text{standard}}} \times 100\%$$

Questions

1 Could this experiment be performed by high performance liquid chromatography? Explain.

2 Explain why absorbance measurements are made at the isosbestic point of the cis and trans isomers.

EXPERIMENT 16-2

Quantitative Analysis of APC Tablets by Liquid-Solid Chromatography and Ultraviolet Spectroscopy

---

Purpose

A common "pain reliever" tablet (APC, Excedrin) is separated into its three active ingredients: aspirin (acetylsalicylic acid), phenacetin (acet-*p*-phenetidine, *p*-ethoxyacetanilide, or acetophenetidin), and caffeine by means of column chromatography. Each of the components is then measured quantitatively with ultraviolet spectroscopy.

References

1 G. Smith, *J. Assoc. Official Agric. Chem., 37,* 677 (1954).

2 J. R. Dyer, "Applications of Absorption Spectroscopy of Organic Compounds," Prentice-Hall, Englewood Cliffs, N.J., 1965.

3 D. R. Browning (ed.), "Spectroscopy," McGraw-Hill, New York, 1969, pp. 8–11.

4 N. E. Heftmann, "Chromatography," 3rd ed., Van Nostrand Reinhold, New York, 1975, chaps. 4 and 5.

5 B. L. Karger, L. R. Snyder, and C. Horvath, "An Introduction to Separation Science," Wiley, New York, 1973, chap. 13.

Apparatus

Chromatography column (Celite 545)
Separatory funnel
Packing rod
Ultraviolet spectrophotometer
Silica cells (2)

Chemicals

1 *M* sodium bicarbonate
9:1 water–sulfuric acid mixture
0.1 *M* acetophenetidin standard
0.1 *M* acetylsalicylic acid standard

0.01 *M* caffeine standard
Acetonitrile
Chloroform
Ether

Theory

Celite (heat-treated siliceous skeletons of dead diatoms) is used as a solid support for the two portions of the column. The first portion of the column has adsorbed sulfuric acid for the stationary phase, which retains the caffeine. The second part of the column is treated with sodium bicarbonate and retains the aspirin. The phenacetin is not held up by the column and is eluted through the column first.

The solvent also plays an important role in column chromatography. The solubilities of the components in various solvents can be used to determine the most effective separation strategy.

| | Solubilities in grams per 100 mL | | | | |
	W(H_2O)	E(Ether)	C($CHCl_3$)	E&W	C&W
Phenacetin	0.077	0.769	6.67	10.0	8.66
Caffeine	2.00	0.167	16.7	0.089	8.35

With just Celite and water, chloroform will elute both phenacetin and caffeine without separation (C&W), while ether will elute phenacetin well before caffeine. Hence, one approach would be to use a mixture of ether and chloroform. To have a reasonable "blank zone" between the two compounds on separation, the ratio of ether to chloroform should be high (e.g., 9:1). The only advantage to the use of chloroform is that ether has a low boiling point and tends to evaporate readily. In this experiment pure ether will be used to elute phenacetin, followed by chloroform to elute the caffeine. Aspirin is eluted last by an acidic mixture of chloroform. In this part of the experiment, one column packed with two portions of solid support is used to separate three components. The same process could be accomplished by repeated extractions in a separatory funnel, but such a method is both tedious and incomplete relative to column chromatography

Ultraviolet
Spectrophotometry

The concentration of the three components which have been separated will be determined by UV spectrophotometry (see Chaps. 6 and 7 for a discussion of the theory and principles).

Procedures

A. Preparation of Ether

The ether and chloroform must be washed prior to use. This is accomplished by putting equal amounts of the ether or chloroform and water into a separatory funnel and shaking well. Discard the first 15 mL and gravity filter the rest.

B. Preparation of Unknown

Weigh accurately a powdered sample of APC (approximately 100 mg) and quantitatively transfer it to a 100-mL volumetric flask. Add about 60 mL of washed chloroform and shake well. Add 0.2 mL of acetic acid (concd.) and dilute to mark with washed chloroform.

C. Preparation of Column

Place the chromatography column in a clamp holder and attach to a ring stand. Put a small piece of rubber tubing with a screw clamp on the base of the column to regulate the flow through the column as necessary. Tightly pack the column with about 1 cm of glass wool. Add an additional 1 cm of glass wool loosely packed to bring the height of the wool to about 2 cm.

Place 6 g of Celite 545 into a 100-mL beaker and add 6.0 mL of the 9:1 water–sulfuric acid solution, and mix until the total amount of Celite looks moist. Add a small amount of ether and stir. Place the Celite in the column (about 1 cm at a time); pack firmly between each addition.

Mix 6.0 g of Celite with 6 mL of 1 M sodium bicarbonate. Pack in the column above the acidified layer; use the same techniques as used with the acidified layer. Add about 15 to 20 mL of washed ether to the column. The flow rate through the column should be about 3 to 4 drops per second without the rubber tube on the end of the column. When the last of the ether has passed through the column, replace the beaker under the column with a clean 150-mL beaker.

D. Separation of Components

Place a 10-mL aliquot of the unknown APC solution in the column. After the solution has passed into the absorbent, pass 5- to 10-mL portions of washed ether through the column, allow each portion to absorb before the next portion is added. Wash the tip of the column with chloroform, and evaporate the eluate to dryness on the steam bath in the fume hood. Dissolve the phenacetin residue in 10 mL of washed chloroform, transfer to a 100-mL volumetric flask, and dilute to the mark with acetonitrile.

Replace the beaker with a 100-mL volumetric flask. Pass about 50 mL of washed chloroform through the column. Wash the tip of the column and dilute to the mark with washed chloroform.

Place another 100-mL volumetric flask under the column. Make a solution of 0.5 mL of acetic acid in 5 mL of chloroform, and pass it through the column. Then pass about 90 mL of 1% acetic acid in washed chloroform and dilute to the mark in the 100-mL volumetric flask.

E. Preparation of Standards

Standards have been prepared for this experiment. 1-mL aliquots of each of the stock standards should be diluted to the mark in 100-mL volumetric flasks. These standard solutions will then be used to prepare the remaining standards. Pipet 10-, 30-, and 50-mL aliquots of the initial standards into 100-mL volumetric flasks and dilute to the mark. This series of dilutions will give four standards that will give a calibration curve that covers one order of magnitude.

When the standards are prepared, use the same solution to dilute them as is used for the unknown. The aspirin should be diluted with 1% acetic acid in washed chloroform, the caffeine in washed chloroform, and the phenacetin in 9:1 acetonitrile-chloroform solution.

F. Spectrophotometric Determinations

Record the absorbance (or $\%T$) for each standard and separated component of the unknown APC tablet.

The wavelengths to be used for the measurements are: aspirin, 286 nm; caffeine, 276 nm; and phenacetin, 286 nm. The solutions of the separated unknown should be measured on the same day that they are separated, if possible. The standards and unknowns must be measured at the same time.

Treatment of Data

The absorbances for each standard can be calculated from the observed $\%T$ values that are read on the instrument. These should be plotted on the ordinate with the concentrations on the abscissa to prepare the calibration curves. If precise quantitative work is required, 15 to 20 points should be used for the calibration. Calculate the amount of each component in the APC tablet.

Question

1 Assume that the manufacturer's analysis of the APC tablets is $3\frac{1}{2}$ grains aspirin, $2\frac{1}{2}$ grains phenacetin, and $\frac{1}{2}$ grain caffeine and determine the percent of recovery that you achieved in your separation experiment.

17

Thin-Layer Chromatography (TLC)

Thin-layer chromatography is based on the same fundamental principles as column chromatography. Both operate as a system of three components: a stationary phase, a mobile phase, and a sample dissolved in a liquid. The advantages of thin-layer chromatography over column and paper chromatography include speed, efficiency of separation, and sensitivity of detection methods.

A thin layer of a stationary solid phase is prepared on a glass or plastic plate, and an organic mobile phase travels up the plate through the thin layer by capillary action (Fig. 17-1). These phases compete for the components of a sample, and separation occurs as components are more or less soluble in the mobile phase and, thus, move rapidly or less rapidly up the plate. The process may be treated in terms of a partition equilibrium between the mobile phase and the surface of the stationary phase (just as for column chromatography). The solvent front is allowed to progress to a predetermined height (sometimes governed by time considerations) and the plate is removed from the elution chamber. The different components appear as spots at different heights on the plate (Fig. 17-2). Their positions are described by an R_f value:

$$R_f = \frac{\text{distance traveled by spot from origin}}{\text{distance traveled by the solvent front from origin}}$$

R_f values are reproducible only under carefully controlled conditions. The R_f value is a function of many parameters; for example, the previous history of the plate (it often adsorbs atmospheric water, which alters the ease of adsorption of organic molecules by the stationary phase) and the distance already traveled by

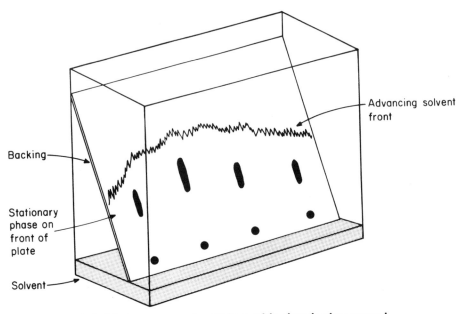

Figure 17-1 / A thin-layer experiment in a wide developing vessel.

the spot. In order to compare R_f's from day to day, or with literature values, it is frequently advantageous to add a standard spot to be run at the side of the unknown. This permits a correction to be made, if desired.

Some common adsorbents that are used in TLC are listed in Table 17-1, and a list of functional groups with their relative adsorption affinity are given in Table 17-2.

Physical detection methods and reagents are used to visualize the substances

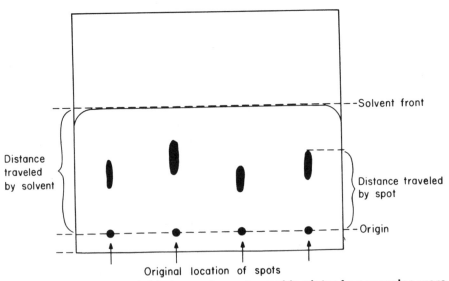

Figure 17-2 / An eluted thin-layer chromatographic plate; four samples were applied.

Table 17-1 / Common Adsorbents Used in TLC (in order of increasing adsorption power)

Cellulose
Kieselguhr (diatomaceous earth)
Hydroxyl apatite
Saccharose, starch
Activated silicic acid (silica gel)
Activated aluminum oxide
Activated charcoal

on the chromatoplates. It is often advisable to apply several indicator methods consecutively to the same chromatoplate to ensure that a substance does not evade detection. Table 17-3 lists some common detection methods that are used in TLC.

Table 17-2 / Functional Groups in Order of Increasing Adsorption Affinity

$-Cl$	Chlorides
$-H$	Hydrocarbons
$-OCH_3$	Ethers
$-NO_2$	Nitro compounds
$-N\begin{smallmatrix}CH_3\\CH_3\end{smallmatrix}$	Tertiary amines
$-C\begin{smallmatrix}O\\OCH_3\end{smallmatrix}$	Esters
$-O-C\begin{smallmatrix}O\\CH_3\end{smallmatrix}$	O-Acetoxy compounds
$-NH_2$	Primary amines
$-NH-C\begin{smallmatrix}O\\CH_3\end{smallmatrix}$	N-Acetoxy compounds
$-OH$	Alcohols
$-C\begin{smallmatrix}O\\NH_2\end{smallmatrix}$	Amides
$-C\begin{smallmatrix}O\\OH\end{smallmatrix}$	Acids

Table 17-3 / Some Common Detection Methods for TLC

Reagent	Color	Components Detected
Daylight	Many colors	Colored compounds
Ultraviolet light (254 and 366 nm)	Fluorescent spots on dark background	Many organic substances
Water	White spots on white background	Hydrophobic substances of high molecular weight
Iodine vapors	Brown spots on yellow background	Most organic substances, especially unsaturated
2',7'-Dichloro-fluorescein	Yellow-green spots on dark purple background in UV light (254 nm)	Most organic compounds
Chromic sulfuric acid solutions, such as 5% $K_2Cr_2O_7$ in 40% H_2SO_4	Black spots on white background after heating (280°C). Color changes during heating	All nonvolatile organic materials
Antimony chlorides 50% $SbCl_3$ in glacial acetic acid 25% $SbCl_6$ in carbon tetrachloride	Many characteristic colors	Steroids, alicyclic vitamins, carotenoids
Ninhydrin, 0.3% in n-butanol containing 3% acetic acid	Pink to purple spots on white background	Amino acids, amines
Diphenylboric acid β-aminoethyl ester 1% in ethanol†	Many characteristic colors	Many natural products

†Fluka/Intern. Chem. Nucl. Corp.

EXPERIMENT 17-1

Thin-Layer Chromatography and Separation of Amino Acid Derivatives†

Purpose

Thin-layer chromatography will be applied to the separation and identification of an amino acid derivative, which will illustrate one of the methods for the determination of protein structure.

References

1 N. E. Heftmann, "Chromatography," 3rd ed., Van Nostrand Reinhold, New York, 1975, chap. 8.
2 D. A. Skoog and D. M. West, "Principles of Instrumental Analysis," 2nd ed., Saunders, Philadelphia, 1980, chap. 25.
3 F. Sanger and H. Tuppy, *Biochem. J., 49,* 463 (1951).
4 F. Sanger and E. O. P. Thompson, *Biochem. J., 53,* 353 (1963).
5 W. A. Schroeder, "The Primary Structure of Proteins," Harper and Row, New York, 1968.

†From a similar experiment used at the University of North Carolina, Chapel Hill.

Apparatus

TLC developing chamber
TLC plates, 20 cm × 20 cm, silica gel (activated at 100°C for 30 min)
Capillary tube
Pipet, disposable (or a 1-μL pipet)
Erlenmeyer flask, 125 mL
Graduated cylinder, 250 mL

Chemicals

Known DNP (dinitrophenyl) derivatives (vial of 1 mg each in 3 mL acetone):
    DNP-β-alanine
    DNP-glutamic acid
    DNP-glycine
    DNP-isoleucine
    DNP-valine
    DNP-serine
    di-DNP-cystine
    2,4-dinitrophenol
    2,4-dinitroaniline
Unknown amino acid, 40 mg (cystine, valine, isoleucine, glycine, serine, alanine, or glutamic acid)
Chloroform
Benzyl alcohol
Methanol
Acetic acid
Sodium carbonate
2,4-dinitrofluorobenzene [DNFB]
HCl, concentrated

Theory

All proteins, including enzymes, are composed of one or more linear sequences (polypeptides) of amino acids. Their action in living systems depends on many factors, but one of the most important is the amino acid content and their sequence. The most elementary form of sequencing is to determine the amino acid at the end of the peptide chain with the free amino group (the N-terminal amino acid). This can be done by treating the enzyme with a tagging reagent followed by hydrolysis. The tagged amino acid usually is a different class of chemical compound from the remaining amino acids, and can be separated and identified. A common method to tag the N-terminal amino acid of a peptide or enzyme is to use 2,4-dinitrofluorobenzene (DNFB) which reacts with the amino group of a terminal amino acid to produce a yellow derivative (a secondary amine). After hydrolysis the end group's properties are such that it behaves like an ordinary carboxylic acid rather than the amino acids which exist in ionic form. After hydrolysis of the enzyme, the yellow derivative can be separated from the other amino acids, concentrated, and spotted on a thin-layer chromatography plate along with derivatives of known amino acids which are called DNP derivatives because they now contain a dinitrophenyl group. The plate is developed and the R_f values of the unknown and the known derivatives are compared and used to identify the unknown.

The actual experiment which will be performed involves the formation of the DNP derivative from an unknown amino acid directly, rather than the formation of one from a protein that is subsequently hydrolyzed. The initial step is to convert the amino acid to an amine

$$\overset{+}{H_3N}-CH-COO^- + CO_3^{2-} \rightarrow H_2N-CH-COO^- + HCO_3^-$$
$$\quad\quad | \quad\quad\quad\quad\quad\quad\quad\quad\quad\quad\quad | $$
$$\quad\quad R \quad\quad\quad\quad\quad\quad\quad\quad\quad\quad\quad R$$

Next the DNP derivative is formed

DNP derivative

By TLC analysis the unknown can be identified from its R_f value, or, at least, the possibilities can be limited.

Procedures

A. Formation of the Derivative

Dissolve a 40-mg sample of the unknown amino acid in 20 mL of water that contains about 1 g sodium carbonate (Na_2CO_3) in a small Erlenmeyer flask. To this mixture add 1.0 mL 2,4-dinitrofluorobenzene (DNFB) with the buret that is provided at the reagent table.

safety caution Do not remove the DNFB from the reagent table. Some individuals are extremely sensitive to this reagent. Care must be taken to prevent contact of skin or clothing with DNFB. In event of such contact, wash immediately with soap and water.

Heat this mixture to 40°C for 30 min (do not let temperature rise much above 40°) with constant swirling. During this time it should be possible to prepare the mobile phase for the experiment (see Procedure B). At the end of the heating period, the aqueous layer should have turned orange. Remove the unreacted DNFB (lower insoluble layer in the form of droplets) with a disposable pipet and discard it. Cautiously add 3 mL concentrated HCl by drops. The derivative should precipitate at this point, usually on the bottom of the flask. If no precipitation occurs, place the flask in an ice bath and allow it to cool. This may take time, but the precipitate will eventually form. Remove the precipitate by decanting the upper layer, or remove solution with a pipet if the precipitate floats on top. Put a small portion of the precipitate in a small test tube, add a drop of acetone, and set it aside briefly.

B. Preparation of the Mobile Phase for TLC

Use a graduated cylinder to prepare a mixture of 140 mL chloroform, 60 mL benzyl alcohol, and 6 mL glacial acetic acid in a large beaker.

C. Preparation of the TLC Plate for Spotting

The stationary phase is silica gel; the plates are already prepared and have been dried in an oven for 30 min at 100°C, so that they have been activated for adsorption chromatography. Take a plate and make a light pencil mark across it, 3 cm above the bottom of the plate.

caution   Do not press so hard on the pencil that you remove the silica gel.

Every 2 cm along this line, put a short vertical mark which will indicate where a known or unknown DMP derivative will be spotted (Fig. 17-3). Keep a record of the order in which the derivatives are spotted in your lab notebook. The best way to do this is to draw a picture of it, labeling the spots.

D.   Spotting the Plate   Use the micropipets (spotting capillaries) to obtain portions of known DNP derivatives that have been dissolved in acetone. When the capillary is placed in the solution, the solution will rise into the tube. Gently touch the capillary to the white side of the TLC plate at the pencil mark, and while blowing on the surface let a tiny bit of solution flow onto the plate. Do not let the flow produce a spot larger than 2 mm in diameter. Once this has dried, build up more solution on the spot while continuing to blow in order to aid evaporation. The idea is to develop a small but visible and built-up spot. The instructor will demonstrate this technique. The spots should be kept as small and concentrated as possible. A major source of error is spotting too much unknown on the plate, which results in an elongated spot from which the R_f value will be difficult to measure. Spot known amino acids in the same manner and make sure to include di-nitro-phenol and di-nitro-aniline because these are breakdown products of the reaction

$$CO_3^{2-} + H_2O + O_2N-\!\!\!\!\!\!\bigcirc\!\!\!\!\!\!\!-\!\!F \rightarrow O_2N-\!\!\!\!\!\!\bigcirc\!\!\!\!\!\!\!-\!\!OH + HCO_3^- + F^-$$

E.   Insertion of the Plate in the Developing Chamber   After spotting all the knowns and the unknown derivative, carefully put the TLC plate between the two glass plates which form the front and back of the chamber. Make sure that the sides are flush and that the bottom of the plate is supported on the lip on the bottom of the glass plates. Now gently attach the clamps so that the alignment is not lost, pour the mobile phase into the tray, and

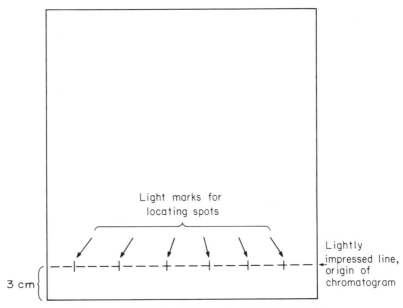

Figure 17-3 / A plate prepared for spotting.

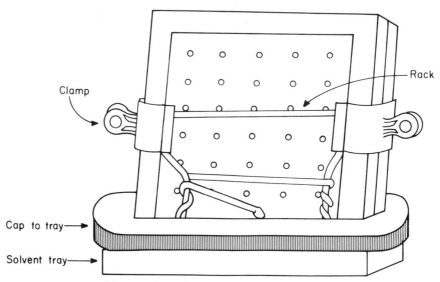

Figure 17-4 / Picture of assembled unit.

gently put the whole assembly into the mobile phase, as shown in Fig. 17-4. Drape your lab towel over the whole assembly to prevent decomposition by light.

F. Development and Documentation of the Chromatogram

The mobile phase will be carried up the plate by capillary action, rapidly at first, and then more slowly as the solvent front rises. Careful observation of the plate will allow observation of advancement of the solvent front; the damp portion of the plate appears darker than the dry portion. If a sufficient amount of material has been spotted at each point at the origin, derivatives will be seen advancing up the plate at different rates. The different spots should have significantly different R_f values when the solvent has advanced about three-fourths the way up the plate.

Remove the plate and immediately score a line with a pencil across the solvent front. The solvent evaporates fairly quickly, and unless marked the front may become undetectable for making R_f measurements. Allow the plates to dry in a hood.

Treatment of Data

In the report, make an illustration of the final appearance of the plate. Measure the distance which the center of the derivative spot has moved from the origin (i.e., the line at which the derivatives were originally spotted) and the distance which the solvent front has moved from the origin for each spot. (Evaporation occurs more readily from the sides of the plate than from the middle, so the solvent may have advanced somewhat more in the middle of the plate than at the edges. This has to be remembered when the distance the solvent has moved is determined; one single value from the center of the plate may not do for spots near the edge.)

Calculate the R_f value for each known amino acid derivative and for the unknown and present these data in a table. Assume that the unknown must be one of the knowns and identify or limit it to two possibilities.

Questions

1 Why do different amino acid derivatives have different R_f values?
2 Would you expect the amino acids to have different R_f values than their DNP derivatives? Why?

3 From the R_f values measured, can you determine whether more polar or less polar molecules have greater R_f values in this particular system? Is this in keeping with what you would predict from analysis of the separation as an adsorption process on silica?

4 Collect about five sets of data from other students. Compare the R_f values for the knowns. How great is the variability from one set of data to the next? Can you suggest reasons for the variability? What effect would different lengths of time for plates in the oven have?

18
Electrophoresis

There are several types of electrophoresis: moving boundary cell, paper, curtain, zone, and immunoelectrophoresis. All depend upon the movement or migration of ions under the influence of an electric field, and separation of ions is based on the different rates of movement or mobilities of ions. For ions that have the same charge, the smaller ions will move faster; for ions that have the same size but different charges, the more highly charged ions move faster. To prevent the separated materials from re-mixing, the operation can be done in several ways: paper strips, cellulose acetate strips, or gels are used in a technique known as *zone* electrophoresis. An experiment consists of supporting a strip of liquid, usually an aqueous buffer solution, on a suitable supporting phase, such as a paper strip. The ends of the strip are contacted with a reservoir of the buffer solution and with two electrodes. A power supply provides a voltage between the two electrodes; this impresses the required electric field, or voltage gradient, along the liquid strip. The sample is placed in a spot or line across the center of the strip (the sample *origin*), the voltage turned on, and the sample allowed to migrate in the electric field for a predetermined period. After this, the voltage is turned off, the sample detected (by chemical or other means if not visibly colored), and its direction of motion and new position recorded.

Moving boundary cells are constructed so that compartments in the cell can be removed after electrophoresis has been carried out for a length of time. Separation is not very complete by this method. Immunoelectrophoresis is usually carried out in an agar tray. A few drops each of antigen and antibody solution are placed separately in small wells cut into the agar. Antigen and antibody diffuse outward toward each other at a rate related to their concentration and their dif-

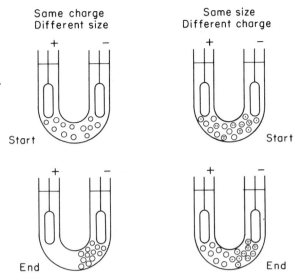

Figure 18-1 / Schematic description of electrophoresis. (Reprinted from "Modern Chemical Technology," vol. 6, rev. ed. figs. 52-1, 52-2, pp. 1149, 1150. Copyright © 1972 by The American Chemical Society and reprinted by permission.)

fusion coefficients. A line of precipitate forms where an antigen encounters its antibody. The line can be identified if either the antigen or the antibody is available in pure form. For example, a precipitate of diphtheria toxin and its antibody can be located.

Figure 18-1 illustrates how electrophoresis works. Figures 18-2 and 18-3 show two ways to set up paper electrophoresis and Fig. 18-4 illustrates a continuous method that combines descending paper chromatography with zone electrophoresis. Electrophoresis tends to move the ionic species horizontally across the paper, while the descending paper chromatography tends to move them straight down. This is an excellent method for the collection of relatively large amounts of pure materials.

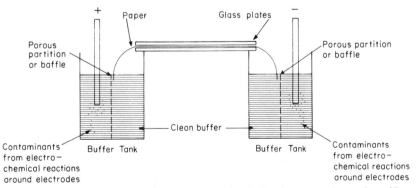

Figure 18-2 / The sandwich technique of paper electrophoresis. (Reprinted from "Modern Chemical Technology," vol. 6, rev. ed. fig. 52-3, p. 1150. Copyright © 1972 by The American Chemical Society and reprinted by permission.)

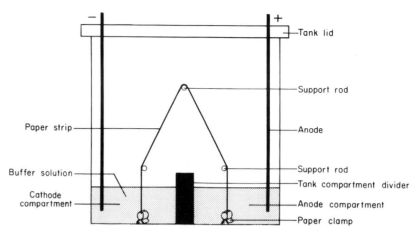

Figure 18-3 / Schematic diagram of apparatus for paper electrophoresis. (Reprinted from "Modern Chemical Technology," vol. 6, rev. ed. fig. 44-2, p. 1001. Copyright © 1972 by The American Chemical Society and reprinted by permission.)

A typical apparatus for this experiment is shown in Fig. 18-5. An overhead view of the extent of migration of various samples during a given period of time is shown in Fig. 18-6. This figure illustrates the dependency of the sample mobility on the molecular charge (compare molecules a and b), that a molecule can have multiple charges but its direction of migration depends on its net charge (compare molecules c, d, and f), that large molecular size generally reduces mobility (compare molecules c and d), and that molecules bearing equal numbers of plus and minus charges are electrically neutral and do not migrate at all (molecule e).

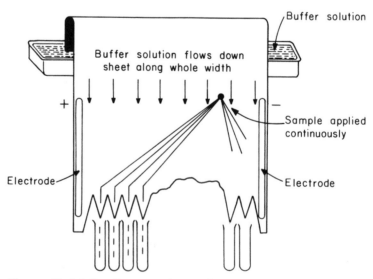

Figure 18-4 / Continuous electrophoresis. (Reprinted from "Modern Chemical Technology," vol. 6, rev. ed. fig. 52-6, p. 1152. Copyright © 1972 by The American Chemical Society and reprinted by permission.)

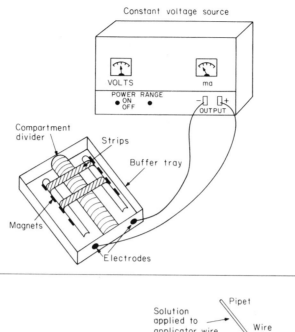

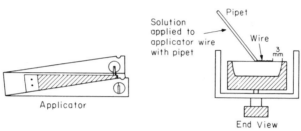

Figure 18-5 / Apparatus for paper electrophoresis.

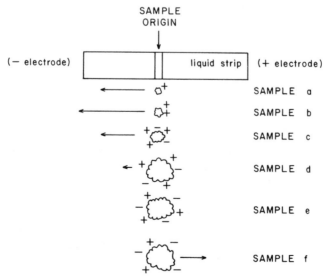

Figure 18-6 / Illustration of factors that influence sample mobility.

Several applications of electrophoresis are evident from the discussion. They include:

1 Determination of net molecular charge, i.e., whether the charge is positive (migration toward the negative electrode), negative (migration toward the positive electrode), or zero (no migration). If the molecular charge is dependent on the pH of the buffer solution, as will be the case if the charged sites on the molecule are weak acid or base groups, then the direction of migration will be pH-dependent and may actually be reversed in experiments conducted over a range of pH values. The pH which leads to electrical neutrality and no migration is called the *isoelectric point* of the molecule.

2 In a mixture of samples that differ in molecular charge and/or size, their difference in mobility can be made the basis of an electrophoretic separation.

EXPERIMENT 18-1

Electrophoresis: Isoelectric Point of a Protein†

Purpose

In this experiment the electrophoretic mobility of a protein, bovine serum albumen (M.W. 70,000), will be measured in a series of aqueous buffers of different pH and the isoelectric point of this substance will be determined.

References

1 J. M. Clark and R. L. Switzer, "Experimental Biochemistry," 2nd ed., Freeman and Co., San Francisco, 1977, pp. 43–55.
2 N. E. Heftmann, "Chromatography," 3rd ed., Van Nostrand Reinhold, New York, 1975, chaps. 10 and 11.
3 B. L. Karger, L. R. Snyder, and C. Horvath, "An Introduction to Separation Science," Wiley, New York, 1973, chap. 17.

Apparatus

Electrophoresis apparatus for paper-strip electrophoresis
Paper strips
Forceps
Paper blotter
Pipets, disposable
Plastic pans for staining baths (2)
Applicators

Chemicals

Buffers, pH 3.0, pH 4.0, pH 5.0, pH 6.0, pH 7.0 (all citrate-phosphate)
(To prepare above buffers, use 1.94 g sodium citrate dihydrate and 3.80 g $Na_3PO_4 \cdot 12H_2O$ per liter; then add the following quantities per liter of a 1:1 concd. HCl:water mixture: pH 3, 5.0 mL; pH 4, 3.7 mL; pH 5, 2.3 mL; pH 6, 1.0 mL; pH 7, 0.20 mL. Adjust to exact pH with a pH meter; use HCl or NaOH.
Bovine serum albumin solution, 5 mg/mL (1 mL per student)
Ponceau S-trichloroacetic acid staining solution
Acetic acid [5% (by volume)]

†From a similar experiment at the University of North Carolina, Chapel Hill.

Theory

The theory for electrophoresis is discussed in the text of the chapter.

Procedure

Check the electrophoresis apparatus first to make sure that the electrical connections to the buffer tray and electrodes are out and that the power supply is OFF. Fill the buffer tray with pH 3.0 citrate-phosphate buffer solution to a depth of about $\frac{1}{2}$ in; make sure that there are equal amounts of buffer in both positive and negative electrode compartments. (Tilt the tray back and forth to accomplish the latter equality.)

Carefully immerse the electrophoresis paper strips in the buffer tray. This is done by initially "floating" the strip on the surface of the liquid, then, when all dry spots have vanished, gently push the strip under the liquid's surface with forceps. After the strip has soaked for a few minutes, carefully remove it with forceps and blot off the excess buffer by laying the strip out flat on a paper blotter on the lab bench.

Now, without delay (to avoid excessive drying out of the strip), apply the bovine serum albumin sample solution. This is done by allowing capillary action to draw a small amount of the sample solution up into a disposable pipet (provided). Touch the pipet tip gently to one end of the parallel wires of the applicator and spread a film of solution to the other end. Make sure that the solution stays suspended between the two wires: you will be able to see this. Place the applicator cross-wise at the middle of the electrophoresis strip and press down hard enough to leave a visible indentation. The point of application is called the origin.

Lay the sample strip in the buffer holder across the mountings in the buffer tray and secure with the magnets at each end. The strip should be straight and level (no sagging) and its ends must contact the buffer solution in each side of the tray. Proper arrangement of the strip in the tray is important to a successful experiment. If you cannot make both ends of the strip contact the liquid surface, add more buffer to the tray. Close the buffer tray.

Prepare a second, identical strip and mount it in the buffer tray in the same fashion.

Now, be sure that the buffer tray is closed and then connect the electrodes (power OFF). Note which electrode is plus and which is minus. Turn on the voltage power supply and adjust the current so that 1.5 mA passes through each strip in the buffer tray. Note the time, and allow migration to proceed for exactly $\frac{1}{2}$ h. At the end of the $\frac{1}{2}$ h, turn off the power supply, remove the electrode connections from the buffer tray, open the tray, and make a mark on each strip to record which end was connected to the plus electrode and which to the minus electrode. Remove the strips from the buffer tray, taking care that no buffer is allowed to flow across them. Now place the strips in the developing or staining bath for 5 min.

safety caution   Use forceps provided for this purpose; the staining bath contains trichloroacetic acid which will rapidly attack your skin.

Wash the strips, by immersion, in water (continue to handle with forceps only), then immerse the strip in 5% acetic acid solution to remove the remaining developing solution. The protein sample will now appear as a red band.

Remove the strips from the acetic acid bath and blot dry with a paper lab towel. Record the direction that the sample moved, the distance that it moved from the origin (measure with a ruler from band center to origin center), and the pH.

Repeat the experiment in identical fashion with buffers of pH 4.0, pH 5.0, pH 6.0, and pH 7.0.

Treatment of Data

The laboratory report should contain:

1. The electrophoretic strips (tape onto paper backing)
2. A table of data for direction and distance migrated as a function of pH
3. The deduction of the isoelectric point of bovine serum albumin
4. A plot of distance migrated vs. pH

Calculation of the
Isoelectric Point

The neutral form of aspartic acid has the formula

$$\overset{\oplus}{N}H_3-\underset{\underset{CH_2-COOH}{|}}{CH}-COO^{\ominus} \ (H_2Asp)$$

with three acidic states; H_3Asp^+ (pK_1, 2), H_2Asp (pK_2, 4), and $HAsp^-$ (pK_3, 10). The acid-dissociation equilibria can be represented by the relations

$$H_3Asp^+ \rightleftharpoons H^+ + H_2Asp \qquad K_1 = 10^{-2}$$
$$H_2Asp \rightleftharpoons H^+ + HAsp^- \qquad K_2 = 10^{-4}$$
$$HAsp^- \rightleftharpoons H^+ + Asp^{2+} \qquad K_3 = 10^{-10}$$

At the isoelectric point the major species is the neutral aspartic acid molecule (H_2Asp), which will disproportionate to a small degree to give the condition $[H_3Asp^+] = [HAsp^-]$. By substition this equality is made dependent on the concentration of H_2Asp and takes the form

$$\frac{[H^+][H_2Asp]}{K_1} = \frac{K_2[H_2Asp]}{[H^+]}$$

This relation can be rearranged to give

$$[H^+]_{iso}^2 = K_1 K_2 \qquad \text{and} \qquad pH_{iso} = \tfrac{1}{2}(pK_1 + pK_2)$$

For aspartic acid, $pH_{iso} = \tfrac{1}{2}(2 + 4) = 3$.

Questions

1. Consider the amino acid alanine. Sketch a qualitative plot of distance and direction migrated against pH for this compound. Explain the plot in terms of the formula of alanine as a function of pH.
2. Why does molecular weight influence electrophoretic mobility?
3. What would be the effect on the distance-migrated values that were measured in this experiment if a higher power supply voltage were used? Would there be any effect on the relative distances at different pH values?

Appendices

1

Special Project Experiments from the *Journal of Chemical Education,* 1972-1982

Part I.   Electrochemical Methods

**Chapter 2.
Potentiometric
Methods**

1   "The Determination of the p*K* of the Dihydrogen Phosphate Ion," I. R. Davies, E. P. Serjeant, and A. G. Warner, *J. Chem. Ed., 54,* 649 (1977).

2   "Determination of the Stability Constants of Nickel(II) Cysteine," T. L. Rose and R. J. Seyse, *J. Chem. Ed., 53,* 728 (1976).

3   "Simultaneous Potentiometric Titration of Cu and Fe in Non-Aqueous Media," R. D. Braun, *J. Chem. Ed., 53,* 463 (1976).

4   "The Carbon Electrode in Potentiometric Titrations," M. Natajaran and A. Ramasubramanian, *J. Chem. Ed., 53,* 663 (1976).

5   "Coated-Wire Ion-Selective Electrodes and Their Application to the Teaching Laboratory," C. R. Martin and H. Freiser, *J. Chem. Ed., 57,* 512 (1980).

6   "Laboratory Experiments with Ion-Selective Electrodes," R. E. Lamb, D. Natusch, J. E. O'Reilly, and C. N. Watkins, *J. Chem. Ed., 50,* 432 (1973).

7   "Ionic Association Ion-Selective Electrode Experiment," M. M. Emara, N. A. Farid, and C. T. Lin, *J. Chem. Ed., 56,* 620 (1979).

8   "Experiments with the PVC Matrix Membrane Calcium Ion-Selective Electrode," B. J. Birth, A. Craggs, G. J. Moody, and J. D. R. Thomas, *J. Chem. Ed., 55,* 740 (1978).

9   "Student Preparation and Analysis of Chloride and Calcium Ion Selective Electrodes," B. W. Lloyd, F. L. O'Brien, and W. D. Wilson, *J. Chem. Ed., 53,* 328 (1976).

10   "Fluoride in Toothpaste Using an Ion-Selective Electrode," T. S. Light and C. C. Cappuccino, *J. Chem. Ed., 52,* 247 (1975).

11 "Construction and Evaluation of a Solid State Iodide-Selective Electrode," D. S. Papastathopoulos and M. I. Karayannis, *J. Chem. Ed.*, *58*, 904 (1981).

12 "Determination of Iodide in Milk with an Ion-Selective Electrode," J. E. O'Reilly, *J. Chem. Ed.*, *56*, 279 (1979).

13 "A Conductometric-Potentiometric Titration for an Advanced Laboratory," J. C. Rosenthal and L. C. Nathan, *J. Chem. Ed.*, *58*, 656 (1981).

Chapter 3. Conductometric Methods

1 "Acetic Acid in Toluene: A Safe Conductivity Experiment," J. W. Hill, *J. Chem. Ed.*, *53*, 778 (1976).

2 "A Non-Aqueous Conductometric Titration for the Analysis of Alkaloids in Cigarette Tobacco," J. M. Hiller, M. S. Mohan, and M. J. Brand, *J. Chem. Ed.*, *56*, 207 (1979).

3 "Study of the Kinetics of the Substitution Reactions of Square Planar Platinum Complexes as Monitored by Conductance Measurements," J. V. Marzik, A. D. Sabatelli, P. J. Fitzgerald, and J. E. Sarneski, *J. Chem. Ed.*, *58*, 589 (1981).

Chapter 4. Controlled Potential Methods

Cyclic Voltammetry

1 "Synthesis and Electrochemistry of Cyclopentadienylcarbonyliron Tetramer," A. J. White and A. J. Cunningham, *J. Chem. Ed.*, *57*, 317 (1980).

Polarography

1 "A Simple Method for Determination of Hemoglobin Oxygen Dissociation Curves," T. Brittain, *J. Chem. Ed.*, *59*, 253 (1982).

Anodic Stripping Voltammetry

1 "Determination of Pb and Cd in Pottery Using Anodic Stripping Voltammetry," M. L. Deanhardt, J. W. Dillard, K. Hanck, and W. L. Switzer, *J. Chem. Ed.*, *54*, 55 (1977).

Spectroelectrochemistry

1 "An Electrochemical Experiment Using an Optically Transparent Thin Layer Electrode," T. P. DeAngelis and W. R. Heineman, *J. Chem. Ed.*, *53*, 594 (1976).

Amperometric Titrations

1 "An Undergraduate Electroanalytical Experiment," J. Janata, *J. Chem. Ed.*, *53*, 399 (1976).

2 "Amperometric Determination of Glucose at Parts per Million Levels with Immobilized Glucose Oxidase," G. Sittampalam and G. C. Wilson, *J. Chem. Ed.*, *59*, 70 (1982).

3 "Amperometric Monitoring of Iron(III) in Ion Exchange Separation from Cobalt," J. T. Stock, *J. Chem. Ed.*, *51*, 491 (1974).

Chapter 5. Electrolytic Methods and Controlled-Current Methods

Coulometric Titrations

1 "An Integrated Circuit Power Source for Constant-Current Coulometry," M. G. Muha, *J. Chem. Ed.*, *53*, 465 (1976).

2 "Coulometry Experiments Using Simple Electronic Devices," E. Grimsrod and J. Amend, *J. Chem. Ed.*, *56*, 131 (1979).

3   "Coulometric Titration Experiment. A Comparison of End Point Methods," S. L. Tackett, *J. Chem. Ed., 49,* 863 (1972).

4   "Coulometric Titrations in Non-Aqueous Media," J. T. Stock, *J. Chem. Ed., 50,* 268 (1973).

5   "Analysis of Hypochlorite in Commercial Liquid Bleaches by Coulometric Titration," *J. Chem. Ed., 52,* 335 (1975).

6   "Analysis of Commercial Vitamin C Tablets by Iodometric and Coulometric Titrimetry," D. G. Marsh, D. L. Jacobs, and H. Vaering, *J. Chem. Ed., 50,* 626 (1973).

Part II.   Methods Based on Electromagnetic Radiation

Chapter 6.   Analytical Ultraviolet-Visible Absorption Spectroscopy

1   "Determination of the Performance Parameters of a Spectrophotometer," V. W. Cope, *J. Chem. Ed., 55,* 680 (1978).

2   "Precision Spectrophotometry Using Modular Instruments," E. H. Piepmeier, *J. Chem. Ed., 50,* 640 (1973).

3   "Computer-Controlled Colorimetry," F. G. Pater and S. P. Perone, *J. Chem. Ed., 50,* 528 (1973).

4   "Spectrophotometric Titrations for Students," S. Amdur and W. J. Levene, *J. Chem. Ed., 51,* 136 (1974).

5   "Principles of Precision Spectrophotometry," F. W. Billmeyer, *J. Chem. Ed., 51,* 530 (1974).

6   "Potential Curves for the I_2 Molecule," R. D'Alterio, R. Mattson, and R. Harris, *J. Chem. Ed., 51,* 282 (1974).

7   "Micro-Determination of Porphyrins," D. Brisbin and J. O. Asgill, *J. Chem. Ed., 51,* 211 (1974).

8   "Laboratory Experiments Using Nicotine," R. M. Navari, *J. Chem. Ed., 51,* 748 (1974).

9   "The Determination of Aspirin by Ultraviolet Absorption," T. L. Hammett and F. H. Klappmeier, *J. Chem. Ed., 55,* 266 (1978).

10   "Nitric Acid in Rain Water," G. I. Gleason, *J. Chem. Ed., 50,* 718 (1973).

11   "Determination of Nitrite in Meat Samples," I. T. Glover and F. T. Johnson, *J. Chem. Ed., 50,* 426 (1973).

12   "Critical Micelle Concentration Using Acridine Orange Dye Probe," M. Rujimethabhas and P. Wilairat, *J. Chem. Ed., 55,* 342 (1978).

13   "Two Convenient Spectrophotometric Enzyme Assays," J. A. Hurlbut, T. N. Ball, H. C. Pound, and J. L. Graves, *J. Chem. Ed., 50,* 149 (1973).

14   "Two Spectrophotometric Experiments with Alpha-Chymotrypsin," E. R. Kantrowitz and G. Eisele, *J. Chem. Ed., 52,* 410 (1975).

Chapter 8.   Infrared Spectroscopy

1   "Characterization of a Coordination Complex Using Infrared Spectroscopy," L. C. Nathan, *J. Chem. Ed., 51,* 285 (1974).

2   "Coffee, Tea, or Cocoa—A Trio of Experiments Including Isolation of Theobromine from Cocoa," D. L. Pavia, *J. Chem. Ed., 50,* 791 (1973).

3   "The Vinyl Acetate Content of Packaging Film," K. N. Allpress and B. J. Cowell, *J. Chem. Ed., 58,* 741 (1981).

Raman Spectroscopy

1   "Raman Spectra of ZXY_3 Compounds. (A Dry-Lab Spectral Analysis Experiment)," F. P. Dehaan, J. C. Thibeault, and D. Kottesen, *J. Chem. Ed., 51,* 263 (1974).

Chapter 9. Atomic Absorption and Atomic Emission Spectroscopy

1 "A Non-Flame Atomic Absorption Attachment for Trace Mercury Determination," V. T. Lieu, A. Cannon, and W. E. Huddleston, *J. Chem. Ed., 51,* 752 (1974).

2 "Determination of the Extraction Constant for Zinc Pyrrolidinecarbodithioate," L. C. Hoskins, P. B. Reichardt, and R. J. Stolzberg, *J. Chem. Ed., 58,* 580 (1981).

3 "Hydrolysis of a Chromium(III) Complex Using Atomic Absorption and Ion Exchange," M. B. Davies and J. W. Lethbridge, *J. Chem. Ed., 50,* 793 (1973).

4 "The Iron Content of Breakfast Cereals," P. H. Laswick, *J. Chem. Ed., 50,* 132 (1973).

5 "Zinc in Hair by Atomic Absorption Spectroscopy," R. K. Pomeroy, N. Drikitis, and Y. Yoga, *J. Chem. Ed., 52,* 544 (1975).

Chapter 10. Fluorescence Spectroscopy

1 "Vibronic Analysis of the Visible Absorption and Fluorescence Spectra of the Fluorescein Dianion," T. Kurucsev, *J. Chem. Ed., 55,* 128 (1978).

2 "The Rate Constant for Fluorescence Quenching," M. W. Legenza and C. J. Marzzacco, *J. Chem. Ed., 54,* 183 (1977).

3 "Quinine Fluorescence Spectra. (An Interpretive Experiment)," J. E. O'Reilly, *J. Chem. Ed., 53,* 191 (1976).

4 "Fluorescence Experiments with Quinine," J. E. O'Reilly, *J. Chem. Ed., 52,* 610 (1975).

5 "Uranyl Luminscence Quenching," H. B. Burrows and S. J. Formosinho, *J. Chem. Ed., 55,* 125 (1978).

Chapter 11. Nuclear Magnetic Resonance Spectroscopy

1 "NMR Spectral Analysis: An Experiment Involving Complete Lineshape Analysis of a Two-Site Exchange Problem," H. M. Bell, *J. Chem. Ed., 53,* 665 (1976).

2 "Interpretation of a ^{13}C Magnetic Resonance Spectrum," H. C. Dorn, D. Kingston, and B. R. Simpers, *J. Chem. Ed., 53,* 584 (1976).

3 "Experimental Determination of pK_a Values by Use of NMR Chemical Shifts," C. S. Handloser, M. R. Chakrabarty, and M. W. Mosher, *J. Chem. Ed., 50,* 510 (1973).

4 "The Fumarase Reaction: NMR Experiment for Biological Chemistry Students," G. J. Kasperek and R. F. Pratt, *J. Chem. Ed., 54,* 515 (1977).

5 "NMR Determination of *n*-Butyllithium," A. Silveira, Jr., and H. D. Bretherick, *J. Chem. Ed., 57,* 560 (1979).

6 "Titration of Alanine Monitored by NMR Spectroscopy," F. C. Waller, I. S. Hartman, and S. T. Kwong, *J. Chem. Ed., 54,* 447 (1977).

Electron Spin Resonance

1 "Two ESR Systems for the Advanced Undergraduate Laboratory," A. Serianz, J. R. Shelton, F. L. Urbach, R. C. Dunbar, and R. F. Kopczewski, *J. Chem. Ed., 53,* 394 (1976).

Part III. Separation Methods

Chapter 12. Gas Chromatography

1 "Illustrating Gas Chromatography and Mass Spectrometry," M. L. Gross, V. K. Olsen, and R. K. Force, *J. Chem. Ed., 52,* 535 (1975).

2 "The Separation and Identification of Straight Chain Hydrocarbons," G. A. Benson, *J. Chem. Ed., 59,* 344 (1982).

3 "Gas Chromatographic Analysis of Gasoline," R. F. Cassidy, Jr., and C. Schuerch, *J. Chem. Ed., 53,* 51 (1976).

4 "Hydrocarbons in Ambient Air," S. R. Dinardi and E. S. Briggs, *J. Chem. Ed.*, *52*, 811 (1975).

5 "Analysis of Chlorinated Hydrocarbon Pesticides," I. T. Glover and A. P. Minter, *J. Chem. Ed.*, *51*, 685 (1974).

6 "DDE Level in Birds," S. Hall and P. B. Reichardt, *J. Chem. Ed.*, *51*, 684 (1974).

7 "Gas Chromatographic Determination of Environmentally Significant Pesticides," W. E. Rudzinski and S. Beu, *J. Chem. Ed.*, *59*, 614 (1982).

8 "Gas-Liquid Chromatography of Derivatized Barbiturates," M. Novotny and K. D. Bartle, *J. Chem. Ed.*, *51*, 333 (1974).

9 "The Fatty Acid Composition of Edible Oils and Fats," D. R. Paulson, J. R. Saranto, and W. A. Forman, *J. Chem. Ed.*, *51*, 406 (1974).

10 "Analysis of Lipid Content and Composition of Ground Beef," G. L. Long, *J. Chem. Ed.*, *52*, 813 (1975).

11 "Egg Yolk Lecithin" (A Combined Thin-Layer, Column, and Gas Chromatography Experiment), B. J. White, C. L. Tipton, and M. Dressel, *J. Chem. Ed.*, *51*, 533 (1974).

Chapter 13. High Performance Liquid Chromatography

1 "Study of the Liquid-Liquid Partitioning Process Using Reverse-Phase Liquid Chromatography," C. H. Lochmuller and D. R. Wilder, *J. Chem. Ed.*, *57*, 381 (1980).

2 "An Introduction to High Performance Liquid Chromatography: Separation of Some FD and C Dyes," H. T. McKone and K. Ivie, *J. Chem. Ed.*, *57*, 321 (1980).

3 "Acylation of Ferrocene," H. T. McKone, *J. Chem. Ed.*, *57*, 380 (1980).

4 "High Performance Liquid Chromatography of Essential Oils," H. T. McKone, *J. Chem. Ed.*, *56*, 698 (1979).

5 "High Performance Liquid Chromatography of Vitamin A," O. Bohman and K. Engdahl, *J. Chem. Ed.*, *59*, 251 (1982).

6 "High Performance Liquid Chromatography of Urinary Compounds," P. T. Kissinger, L. J. Felice, W. P. King, L. A. Pachla, R. M. Riggin, and R. E. Shoup, *J. Chem. Ed.*, *54*, 766 (1977).

Chapters 14–18. Other Separation Methods

1 "A Laboratory Introduction to Quantitative Column Chromatography," J. F. Rubinson and K. A. Rubinson, *J. Chem. Ed.*, *57*, 909 (1980).

2 "Ion Exchange Separation of the Oxidation States of Vanadium," R. Cornelius, *J. Chem. Ed.*, *57*, 316 (1980).

3 "Mini-Column Ion Exchange Separation and Atomic Absorption Quantitation of Nickel, Cobalt, and Iron," J. L. Anderson, J. Grohs, and D. Frick, *J. Chem. Ed.*, *57*, 521 (1980).

4 "Separation and Determination of Hydroxyacetophenones" (An Experiment in Preparative TLC and UV Spectroscopy), I. R. C. Bick and A. J. Blackman, *J. Chem. Ed.*, *59*, 618 (1982).

5 "Isolation, Separation, and Identification of Synthetic Food Colors," E. A. Dixon and G. Renyk, *J. Chem. Ed.*, *59*, 67 (1982).

6 "Thin Layer Chromatography of Darvon Compound-65," D. W. Chasar and G. B. Toth, *J. Chem. Ed.*, *51*, 487 (1974).

7 "An Analytical Procedure for Determination of Pesticides in Food," D. J. Subach and M. E. Butwill Ball, *J. Chem. Ed.*, *50*, 855 (1973).

8 "Polyacrylamide Gel Electrophoresis of Yeast Invertase," C. A. Roberts, C. Jones, E. J. Spencer, G. Bowman, and D. Blackman, *J. Chem. Ed.*, *53*, 62 (1976).

2
Reference Tables

Table A-1 / Physical Constants and Units

| Quantity | Symbol | Value | DECIMAL AND UNITS | |
			SI	cgs
Velocity of light	c	2.9979	10^8 m/s	10^{10} cm/s
Electron charge	e	1.602	10^{-19} C	10^{-20} emu
		4.803	—	10^{-10} esu
Planck's constant	h	6.6256	10^{-34} J · s	10^{-27} erg · s
Electron volt	eV	1.6021	10^{-19} J	10^{-12} erg
		3.837	—	10^{-20} cal
Avogadro's number	N	6.022	10^{23} mol^{-1}	10^{23} mol^{-1}
Atomic mass unit	amu	1.6604	10^{-27} kg	10^{-24} g
Faraday constant	F	9.6491	10^4 C/mol	10^4 esu/mol
Gas constant	R	1.9872	—	cal/K · mol
		8.3143	J/K · mol	10^7 erg/K · mol
		8.2054	—	10^{-2} L · atm/K · mol
Rydberg constant	R_∞	1.0974	10^7 m^{-1}	10^5 cm^{-1}
Bohr magneton	μ_B	9.2741	10^{-24} J/T	10^{-21} erg/G
Boltzmann constant	k	1.3805	10^{-23} J/K	10^{-16} erg/K

Table A-2 / SI Derived Units

Quantity	Name	Symbol	Units	Special Multiples
Frequency	hertz	Hz	s^{-1}	
Force	newton	N	$kg \cdot m/s^2$	10^{-5} N / 1 dyne (dyn)
Pressure†	pascal	Pa	$kg/m \cdot s^2 =$ N/m^2	10^5 Pa = 1 bar
Power, radiant flux	watt	W	$kg \cdot m^2/s^3 = J/s$	
Electric charge, quantity of electricity	coulomb	C	$A \cdot s$	
Electric potential, potential difference, electromotive force	volt	V	$kg \cdot m^2/s^3 \cdot A$	
Electric resistance	ohm	Ω	$kg \cdot m^2/s^3 \cdot A^2$	
Electrical capacitance	farad	F	$A^2 \cdot s^4/kg \cdot m^2$	
Conductance	siemens	S	$kg^{-1} \cdot m^{-2} \cdot s^3 \cdot A^2 =$ Ω^{-1}	
Energy, work, quantity of heat‡	joule	J	$kg \cdot m^2/s^2 = V \cdot C$	10^{-7} J = 1 erg
Magnetic flux	weber	Wb	$kg \cdot m^2/s^2 \cdot A$	10^{-8} Wb = 1 maxwell (Mx)
Inductance	henry	H	$kg \cdot m^2/s^2 \cdot A^2$	
Magnetic flux density	tesla	T	$kg/s^2 \cdot A$	10^{-4} T = 1 gauss (G)
Temperature	kelvin	K		

†101,325 Pa = 1 atmosphere (atm) = 760 millimeters of mercury (mmHg)
133.322 Pa = 1 torr = 1 millimeter of mercury (mmHg)
‡3.6 × 10^6 J = 1 kilowatthour (kWh)
1055.056 J = 1 British thermal unit (Btu)
4.184 J = 1 thermochemical calorie (cal_{th})

Table A-3 / Acidity Constants of Inorganic Substances

Acid	Equilibrium Equation	pK
Aluminum hydroxide	$Al(OH)_3 \rightarrow H_3O^+ + AlO_2^-$	12.4
Aluminum ion	$Al^{3+} + 2H_2O \rightarrow H_3O^+ + AlOH^{2+}$	4.9
Ammonium ion	$NH_4^+ + H_2O \rightarrow H_3O^+ + NH_3$	9.3
Antimony(III) hydroxide	$SbOOH + H_2O \rightarrow H_3O^+ + SbO_2^-$	11.0
Arsenic (ortho)	$H_3AsO_4 + H_2O \rightarrow H_3O^+ + H_2AsO_4^-$	3.6
	$H_2AsO_4^- + H_2O \rightarrow H_3O^+ + HAsO_4^{2-}$	7.3
	$HAsO_4^{2-} + H_2O \rightarrow H_3O^+ + AsO_4^{3-}$	12.5
Arsenous (meta)	$HAsO_2 + H_2O \rightarrow H_3O^+ + AsO_2^-$	9.2
Bismuth(III) ion	$Bi^{3+} + 2H_2O \rightarrow H_3O^+ + BiOH^{2+}$	2.0
Boric (ortho)	$H_3BO_3 + H_2O \rightarrow H_3O^+ + H_2BO_3^-$	9.2
Carbonic	$H_2CO_3 + H_2O \rightarrow H_3O^+ + HCO_3^-$	6.4
	$HCO_3^- + H_2O \rightarrow H_3O^+ + CO_3^{2-}$	10.3
Chromic	$H_2CrO_4 + H_2O \rightarrow H_3O^+ + HCrO_4^-$	ca. 1.0
	$HCrO_4^- + H_2O \rightarrow H_3O^+ + CrO_4^{2-}$	6.5
	$2HCrO_4^- \rightarrow C_2O_7^{2-} + H_2O$	ca. -0.4
Chromium(III) hydroxide	$Cr(OH)_3 \rightarrow H_3O^+ + CrO_2^-$	16.1
Chromium(III) ion	$Cr^{3+} + 2H_2O \rightarrow H_3O^+ + CrOH^{2+}$	4.0
Copper(II) hydroxide	$Cu(OH)_2 + H_2O \rightarrow H_3O^+ + HCuO_2^-$	19.0
	$HCuO_2^- + H_2O \rightarrow H_3O^+ + CuO_2^{2-}$	13.1
Copper(II) ion	$Cu^{2+} + 2H_2O \rightarrow H_3O^+ + CuOH^+$	8.0
Hexacyanoferric(III)	$H_3[Fe(CN)_6] + H_2O \rightarrow H_3O^+ + H_2[Fe(CN)_6]^-$	Neg.
	$H_2[Fe(CN)_6]^- + H_2O \rightarrow H_3O^+ + H[Fe(CN)_6]^{2-}$	Neg.
	$H[Fe(CN)_6]^{2-} + H_2O \rightarrow H_3O^+ + [Fe(CN)_6]^{3-}$	Neg.
Hexacyanoferric(II)	$H_4[Fe(CN)_6] + H_2O \rightarrow H_3O^+ + H_3[Fe(CN)_6]^-$	Neg.
	$H_3[Fe(CN)_6]^- + H_2O \rightarrow H_3O^+ + H_2[Fe(CN)_6]^{2-}$	Neg.
	$H_2[Fe(CN)_6]^{2-} + H_2O \rightarrow H_3O^+ + H[Fe(CN)_6]^{3-}$	ca. 3
	$H[Fe(CN)_6]^{3-} + H_2O \rightarrow H_3O^+ + [Fe(CN)_6]^{4-}$	4.2
Hydriodic	$HI + H_2O \rightarrow H_3O^+ + I^-$	Neg.
Hydrobromic	$HBr + H_2O \rightarrow H_3O^+ + Br^-$	Neg.
Hydrochloric	$HCl + H_2O \rightarrow H_3O^+ + Cl^-$	Neg.

Acid	Equilibrium Equation	pK
Hydrocyanic	$HCN + H_2O \rightarrow H_3O^+ + CN^-$	9.4
Hydrofluoric	$HF + H_2O \rightarrow H_3O^+ + F^-$	3.2
Hydrogen peroxide	$H_2O_2 + H_2O \rightarrow H_3O^+ + HO_2^-$	11.6
Hydrosulfuric	$H_2S + H_2O \rightarrow H_3O^+ + HS^-$	7.0
	$HS^- + H_2O \rightarrow H_3O^+ + S^{2-}$	12.9
Hypochlorous	$HClO + H_2O \rightarrow H_3O^+ + ClO^-$	7.5
Iron(III) ion	$Fe^{3+} + 2H_2O \rightarrow H_3O^+ + FeOH^{2+}$	2.4
Iron(II) ion	$Fe^{2+} + 2H_2O \rightarrow H_3O^+ + FeOH^+$	5.9
Lead(II) hydroxide	$Pb(OH)_2 + H_2O \rightarrow H_3O^+ + HPbO_2^-$	15.3
Magnesium ion	$Mg^{2+} + 2H_2O \rightarrow H_3O^+ + MgOH^+$	11.7
Mercury(II) ion	$Hg^{2+} + 2H_2O \rightarrow H_3O^+ + HgOH^+$	2.7
Nitric	$HNO_3 + H_2O \rightarrow H_3O^+ + NO_3^-$	Neg.
Nitrous	$HNO_2 + H_2O \rightarrow H_3O^+ + NO_2^-$	3.4
Perchloric	$HClO_4 + H_2O \rightarrow H_3O^+ + ClO_4^-$	Neg.
Permanganic	$HMnO_4 + H_2O \rightarrow H_3O^+ + MnO_4^-$	Neg.
Phosphoric (ortho)	$H_3PO_4 + H_2O \rightarrow H_3O^+ + H_2PO_4^-$	2.1
	$H_2PO_4^- + H_2O \rightarrow H_3O^+ + HPO_4^{2-}$	7.2
	$HPO_4^{2-} + H_2O \rightarrow H_3O^+ + PO_4^{3-}$	ca. 12
Silicic (meta)	$H_2SiO_3 + H_2O \rightarrow H_3O^+ + HSiO_3^-$	9.5
	$HSiO_3^- + H_2O \rightarrow H_3O^+ + SiO_3^{2-}$	11.8
Sulfuric	$H_2SO_4 + H_2O \rightarrow H_3O^+ + HSO_4^-$	Neg.
	$HSO_4^- + H_2O \rightarrow H_3O^+ + SO_4^{2-}$	1.9
Sulfurous	$H_2SO_3 + H_2O \rightarrow H_3O^+ + HSO_3^-$	1.9
	$HSO_3^- + H_2O \rightarrow H_3O^+ + SO_3^{2-}$	7.3
Thiocyanic	$HNCS + H_2O \rightarrow H_3O^+ + NCS^-$	Neg.
Thiosulfuric	$H_2S_2O_3 + H_2O \rightarrow H_3O^+ + HS_2O_3^-$	1.7
	$HS_2O_3^- + H_2O \rightarrow H_3O^+ + S_2O_3^{2-}$	2.5
Tin(IV) hydroxide	$Sn(OH)_4 + 4H_2O \rightarrow 2H_3O^+ + [Sn(OH)_6]^{2-}$	Neg.
Tin(II) hydroxide	$Sn(OH)_2 + H_2O \rightarrow H_3O^+ + HSnO_2^-$	14.4
Zinc hydroxide	$Zn(OH)_2 + 2H_2O \rightarrow 2H_3O^+ + ZnO_2^{2-}$	29.0
Zinc ion	$Zn^{2+} + 2H_2O \rightarrow H_3O^+ + ZnOH^+$	9.6

Table A-4 / Acidity Constants of Organic Substances

Substance	pK₁	pK₂	pK₃	pK₄	pK₅
Acetic acid	4.7				
Acetylacetone	8.9				
Benzoic acid	4.7				
Bromoacetic acid	2.9				
n-Butyric acid	4.8				
Chloroacetic acid	2.9				
Citric acid	3.1	4.8	6.7		
Dichloroacetic acid	1.3				
Diethanolamine	9.0				
Diethylamine	11.1				
Diethylenetriamine-pentaacetic acid	2.08	2.41	4.27	8.60	10.55
Dimethylamine	11.1				
Dipropylamine	11.1				
Ethanolamine	9.5				
Ethylamine	10.8				
Ethylenediamine	7.5	10.2			
Ethylenediamine-tetraacetic acid	2.0	2.76	6.16	10.26	
Formic acid	3.7				
Fumaric acid	3.02	4.39			
Glycine	2.35	9.78			
Hexamethylenetet-ramine	4.9				
Hydroquinone	9.9				
8-hydroxyquinoline-5-sulfonic acid	1.3	4.2	8.7		
Iminodiacetic acid	2.6	9.3			
Isopropylamine	10.8				
Lactic acid	3.8				
Maleic acid	1.92	6.22			
Malonic acid	2.7	5.3			
Methylamine	10.7				
Nitrilotriacetic acid	1.9	2.49	9.73		
Oxalic acid	1.2	4.3			
Phenol	9.9				
Phthalic acid	2.9	5.1			
Picric acid	0.4				
n-Propylamine	10.7				
Pyridine	5.15				
Salicylic acid	3.0	14.4			
Succinic acid	4.1	5.3			
Sulfanilic acid	3.2				
Tartaric acid	3.0	4.5			
Triethanolamine	7.8				
Triethylamine	10.8				
Triethylenetetramine	3.32	6.67	9.20	9.92	
Trishydroxymethyl-aminomethane	8.1				
Trimethylamine	9.9				
Urea	0.8				

Table A-5 / Acid-Base Indicators

pH Range		pH Range	
0.0–2.0	Malachite green hydrochloride, yellow-bluish green	5.0–6.0	Hematoxylin C. P., yellow-red
0.0–3.0	Eosin Y, orange-green	5.0–7.6	p-Nitrophenol, colorless-yellow
0.0–3.6	Erythrosin B, orange-red	5.2–6.8	Brom cresol purple (dibromo-o-cresolsulfonphthalein), yellow-purple
0.1–3.2	Methyl violet 2B, yellow-violet	5.2–7.0	Brom phenol red (dibromophenolsulfonphthalein), yellow-red
1–2	Tetranitrophenolsulfonphthalein, yellow-red	5.5–5.8	Sodium alizarinsulfonate (alizarine red S), yellow-purple
1.2–2.1	Diphenylaminoazobenzene, red-yellow	6.0–7.6	Brom thymol blue (dibromothymolsulfonphthalein), yellow-blue
1.2–2.3	Metanil yellow, red-yellow	6.0–8.0	Brilliant yellow, yellow-red
1.2–2.8	Meta cresol purple (m-cresolsulfonphthalein), red-yellow	6.6–8.0	Corallin Na salt, water soluble, yellow-red
1.2–2.8	Thymol blue (thymolsulfonphthalein), red-yellow	6.8–8.0	Neutral red, red-yellowish orange
1.2–3.0	Basic fuchsin, purple-red	6.8–8.4	Phenol red (phenolsulfonphthalein), red-yellow
1.3–3.0	Orange IV, red-yellowish orange	6.9–8.0	Rosolic acid, yellow-red
1.3–4.0	Benzopurpurin 4B, bluish violet-red	7.2–8.8	Cresol red (o-cresolsulfonphthalein), yellow-red
1.4–2.6	Tropacolin 00, red-yellow	7.3–8.7	α-Naphtholphthalein, rose-green
2.0–3.0	Cresol red (o-cresolsulfonphthalein), orange-yellow	7.4–8.6	Turmeric, yellow-brown
2.9–4.0	Dimethylaminoazobenzene, red-yellow	7.4–8.8	Orange I, yellow-red
2.9–4.0	Ethyl red, rose red-yellowish orange	7.4–8.8	Tropacolin 000 No. 1, yellow-red
3.0–4.0	Naphthol green B	7.4–9.0	Meta cresol purple (m-cresolsulfonphthalein), red-yellow
3.0–4.6	Brom phenol blue (tetrabromophenolsulfonphthalein), yellow-purple	7.6–9.4	Tetrabromophenolphthalein, colorless-violet
3.0–5.2	Congo red, blue-red	7.8–9.2	Curcumin, yellow-reddish brown
3.0–6.0	Phenacetolin, yellow-red	8.0–9.6	Thymol blue (thymolsulfonphthalein), yellow-blue
3.1–4.4	Methyl orange, red-yellowish orange	8.3–10	Phenolphthalein, colorless-red
3.2–4.8	Brom chlor phenol blue (dibromodichlorophenolsulfonphthalein), yellow-purple	9.3–10.5	Thymolphthalein, colorless-blue
3.3–4.8	Ethyl orange, rose red-yellow	9.4–14.0	Alkali blue 6B, light blue-rose
3.4–4.9	Propyl orange	9.8–11.6	α-Naphtholbenzein, yellow-blue
3.5–5.7	α-Naphthylamineazobenzene-p-sodium sulfonate, red-yellow	10.0–12.0	Tetranitrophenolsulfonphthalein, red-yellow
3.7–5.2	Alizarin sulfonate, yellow-violet	10.0–13.0	Phenacetolin, red-colorless
3.8–4.3	Fluorescein (Na salt), bluish green fluorescence-green fluorescence	10.0–13.0	Poirrier's blue, blue-purple
3.8–6.6	Gallein, light brownish yellow-rose	10.1–12.1	Sodium alizarin sulfonate (alizarin red S), violet-purple
4.0–5.6	Brom cresol green (tetrabromo-m-cresolsulfonphthalein), yellow-blue	11.0–13.0	Alizarin blue S, yellowish red-blue
4.4–6.2	Methyl red (dimethylaminoazobenzene-o-carboxylic acid), red-yellow	11.1–12.7	Tropaeolin 0, yellow-orange brown
4.5–8.3	Azolitmin, red-blue	11.5–13	Trinitrobenzene, colorless-reddish brown
4.7–6.2	Cochineal-powder, red-violet	11.5–14.0	Malachite green hydrochloride, blue-colorless
4.8–6.1	Carminic acid, yellow-violet	11.6–14.0	Basic fuchsin, red-colorless
4.8–6.4	Chlor phenol red (dichlorophenolsulfonphthalein), yellow-red	12.0–14.0	Acid fuchsin, red-colorless
4.8–6.4	Propyl red, red-yellow	12.1–14.0	Clayton yellow (titan yellow), yellow-red

Table A-6 / Solubility-Product Constants

Anion	Substance	pK_{sp}	Anion	Substance	pK_{sp}
Acetate	$AgC_2H_3O_2$	2.4	Iodide	PbI_2	8.1
	$Hg_2(C_2H_3O_2)_2$	9.4		CuI	12.0
Arsenate	Ag_3AsO_4	22.0		AgI	16.1
Bromide	$PbBr_2$	5.3		HgI_2	25.6
	$CuBr$	8.2		Hg_2I_2	28.3
	$AgBr$	12.3	Nitrite	$AgNO_2$	3.9
	Hg_2Br_2	21.9		$K_2Na[Co(NO_2)_6]$	10.7
Carbonate	Li_2CO_3	0.5		$(NH_4)_2Na[Co(NO_2)_6]$	12.
	$MgCO_3$	4.4	Oxalate	MgC_2O_4	4.1
	$NiCO_3$	6.9		CoC_2O_4	5.4
	$CaCO_3$	8.3		FeC_2O_4	6.7
	$BaCO_3$	8.8		NiC_2O_4	7.0
	$SrCO_3$	9.2		SrC_2O_4	7.3
	$CuCO_3$	9.6		CuC_2O_4	7.5
	$ZnCO_3$	9.7		BaC_2O_4	7.8
	$MnCO_3$	10.1		CdC_2O_4	7.8
	$FeCO_3$	10.7		ZnC_2O_4	8.8
	Ag_2CO_3	11.1		CaC_2O_4	8.9
	$CdCO_3$	11.3		$Ag_2C_2O_4$	11.0
	$CoCO_3$	12.1		PbC_2O_4	11.1
	$PbCO_3$	12.8		$Hg_2C_2O_4$	13.0
Chloride	$PbCl_2$	4.8		MnC_2O_4	15.0
	$CuCl$	6.5		$La_2(C_2O_4)_3$	27.7
	$AgCl$	9.6	Phosphate	$AlPO_4$	6.
	Hg_2Cl_2	17.9		Li_3PO_4	12.5
Chromate	$CaCrO_4$	3.2		$Mg_3(PO_4)_2$	12.6
	$SrCrO_4$	4.4		$MgNH_4PO_4$	12.6
	Hg_2CrO_4	8.7		$BiPO_4$	20.0
	$BaCrO_4$	10.1		Ag_3PO_4	20.8
	Ag_2CrO_4	11.7		$Mn_3(PO_4)_2$	22.0
	$PbCrO_4$	15.7		$Ca_3(PO_4)_2$	25.0
Cyanide	$Ag[Ag(CN)_2]$	11.4		$Pb_3(PO_4)_2$	32.0
Fluoride	LiF	2.3	Sulfate	Ag_2SO_4	4.2
	BaF_2	4.6		$CaSO_4$	4.4
	MgF_2	7.1		Hg_2SO_4	6.0
	PbF_2	7.4		$SrSO_4$	6.1
	SrF_2	9.1		$PbSO_4$	7.9
	CaF_2	9.7		$BaSO_4$	8.8
	ThF_4	27.4	Sulfite	$CaSO_3$	4.0
Hexacyanoferrate (II)	$Ag_4[Fe(CN)_6]$	40.8		$BaSO_3$	8.0
	$K_2Zn_3[Fe(CN)_6]_2$	95.0	Sulfide	MnS	15.2
Hydroxyl	Ag_2O	7.7		FeS	18.4
	$Mg(OH)_2$	11.1		NiS	20.5
	$BiO(OH)$	12.0		CoS	21.3
	$Mn(OH)_2$	12.7		ZnS	22.8
	$Cd(OH)_2$	13.7		SnS	26.0
	$Pb(OH)_2$	14.4		CdS	28.0
	$Fe(OH)_2$	14.7		PbS	28.2
	$Co(OH)_2$	15.6		CuS	36.1
	$Ni(OH)_2$	15.8		Cu_2S	48.9
	$Zn(OH)_2$	16.3		Ag_2S	50.3
	$SbO(OH)$	17.0		HgS	53.8
	$Cu(OH)_2$	18.8		Fe_2S_3	88.0
	$Hg(OH)_2$	21.9	Thiocyanate	$AgSCN$	12.0
	$Sn(OH)_2$	26.5		$CuSCN$	13.4
	$Cr(OH)_3$	30.2		$Hg_2(SCN)_2$	19.5
	$Al(OH)_3$	32.3	Thiosulfate	BaS_2O_3	4.0
	$Fe(OH)_3$	37.2			
	$Sn(OH)_4$	57.			

Table A-7 / Formation Constants (log *K*) of Some Metal Chelonates

Chelon	Abbrev.	Formula
1. Triethylenetetramine	Trien	CH_2—NH—CH_2—CH_2—NH_2 | CH_2—NH—CH_2—CH_2—NH_2
2. Tetraethylenepentamine	Tetren	CH_2—CH_2—NH—CH_2—CH_2—NH_2 H—N CH_2—CH_2—NH—CH_2—CH_2—NH_2
3. Nitrilotriacetic acid Ammoniatriacetic acid	NTA	CH_2COO^- $^-OOCCH_2$—N CH_2COO^-
4. Ethylenediaminetetraacetic acid	EDTA	$^-OOCH_2C$ CH_2COO^- N—CH_2—CH_2—N $^-OOCH_2C$ CH_2COO^-
5. *N*-Hydroxyethylenediamine-triacetic acid	HEDTA	$^-OOCH_2C$ CH_2COO^- N—CH_2—CH_2—N HOH_2C—H_2C CH_2COO^-
6. Ethyletherdiaminetetraacetic acid	EEDTA	$^-OOCH_2C$ CH_2COO^- N—CH_2—CH_2—O—CH_2—CH_2—N $^-OOCH_2C$ CH_2COO^-
7. Ethylene glycol-bis-(β-amino-ethyl ether)-*N,N'*-tetraacetic acid	EGTA	$^-OOCH_2C$ CH_2COO^- N—H_2C—H_2C—O—CH_2—CH_2—O—CH_2—CH_2—N $^-OOCH_2C$ CH_2COO^-
8. Diethylenetriamine-pentaacetic acid	DTPA	$^-OOCH_2C$ CH_2COO^- CH_2COO^- N—H_2C—H_2C—N—CH_2—CH_2—N $^-OOCH_2C$ CH_2COO^-
9. Cyclohexanediaminetetraacetic acid	CyDTA	$^-OOCH_2C$ CH_2COO^- N N $^-OOCH_2C$ HC—CH CH_2COO^- H_2C CH_2 H_2C—CH_2

†R.E. = rare earths
‡Negl = negligible reaction

PROTON FORMATION CONSTANTS (LOG)														
H_1	H_2	H_3	H_4	H_5	Mg^{2+}	Ca^{2+}	Sr^{2+}	Ba^{2+}	La^{3+}	$R.E.^{3+}†$	Mn^{2+}	Fe^{2+}	Fe^{3+}	Co^{2+}
9.92	9.20	6.67	3.32	—	Negl‡	Negl	Negl	Negl	Negl	—	4.9	7.8	—	11.0
10.0	9.2	8.2	4.1	2.6	Negl	Negl	Negl	Negl	Negl	—	7.0	—	—	15.1
9.73	2.49	1.9	—	—	5.4	6.4	5.00	4.8	10.4	10.4–12.2	7.4	8.8	15.8	10.4
10.26	6.16	2.76	2.0	—	8.7	10.7	8.7	7.9	15.5	15.8–19.8	13.8	14.4	25.1	16.3
9.73	5.33	2.64	—	—	7.0	8.0	6.8	6.2	13.2	14.1–15.8	10.7	11.6	—	14.4
9.49	8.82	2.67	1.90	—	8.3	10.0	8.6	8.2	—	—	13.2	—	—	14.7
9.43	8.85	2.68	2.0	—	5.4	10.9	8.5	8.4	—	—	12.3	—	—	12.3
10.55	8.60	4.27	2.41	2.08	9.0	10.7	9.7	8.6	—	19.1 (La)	15.5	16.7	27.5	19.0
11.70	6.12	3.52	2.40	—	10.3	12.3	10.0	8.0	—	16.8–21.5	16.8	—	—	18.9

Table A-7 (cont.) / Formation Constants (log K) of Some Metal Chelonates

Chelon	Ni^{2+}	Cu^{2+}	Zn^{2+}	Cd^{2+}	Hg^{2+}	Al^{3+}	Pb^{2+}
1. Triethylenetetramine	14.0	20.1	11.9	10.8	25.0	Negl	10.4
2. Tetraethylenepentamine	17.8	22.9	15.4	14.0	27.7	Negl	10–11
3. Nitrilotriacetic acid Ammoniatriacetic Acid	11.5	12.6	10.5	9.8	—	—	11.1
4. Ethylenediaminetetraacetic acid	18.6	18.8	16.5	16.5	22.1	16.1	17.9
5. N-Hydroxyethylenediamine-triacetic acid	17.0	17.4	14.5	13.0	20.1	—	15.5
6. Ethyletherdiaminetetraacetic acid	14.7	17.8	15.3	16.3	23.1	—	14.4
7. Ethylene glycol-bis-(β-amino-ethyl ether)-N,N'-tetraacetic acid	13.6	17.8	13.0	16.7	23.8	—	14.6
8. Diethylenetriamine-pentaacetic acid	20.2	21.0	18.8	19.0	27.0	—	18.6
9. Cyclohexanediaminetetraacetic acid	19.4	21.3	18.6	19.2	24.4	17.6	19.7

Table A-8 / Formation Constants of Some Metal Complexes

	log K_1	log K_2	log K_3	log K_4	log K_5	log K_6
Ammonia						
Co^{2+}	2.11	1.63	1.05	0.76	0.18	−0.62
Ni^{2+}	2.80	2.24	1.73	1.19	0.75	0.03
Cu^+	5.93	4.93				
Cu^{2+}	4.31	3.67	3.04	2.30	−0.46	
Ag^+	3.24	3.81				
Zn^{2+}	2.37	2.44	2.50	2.15		
Cd^{2+}	2.65	2.10	1.44	0.93	−0.32	−1.66
Hg^{2+}	8.8	8.7	1.00	0.78		
Ethylenediamine						
Ag^+	4.70	3.00				
Cd^{2+}	5.47	4.55	2.07			
Co^{2+}	5.93	4.73	3.30			
Cu^{2+}	10.72	9.31				
Fe^{2+}	4.28	3.25	1.99			
Hg^{2+}	log $K_1 \cdot K_2 = 23.18$					
Mn^{2+}	2.73	2.06	0.88			
Ni^{2+}	7.60	6.48	5.03			
Zn^{2+}	6.00	4.81	2.17			
Glycine						
Ag^+	3.51	3.38				
Cd^{2+}	4.74	3.86				
Co^{2+}	4.66	3.80	2.29			
Cu^{2+}	8.38	6.87				
Hg^{2+}	10.3	8.9				
Ni^{2+}	5.77	4.80	3.61			
Pb^{2+}	5.53	4.45				
Zn^{2+}	5.42	4.59				
Oxalate						
Al^{3+}	log $K_1 \cdot K_2 = 13$		3.8			
Cd^{2+}	3.5	2.2				
Ce^{3+}	6.5	4.0	0.8			
Co^{2+}	4.7	2.0				
Cu^{2+}	log $K_1 \cdot K_2 = 8.3$					
Fe^{3+}	9.4	6.8	4.0			
Mn^{2+}	log $K_1 \cdot K_2 \cdot K_3 = 2.4$					
Ni^{2+}	3.9	2.6				
Zn^{2+}	4.7	2.3	2.0			
Cyanide						
Fe^{3+}	log $K_1 \cdot K_2 \cdot K_3 \cdot K_4 \cdot K_5 \cdot K_6 = 31$					
Co^{2+}	log $K_1 \cdot K_2 \cdots K_6 = 19.09$					
Ni^{2+}	log $K_1 \cdot K_2 \cdot K_3 \cdot K_4 = 22$					
Cu^+	log $K_1 \cdot K_2 = 24$	4.59	1.70			
Ag^+	log $K_1 \cdot K_2 = 21.1$	0.89				
Zn^{2+}	log $K_1 \cdot K_2 \cdot K_3 = 17.5$	2.7				
Cd^{2+}	5.48	5.12	4.63	3.55		
Hg^{2+}	18.00	16.70	3.83	2.98		

Table A-9 / Standard and Formal Electrode Potentials at 25°C

Electrode	Reaction	$E°$ (volts) vs. N.H.E.
Li^+, Li	$Li^+ + e = Li$	−3.045
K^+, K	$K^+ + e = K$	−2.925
Ba^{2+}, Ba	$Ba^{2+} + 2e = Ba$	−2.90
Sr^{2+}, Sr	$Sr^{2+} + 2e = Sr$	−2.89
Ca^{2+}, Ca	$Ca^{2+} + 2e = Ca$	−2.87
Na^+, Na	$Na^+ + e = Na$	−2.714
La^{3+}, La	$La^{3+} + 3e = La$	−2.52
Ce^{3+}, Ce	$Ce^{3+} + 3e = Ce$	−2.48
Mg^{2+}, Mg	$Mg^{2+} + 2e = Mg$	−2.37
AlF_6^{3-}, Al	$AlF_6^{3-} + 3e = Al + 6F^-$	−2.07
Pu^{3+}, Pu	$Pu^{3+} + 3e = Pu$	−2.07
Th^{4+}, Th	$Th^{4+} + 4e = Th$	−1.90
Np^{3+}, Np	$Np^{3+} + 3e = Np$	−1.86
Be^{2+}, Be	$Be^{2+} + 2e = Be$	−1.85
U^{3+}, U	$U^{3+} + 3e = U$	−1.80
Al^{3+}, Al	$Al^{3+} + 3e = Al$	−1.66
Ti^{2+}, Ti	$Ti^{2+} + 2e = Ti$	−1.63
V^{2+}, V	$V^{2+} + 2e = V$	−1.18
Mn^{2+}, Mn	$Mn^{2+} + 2e = Mn$	−1.18
Cr^{2+}, Cr	$Cr^{2+} + 2e = Cr$	−0.91
TiO^{2+}, Ti	$TiO^{2+} + 2H^+ + 4e = Ti + H_2O$	−0.89
H_2O, H_2, Pt	$2H_2O + 2e = H_2 + 2OH^-$	−0.828
$Cd(OH)_2$, Cd	$Cd(OH)_2 + 2e = Cd + 2OH^-$	−0.809
Zn^{2+}, Zn	$Zn^{2+} + 2e = Zn$	−0.763
TlI, Tl	$TlI + e = Tl + I^-$	−0.753
Cr^{3+}, Cr	$Cr^{3+} + 3e = Cr$	−0.74
TlBr, Tl	$TlBr + e = Tl + Br^-$	−0.658
U^{4+}, U^{3+}, Pt	$U^{4+} + e = U^{3+}$	−0.61
TlCl, Tl	$TlCl + e = Tl + Cl^-$	−0.557
Ga^{3+}, Ga	$Ga^{3+} + 3e = Ga$	−0.53
CO_2, $H_2C_2O_4$, Pt	$2CO_2 + 2H^+ + 2e = H_2C_2O_4$	−0.49
Fe^{2+}, Fe	$Fe^{2+} + 2e = Fe$	−0.440
Cr^{3+}, Cr^{2+}, Pt	$Cr^{3+} + e = Cr^{2+}$	−0.41
Cd^{2+}, Cd	$Cd^{2+} + 2e = Cd$	−0.403
Ti^{3+}, Ti^{2+}, Pt	$Ti^{3+} + e = Ti^{2+}$	−0.37
PbI_2, Pb	$PbI_2 + 2e = Pb + 2I^-$	−0.365
$PbSO_4$, Pb	$PbSO_4 + 2e = Pb + SO_4^{2-}$	−0.356
Cu_2O, Cu	$Cu_2O + H_2O + 2e = 2Cu + 2OH^-$	−0.34
Tl^+, Tl	$Tl^+ + e = Tl$	−0.336
$Ag(CN)_2^-$, Ag	$Ag(CN)_2^- + e = Ag + 2CN^-$	−0.31
$PbBr_2$, Pb	$PbBr_2 + 2e = Pb + 2Br^-$	−0.280
Co^{2+}, Co	$Co^{2+} + 2e = Co$	−0.277
$PbCl_2$, Pb	$PbCl_2 + 2e = Pb + 2Cl^-$	−0.268
V^{3+}, V^{2+}, Pt	$V^{3+} + e = V^{2+}$	−0.255
Ni^{2+}, Ni	$Ni^{2+} + 2e = Ni$	−0.250
N_2, $N_2H_5^+$, Pt	$N_2 + 5H^+ + 4e = N_2H_5^+$	−0.23
Mo^{3+}, Mo	$Mo^{3+} + 3e = Mo$	−0.2
Cu_2I_2, Cu	$Cu_2I_2 + 2e = 2Cu + 2I^-$	−0.185
AgI, Ag	$AgI + e = Ag + I^-$	−0.151
Sn^{2+}, Sn	$Sn^{2+} + 2e = Sn$	−0.136
Pb^{2+}, Pb	$Pb^{2+} + 2e = Pb$	−0.126
HgI_4^{2-}, Hg	$HgI_4^{2-} + 2e = Hg + 4I^-$	−0.04
Hg_2I_2, Hg	$Hg_2I_2 + 2e = 2Hg + 2I^-$	−0.04
H^+, H_2, Pt	$2H^+ + 2e = H_2$	0.000
CuBr, Cu	$CuBr + e = Cu + Br^-$	+0.033
UO_2^{2+}, UO_2^+, Pt	$UO_2^{2+} + e = UO_2^+$	+0.05
$S_4O_6^{2-}$, $S_2O_3^{2-}$, Pt	$S_4O_6^{2-} + 2e = 2S_2O_3^{2-}$	+0.08
AgBr, Ag	$AgBr + e = Ag + Br^-$	+0.095
HgO, Hg	$H_2O + HgO + 2e = Hg + 2OH^-$	+0.098
TiO^{2+}, Ti^{3+}, Pt	$TiO^{2+} + 2H^+ + e = Ti^{3+} + H_2O$	+0.1

Note: An alphabetical listing is presented in D. A. Skoog and D. M. West, "Fundamentals of Analytical Chemistry," 4th ed., Saunders, Philadelphia, 1982, pp. 828–831.

$E^{\circ\prime}$ (VOLTS) VS. N.H.E.			
1 M HClO$_4$	1 M H$_2$SO$_4$	1 M HCl	Other Media
		−0.551	
	−0.29		
−0.33	−0.33	−0.551	
−0.21			
			−0.137 (1 M KI)
−0.14			−0.32 (1 M NaOAc)
+0.005		+0.005	+0.005 (1 M HNO$_3$)
	+0.04		

Table A-9 (cont.) / Standard and Formal Electrode Potentials at 25°C

Electrode	Reaction	$E°$ (volts) vs. N.H.E.
CuCl, Cu	$CuCl + e = Cu + Cl^-$	+0.137
Hg_2Br_2, Hg	$Hg_2Br_2 + 2e = 2Hg + 2Br^-$	+0.140
Sn^{4+}, Sn^{2+}	$Sn^{4+} + 2e = Sn^{2+}$	+0.15
Sb_2O_3, Sb	$Sb_2O_3 + 6H^+ + 6e = Sb + 3H_2O$	+0.152
Cu^{2+}, Cu^+, Pt	$Cu^{2+} + e = Cu^+$	+0.153
SO_4^{2-}, H_2SO_3, Pt	$SO_4^{2-} + 4H^+ + 2e = H_2SO_3 + H_2O$	+0.17
$HgBr_4^{2-}$, Hg	$HgBr_4^{2-} + 2e = Hg + 4Br^-$	+0.21
AgCl, Ag	$AgCl + e = Ag + Cl^-$	+0.222
H_3AsO_3, As	$H_3AsO_3 + 3H^+ + 3e = As + 3H_2O$	+0.24
Hg_2Cl_2, Hg	$Hg_2Cl_2 + 2e = 2Hg + 2Cl^-$	+0.268
BiO^+, Bi	$BiO^+ + 2H^+ + 3e = Bi + H_2O$	+0.31
UO_2^{2+}, U^{4+}, Pt	$UO_2^{2+} + 4H^+ + 2e = U^{4+} + 2H_2O$	+0.334
Cu^{2+}, Cu	$Cu^{2+} + 2e = Cu$	+0.337
Ag_2O, Ag	$Ag_2O + H_2O + 2e = 2Ag + 2OH^-$	+0.344
$Fe(CN)_6^{3-}$, $Fe(CN)_6^{4-}$, Pt	$Fe(CN)_6^{3-} + e = Fe(CN)_6^{4-}$	+0.36
VO^{2+}, V^{3+}, Pt	$VO^{2+} + 2H^+ + e = V^{3+} + H_2O$	+0.361
Ag_2CrO_4, Ag	$Ag_2CrO_4 + 2e = 2Ag + CrO_4^{2-}$	+0.446
Cu^+, Cu	$Cu^+ + e = Cu$	+0.521
I_2, I^-	$I_2 + 2e = 2I^-$	+0.536
I_3^-, I^-, Pt	$I_3^- + 2e = 3I^-$	+0.536
Cu^{2+}, Cu_2Cl_2, Pt	$2Cu^{2+} + 2Cl^- + 2e = Cu_2Cl_2$	+0.538
H_3AsO_4, $HAsO_2$, Pt	$H_3AsO_4 + 2H^+ + 2e = HAsO_2 + 2H_2O$	+0.559
MnO_4^-, MnO_4^{2-}, Pt	$MnO_4^- + e = MnO_4^{2-}$	+0.564
UO_2^+, U^{4+}, Pt	$UO_2^+ + 4H^+ + e = U^{4+} + 2H_2O$	+0.62
AgOAc, Ag	$AgOAc + e = Ag + OAc^-$	+0.643
Ag_2SO_4, Ag	$Ag_2SO_4 + 2e = 2Ag + SO_4^{2-}$	+0.653
O_2, H_2O_2, Pt	$O_2 + 2H^+ + 2e = H_2O_2$	+0.682
Q, H_2Q, Pt	$Q + 2H^+ + 2e = H_2Q$	+0.699
Fe^{3+}, Fe^{2+}, Pt	$Fe^{3+} + e = Fe^{2+}$	+0.771
Hg_2^{2+}, Hg	$Hg_2^{2+} + 2e = 2Hg$	+0.789
Ag^+, Ag	$Ag^+ + e = Ag$	+0.799
Hg^{2+}, Hg	$Hg^{2+} + 2e = Hg$	+0.854
Cu^{2+}, Cu_2I_2, Pt	$2Cu^{2+} + 2I^- + 2e = Cu_2I_2$	+0.86
Hg^{2+}, Hg_2^{2+}, Pt	$2Hg^{2+} + 2e = Hg_2^{2+}$	+0.920
NO_3^-, HNO_2, Pt	$NO_3^- + 3H^+ + 2e = HNO_2 + H_2O$	+0.94
VO_2^+, VO^{2+}, Pt	$VO_2^+ + 2H^+ + e = VO^{2+} + H_2O$	+1.00
Br_2, Br^-	$Br_2(l) + 2e = 2Br^-$	+1.065
ClO_4^-, ClO_3^-, Pt	$ClO_4^- + 2H^+ + 2e = ClO_3^- + H_2O$	+1.19
IO_3^-, I_2	$2IO_3^- + 12H^+ + 10e = I_2 + 6H_2O$	+1.195
O_2, H_2O, Pt	$O_2 + 4H^+ + 4e = 2H_2O$	+1.229
MnO_2, Mn^{2+}, Pt	$MnO_2 + 4H^+ + 2e = Mn^{2+} + 2H_2O$	+1.23
Tl^{3+}, Tl^+, Pt	$Tl^{3+} + 2e = Tl^+$	+1.25
$Cr_2O_7^{2-}$, Cr^{3+}, Pt	$Cr_2O_7^{2-} + 14H^+ + 6e = 2Cr^{3+} + 7H_2O$	+1.33
Cl_2, Cl^-	$Cl_2 + 2e = 2Cl^-$	+1.360
PbO_2, Pb^{2+}, Pt	$PbO_2 + 4H^+ + 2e = Pb^{2+} + 2H_2O$	+1.455
MnO_4^-, Mn^{2+}, Pt	$MnO_4^- + 8H^+ + 5e = Mn^{2+} + 4H_2O$	+1.51
BrO_3^-, Br_2	$2BrO_3^- + 12H^+ + 10e = Br_2 + 6H_2O$	+1.52
$NaBiO_3$, Bi^{3+}, Pt	$NaBiO_3 + 6H^+ + 3e = Bi^{3+} + Na^+ + 3H_2O$	+1.59
Ce^{4+}, Ce^{3+}, Pt	$Ce^{4+} + e = Ce^{3+}$	+1.61
NiO_2, Ni^{2+}, Pt	$NiO_2 + 4H^+ + 2e = Ni^{2+} + 2H_2O$	+1.68
PbO_2, $PbSO_4$, Pt	$PbO_2 + 4H^+ + SO_4^{2-} + 2e = PbSO_4 + 2H_2O$	+1.685
MnO_4^-, MnO_2, Pt	$MnO_4^- + 4H^+ + 3e = MnO_2 + 2H_2O$	+1.695
IO_4^-, IO_3^-, Pt	$IO_4^- + 2H^+ + 2e = IO_3^- + H_2O$	+1.70
H_2O_2, H_2O, Pt	$H_2O_2 + 2H^+ + 2e = 2H_2O$	+1.77
Co^{3+}, Co^{2+}, Pt	$Co^{3+} + e = Co^{2+}$	+1.842
Ag^{2+}, Ag^+, Pt	$Ag^{2+} + e = Ag^+$	+1.98
$S_2O_8^{2-}$, SO_4^{2-}, Pt	$S_2O_8^{2-} + 2e = 2SO_4^{2-}$	+2.01
F_2, HF	$F_2 + 2H^+ + 2e = 2HF$	+3.06

$E^{\circ\prime}$ (VOLTS) VS. N.H.E.			
1 M HClO$_4$	**1 M H$_2$SO$_4$**	**1 M HCl**	**Other Media**
		+0.14	
	+0.07		
+0.249	+0.25		+0.228 (1 M KCl)
			+0.282 (1 M KCl)
			+0.34 (0.02 M HCl)
	+0.4		
+0.72	+0.72	+0.71	+0.48 (0.01 M HCl)
	+0.360		
+0.577		+0.45	
		+0.577	
+0.696	+0.696	+0.696	(quinhydrone)
+0.732	+0.68	+0.700	+0.61 (1 M H$_2$SO$_4$ & 0.5 M H$_3$PO$_4$)
+0.776	+0.674	+0.274	+0.282 (1 M KCl)
+0.792	+0.77	+0.228	
+0.907			+0.92 (1 M HNO$_3$)
+1.02	+1.0	+1.02	
			+1.05 (4 M HCl)
+1.24	+1.22	+0.77	+1.23 (1 M HNO$_3$)
+1.26		+1.09	+1.10 (2 M H$_2$SO$_4$)
+1.47	+1.628		
+1.70	+1.44	+1.28	+1.61 (1 M HNO$_3$)
	+1.82		+1.83 (1 M HNO$_3$)
			+1.91 (1 M HNO$_3$)

Table A-10 / Potentials of the Common Reference Electrodes†

Electrode	Potential at 25°C (volts) vs. N.H.E.	Electrode	Potential at 25°C (volts) vs. N.H.E.
Hg|Hg$_2$Cl$_2$(s), KCl(s) (S.C.E.)	+0.2444	Hg|HgO(s), 1.0 M NaOH	+0.140
Hg|Hg$_2$Cl$_2$(s), 1.0 M KCl (N.C.E.)	+0.2810	Ag|AgCl(s), KCl(s)	+0.199
Hg|Hg$_2$Cl$_2$(s), 0.10 M KCl	+0.3356	Ag|AgCl(s), 1.0 M KCl	+0.237
Hg|Hg$_2$SO$_4$(s), K$_2$SO$_4$(s)	+0.64	Ag|AgCl(s), 0.10 M KCl	+0.290
Hg|Hg$_2$SO$_4$(s), 1.0 M H$_2$SO$_4$	+0.682		

†With well buffered solutions a glass electrode serves as a satisfactory reference electrode for potentiometric titrations.

Table A-11 / Equivalent Ionic Conductances at Infinite Dilution
(The values are for aqueous solutions at 25°C)

Cations	λ, mho · cm^2/g equiv	Anions	λ, mho · cm^2/g equiv
H$^+$	349.82	OH$^-$	197.6
Co(NH$_3$)$_6$$^{3+}$	102.3	Fe(CN)$_6$$^{4-}$	110.5
Tl$^+$	74.7	Fe(CN)$_6$$^{3-}$	101.0
K$^+$	73.52	SO$_4$$^{2-}$	79.8
NH$_4$$^+$	73.4	Br$^-$	78.4
Pb^{2+}	73.	I$^-$	76.8
La^{3+}	69.5	Cl$^-$	76.34
Ba^{2+}	63.64	C$_2$O$_4$$^{2-}$	74.2
Ag$^+$	61.92	NO$_3$$^-$	71.44
Ca^{2+}	59.50	CO$_3$$^{2-}$	69.3
Sr^{2+}	59.46	ClO$_4$$^-$	67.32
Cu^{2+}	54.	ClO$_3$$^-$	64.58
Fe^{2+}	54.	BrO$_3$$^-$	55.78
Mg^{2+}	53.06	F$^-$	55.
Zn^{2+}	52.8	HCOO$^-$	54.6
UO$_2$$^{2+}$	51.	IO$_4$$^-$	54.4
Na$^+$	50.11	HCO$_3$$^-$	44.48
Li$^+$	38.69	CH$_3$COO$^-$	40.9
		IO$_3$$^-$	40.8
		HC$_2$O$_4$$^-$	40.2
		C$_6$H$_5$COO$^-$	32.3
		B(C$_6$H$_5$)$_4$$^-$	21.

Table A-12 / Polarographic Half-Wave Potentials

Element and Oxidation State	SUPPORTING ELECTROLYTE				
	1 M NH$_3$, 1 M NH$_4$Cl	1 M KCl	7.3 M H$_3$PO$_4$	1 M NaOH	1 M KSCN
Ag(I)	>0 (0)				
Al(III)	—	−1.75 w (0)	NR	NR	
As(III)	−1.46 w (0)	—	−0.46 w (0)	(−0.27) w (V)	−0.68 w (0)
	−1.64 w (−III)	—	−0.71 i (−III)	—	−0.09 i (−III)
					−1.56 i
As(V)	NR	NR	NR	NR	NR
Bi(III)	—	−0.09 w (0, R)†	−0.15 w (0)	−0.6 (0)	>0 w (0)
Cd(II)	−0.81 w (0, R)	−0.64 w (0, R)	−0.71 w (0, R)	−0.76 (0)	−0.65 w (0, R)
Co(II)	−1.29 w (0)	−1.20 w (0)	−1.20 i (0)	−1.43 (0)	−1.06 (0)
Cr(III)	−1.43 (II)	−0.61 (II)	−1.02 w (II)		−1.05 i (II)
	−1.71 (0)	−0.85 (II)			
		−1.47 (0)			
Cr(VI)	−0.2 (III)	−0.3 (III)	>0 i (III)	−0.85 w (III)	>0 (III)
	−1.6 (0)	−1.0 (III)	—	—	−0.46 (III)
		−1.55 (II)	—	—	−0.95 (II)
		−1.8 (0)			
Cu(II)	−0.24 w (I, R)	+0.04 (I)	−0.09 w (0, R)	−0.41 w (0)	>0 w
	−0.51 w (0, R)	−0.22 w (0)	—	—	−0.54 w (0)
Fe(II)	(−0.34) (III)	−1.3 (0)	—	(−0.9) (III, R)	−1.52 i (0)
				−1.46 (II)	
Fe(III)	—	>0 (II)	+0.06 (II, R)	−0.9 (II, R)	>0 w (II)
Mn(II)	−1.66 w (0)	−1.51 w (0)	—	−1.70 (0)	−1.54 w (0)
Mo(VI)	−1.71 w (V ?)	—	±0.0 i	—	NR
			−0.49 w		
Ni(II)	−1.10 w (0)	−1.1 (0)	−1.18 i (0)	—	−0.68 w (0)
Pb(II)	—	−0.44 w (0, R)	−0.53 w (0, R)	−0.76 w (0, R)	−0.44 w (0, R)
Sb(III)	—	−0.15 w (0, R)†	−0.29 w (0)	(−0.45) w (V)	>0 w (0)
Sb(V)	NR	—	NR	—	NR
Sn(II)	—	(−0.1) i (IV)	−0.58 (0)	(−0.73) w (IV)	−0.46 w (0, R)
		−0.47 (0)	—	−1.22 w (0)	
Sn(IV)	—	−0.1 (II)	−0.65 i	NR	−0.5 i (0)
Tl(I)	−0.48 w (0, R)	−0.48 w (0, R)	−0.63 w (0, R)	−0.48 w (0, R)	−0.52 w (0, R)
U(VI)	−0.8 (V)	−0.2 (IV +V)	−0.12	−0.95 w	−0.24 i (V, R)
	−1.4 (III)	−0.9 (III)	−0.58	—	−0.54 w (IV ±V)
					−1.21 w (IV)
V(IV)	(−0.32) w (V)	Depends on pH	−0.6 i	(−0.43) w (V)	Depends on pH
	−1.28 w (II)		−0.93 i (II)		
V(V)	−0.96 w (IV)	Depends on pH	>0 (IV)	−1.7 i (II)	Depends on pH
	−1.26 w (II)	—	−0.54 i		
			−0.91 (II)		
Zn(II)	−1.35 w (0)	−1.00 w (0, R)	−1.13 i (0)	−1.53 w (0, R)	−1.06 w (0, R)

†In 1 M HCl.

Note: All values are referred to the saturated calomel electrode at 25°C. The symbol >0 indicates that the wave starts from zero applied potential, merging with the anodic wave due to oxidation of the mercury. NR indicates that the substance in question gives no wave. A half-wave potential enclosed in parentheses corresponds to an anodic (oxidation) wave. The appearance of a w following the half-wave potential indicates that the wave is well-defined and that its height is easily and accurately measurable, whereas i means that the wave is ill-defined and that its diffusion current cannot be measured accurately. The roman numeral in parentheses gives the oxidation state of the product of the half-reaction which occurs at the dropping electrode when this is known with reasonable certainty, and an R following this signifies that the oxidation or reduction proceeds with thermodynamic reversibility.

Table A-13 / Spectrophotometric Solvents for UV and IR Spectra

SOLVENT	ULTRAVIOLET CUTOFF (nm)
ACETONE	330
ACETONITRILE	210
BENZENE	280
BROMOFORM (stabilized with Diphenylamine)	360
BUTYL ALCOHOL	210
CARBON TETRACHLORIDE	265
CHLOROFORM (stabilized with alcohol)	245
CYCLOHEXANE	210
1, 2-DICHLOROETHANE	235
DICHLOROMETHANE	235
N,N-DIMETHYLFORMAMIDE	270
ETHYL ETHER	210
METHANOL	210
METHYLCYCLOHEXANE	210
METHYL FORMATE	260
NITROMETHANE	380
ISO-PROPYL ALCOHOL	210
PYRIDINE	305
TETRACHLOROETHYLENE (stabilized with Thymol)	290
2,2,4-TRIMETHYLPENTANE	210

WAVELENGTH (micrometers): 2, 4, 6, 8, 10, 12, 14, 16

WAVE NUMBERS (cm^{-1}): 5000, 2500, 1500, 1200, 1000, 800, 700, 625

Note: The ultraviolet cutoff is the wavelength at which the absorbance (optical density) of a 1-cm layer is about 1. Below 220 nm, cutoff values should be determined in the user's own equipment because of instrumental variations.

The shaded areas in the chart show the infrared regions for which these liquids have been found useful in the cell thicknesses commonly employed.

Table A-14 / Useful Spectral Regions for Absorption Cells and Instruments

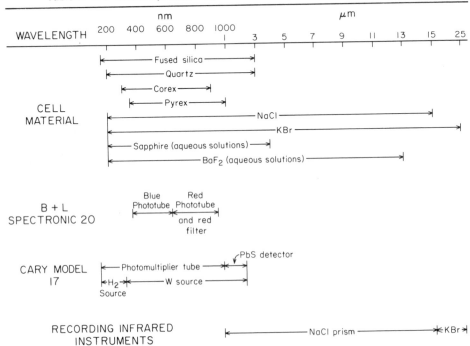

Table A-15 / Emission Spectra; Principal Analysis Lines of the Elements

Strong, medium, and *weak* analysis lines are indicated by **bold face,** normal, and *italic* type. Important lines outside this spectral region are given in parentheses. A semi-quantitative estimate of the elemental concentration can be made from the presence or absence of lines of different intensity values.

In the Region of 2400 to 3300 Å

Ag	2375, 2437, 2448, 2824, 2938, **3281, (3383)**
Al	2568, 2575, 3054, 3064, 3066, **3093, (3962)**
As	**2350,** 2780, 2860, 2899
Au	**2428,** 2590, **2676,** 2748, 3123
B	**2497, 2498**
Ba	2634, 3071, **(3501)**
Be	**2349,** 2651, 3019, **3131,** 3321
Bi	2697, 2780, 2898, 2938, **3068**
C	**2479**
Ca	**2398,** 2721, 2995, 3159, 3179, **(3934)**
Cd	**2288,** 2763, 2775, 2837, **2981, 3261**
Co	2511, 2521, 2649, 2731, 2987, **3044, 3072, (3405, 3454)**
Cr	2383, 2409, **2781, 2836, 2986, 3015, (4254)**
Cu	2393, 2442, 2618, 3108, **3247, 3274**
Fe	2396, 2483, 2599, 2749, 2756, **3021,** 3101
Ga	2874, 2943, **(4033, 4172)**
Ge	2651, 2754, **3039, 3269**
Hg	**2536,** 2752, 2967, 3021
Ir	2544, 2850, 2925, 3133, **3221**
Mg	2780, 2782, 2795, 2802, **2852,** 2915, 3074, 3097
Mn	**2576, 2594, 2606, 2795,** 2914, 2933, 2949, 3212
Mo	**2816,** 2848, 2871, 2891, **3132,** 3158
Na	2680, 2853, **3302, (5890)**
Ni	2907, 2943, **3002,** 3051, 3081, 3129
Os	2637, 2839, **2909, 3059, 3302**
P	2534, 2553
Pb	**2393,** 2577, **2614,** 2650, 2663, 2802, 2823, **2833,** 2873
Pt	2651, **2659,** 2702, **2830, 2998,** 3042, **3065**
Sb	2474, **2528,** 2598, 2718, 2727, **2770, 2878**
Si	2435, 2507, 2516, 2519, **2882,** 2988
Sn	2422, 2547, 2706, 2814, **2840, 2863,** 3009, 3034, 3142, **3175**
Ta	2647, 2653, 2675, 2714, 2892, 2902, **(3311)**
Tl	2580, 2609, 2709, 2721, **3230**
Zn	**(2139),** 2608, 2771, **2801,** 3018, 3035, 3075, 3282

In the Region of 3200 to 5100 Å

Ag	3281, **3383,** 3933, 3968, 3981, 4055, 4210, **(5209)**
Al	(3082) **3944, 3962**
Ba	3501, 3910, 4130, **4554,** 4934
Be	**3321**
Ca	3350, 3487, 3706, **3934, 3969, 4227**
Cd	3261, 3404, **3466,** 3500, **3611,** 4800, **5086,** (6438)
Co	**3405,** 3412, 3449, **3453, 3502, 3894, 3995, 4118, 4121**
Cr	3403, 3408, 3605, 3928, 3941, 3980, 3991, 3992, 3994, **4254, 4275, 4290,** 4496
Cs	3876, **4555,** 4593
Cu	**3247, 3274,** 4022, 4104, 4249
Fe	**3581, 3720, 3735, 3749,** 3753, 4045, 4063, **4271**
Ga	**4033,** 4172
Hg	3341, 3650, 3663, **4358**
In	(3039), 3256, 3259, 4057, **4101, 4511**
Ir	**3513,** 3800, 4400
K	3217, 3447, **4044, 4047,** (7665)
Li	**3233,** 3915, 4273, 4603, **(6104), (6708)**
Mg	3332, 3336, **3829, 3832, 3838**
Mn	3212, 3531, **4031, 4033, 4034,** 4041, 4048
Mo	**3194,** 3208, **3798, 3864, 3903,** 4062, 4070, 4128
Na	3302, **(5890)**
Ni	3310, 3320, **3415, 3446, 3493, 3524, 3619,** 3674, 3994
Pb	3640, 3683, 3740, **4058**
Pd	**3243,** 3373, **3405, 3421,** 3460, 3609, 3634
Rb	**4202,** 4215
Rh	3397, **3435,** 3503, 3528, 3658, 3692, 4375
Ru	2651, **3499,** 3727, 3799, **4554**
Sc	3573, 3614, 3631, 4024, 4247, 4314
Sn	3262, 3331, 3656, **4525**
Sr	3464, **4078,** 4216, 4337, **4607**
Th	3291, 3539, 3601, 3741, 3836, 4019, 4069, 4391, 4919
Ti	3234, 3261, **3349,** 3361, 3372, 3653, 3729
Tl	3230, **3519,** 3529, **3776, (5351)**
U	3552, 3670, 3859, 4090, 4171, 4242
V	**3184,** 3202, 4112, 4379, 4385, 4390, 4409
W	(2944), **4009,** 4074, 4295, 4302
Zn	3282, **3303, 3345,** 4298, 4680, 4722, 4811
Zr	**3392,** 3438, 3496